EINSTEIN

EINSTEIN

A BIOGRAPHY

JÜRGEN NEFFE

TRANSATED BY SHELLEY FRISCH

THE JOHNS HOPKINS UNIVERSITY PRESS

BALTIMORE

© 2005 Rowohlt Verlag
Translation copyright © 2007 by Shelley Frisch
All rights reserved
Published by arrangement with Farrar, Straus and Girous, LLC
Printed in the United States of America on acid-free paper

First American edition published in 2007 by Farrar, Straus and Giroux
Johns Hopkins Paperback edition, 2009
2 4 6 8 9 7 5 3 1

The Johns Hopkins University Press
2715 North Charles Street
Baltimore, Maryland 21218-4363
www.press.jhu.edu

Library of Congress Control Number: 2008942665

ISBN 13: 978-0-8018-9310-0
ISBN 10: 0-8018-9310-0

A catalog record for this book is available from the British Library.

Designed by Jonathan D. Lippincott

Special discounts are available for bulk purchases of this book.
For more information, please contact Special Sales at 410-516-6936
or specialsales@press.jhu.edu.

The Johns Hopkins University Press uses environmentally friendly
book materials, including recycled text paper that is composed of at
least 30 percent post-consumer waste, whenever possible. All of our
book papers are acid-free, and our jackets and covers are printed on
paper with recycled content.

FOR YOU

CONTENTS

TRANSLATOR'S PREFACE

This book is an expanded version of the original German text, which was published in 2005 in Germany to great acclaim. The biography spent the better part of the Einstein Year on the bestseller list. Since the book's original publication, the author has come across newly available information on Einstein, notably in the diaries of János Plesch, who was Einstein's physician. Plesch's diary entries provide intriguing insights into Einstein's daily routines, the ups and downs of his health, and his scientific approach. This past year has also seen the publication of Volume Ten of Einstein's *Collected Papers*, which offers fresh perspectives on Einstein's relationships with family, friends, colleagues, and the public at large, as well as new information concerning his political outlook and how his scientific innovations were received in the academic community. This English-language edition of Jürgen Neffe's biography of Einstein is one of the first to incorporate these newly published materials.

In translating Einstein's words, I cite the *Collected Papers* wherever they contain the pertinent passages. The *Collected Papers*, published by Princeton University Press, are a work in progress; at the time of this writing, they comprise ten hardbound volumes in the original German, along with English translations of selected passages in accompanying softbound volumes. These translations are the work of several individuals, and since their accuracy fluctuates, I have revised them where they were in need of updating or correction. I similarly refer the reader to other extant volumes of Einstein's correspondence, and to anthologies of Einstein's writings that have appeared in English translation, but again, I have corrected inaccuracies and reworked infelicitous phrasings in the quoted texts.

I am fortunate to live in Princeton, within walking distance of Mercer Street, where Einstein lived, and the Institute for Advanced Study, where he worked. Princeton abounds in physicists and Einstein scholars, several of whom generously shared with me their books and their specialized knowledge of physics and the workings of the Institute for Advanced Study, and supplied helpful leads in locating the sources and terminology this translation required. I am especially grateful to my neighbor and friend Alice Calaprice, author of several books on Einstein and long-time editor of the *Collected Papers,* and to Linny Schenk at Princeton University Press, who made materials on Einstein available to me before they reached bookstores. I would also like to express my appreciation for the invaluable advice I received from Princeton University physicists Bill Bialek and Chiara Nappi, who took it in their stride when I came at them in cafés with terminological queries. And Ed Tenner's encyclopedic knowledge proved invaluable at several critical junctures.

Aaron Wiener, a student of history at Yale University, served as my research assistant, tirelessly tracking down bibliographical references and often-elusive passages in writings by Einstein and his contemporaries. He also supplied sparkling translations of Einstein's poetry for use in this text. Physicist and Einstein translator Josef Eisinger, who has produced an inspired rendering of Einstein's travel diaries, kindly granted me permission to quote from his translations of these texts. Thanks also go to Eric Chinski, my editor at Farrar, Straus and Giroux, and his able assistant, Gena Hamshaw.

Author Jürgen Neffe and I enjoyed a productive correspondence throughout the project, and we met in person on several occasions, in Germany and in the United States. He was most generous with his time and his knowledge.

Shelley Frisch
Princeton, New Jersey

EINSTEIN

THE IMMORTAL

EINSTEIN'S SECRET

Princeton, New Jersey, April 18, 1955. A sunny Monday morning. Pathologist Thomas Harvey's shift begins at the hospital in this university town. The dissection table in the autopsy room holds a dead man whose presence offers Harvey the opportunity of a lifetime. The forty-two-year-old starts in as he would on any other workday, picking up a hospital form and entering the requisite data in the spaces provided. Name: Albert . . . Family Name: Einstein . . . Gender: Male . . . Age: 76 . . . Year: 55 . . . Postmortem Serial Number for the Year 1955: 33. Then the medical examiner begins the autopsy.

He places his scalpel behind one of the dead man's ears and pulls it hard over the neck and thorax through the cold, pale skin down to the abdomen. Then he repeats this cut beginning with the other ear. The result is the Y incision that Rudolf Virchow, a Berlin doctor, had introduced to pathology 150 years earlier.

Blood trickles out of the abdominal cavity. Harvey suspects that a ruptured aorta is the cause of death. It soon becomes apparent that his hypothesis is correct. Einstein had been suffering from an aneurysm for years, a blood-filled protrusion of his abdominal artery, and it had burst during the night, evidently owing to a weakness in the vascular wall. The inevitable result was internal bleeding and death. The doctor announces these findings to the journalists eagerly assembled in front of the clinic to report every detail to the world.

The pathologist has run into the physicist now lying on the autopsy table several times in the past, which is nothing out of the ordinary in a small town like Princeton, where Einstein spent the final twenty-two years of his life. The only time the doctor came into direct contact with his prominent fellow Princetonian, however, was during a house call, when he was standing in for a female colleague.

"I see you've switched genders," Einstein quipped when the doctor entered his room for that visit. Evidently he preferred the female variety of medical care. He was lying in his bed, which took up nearly half of his room. A feather quilt covered his stocky body, and his famous shock of hair was spread out on the pillow. The patient was again suffering from an upset stomach, as he had off and on since his childhood.

Harvey asked him to hold out one of his arms. He looked for a suitable vein, stuck a needle into the skin, and drew blood into a syringe. While doing so, he told Einstein how he had bicycled through Europe with friends for a few weeks before the war and had seen something of Germany along the way. The emigrant listened attentively. Finally the doctor gave him a glass and asked him for a urine sample. When Einstein returned from the bathroom and handed him the warm container, Harvey kept thinking, *This is from the greatest genius of all time.*

And now Einstein's cold corpse is lying cut open before him. It is Harvey's last chance to take something from the body before it goes to the crematorium. Suddenly the pathologist sees, and seizes, his once-in-a-lifetime opportunity. Case 55-33 will change his life.

Removing and examining the brain of a dead person does not go beyond the purview of standard autopsy procedure. Harvey, however, has been neither asked nor authorized to do what he does next to Einstein's body, nor does the Hippocratic oath endorse his actions. He saws off the head of the dead man and scoops out its contents. He holds the brain in his hand the way Hamlet held Yorick's skull. In these two and a half pounds of nerve tissue, he is certain, lies the key to understanding the greatest intellectual creative power. If it were possible to elicit the trade secret from this organ, he, the pathologist, would gain fame and honor. He decides to walk off with it and never give it back.

Princeton Hospital, half a century later. Like a criminal returning to the scene of the crime, Harvey heads straight to the former autopsy room. A windowless, fluorescent-lit back room, part office, part laboratory, full of beakers, retorts, refrigerators, bins, files, and discarded furnishings. The table, made of shiny high-grade steel, still dominates the center of the room. Harvey, now white-haired and ninety, his back bent by the trials and tribulations of his life, waits in front of it. He is wearing a sleeveless vest over his sport shirt.

A young doctor in a white coat enters the room and places a cardboard box on the steel table. Harvey opens the box like someone who has performed this movement a thousand times. He pulls out crumpled rags, then heaves two heavy glass containers in the shape of big cookie jars onto the table. Both are filled to the top with a yellowish, translucent, somewhat cloudy liquid. In this liquid is a stack of rosy gray chunks, wrapped in fine gauze and marked with tiny numbers—Einstein's brain, sectioned and stored in an alcohol solution.

"Is everything all right, Dr. Harvey?" the young doctor inquires. "Thanks, Elliott, everything's fine. Let's have a look, shall we?" Harvey carefully holds one of the jars up to the light and rotates it with both hands. "My treasure," he exults. Eyes fixed on the shimmering pallid cubes, he describes his venturesome life starting with that fateful Monday morning when he took the gem into his possession.

How he meticulously prepared the brain, sectioned it into about two hundred cubes, and divided them between the two jars. How he lost his job in the aftermath of walking off with the brain. How the jars, packed away in rags in the carton, accompanied him on his travels all over the country. How he had to keep hiding the brain in unlikely locations, underneath a beer cooler or in a closet of a student apartment, when his impoverishment after having lost his medical privileges drove him to seek employment in a factory in Kansas. And how, after more than forty years, he ruefully returned the infamous stolen goods to the safekeeping of his former workplace.

Elliott Krauss, the successor to his successor in pathology, knows the story by heart. "It all happened right here in this room, didn't it?" "Yes, it did, Elliott." The aged doctor continues to regard his actions as a kind of peccadillo.

Einstein would certainly have condemned what Harvey had done under the white-draped guise of medicine—even though in principle he was not against having his brain examined. However, Harvey was unaware of that. In his will, Einstein had stipulated precisely what was to happen with his body after his death. His mortal remains were to be burned on the day of his death and the ashes scattered in a secret location; these wishes were respected. He did not want to leave behind anything that could be used as a place of pilgrimage or worship. He himself was the monument. Gods have no graves.

But who could blame Harvey? After all, when Einstein's ophthalmologist and longtime friend Henry Abrams heard that the autopsy had just been completed, he too seized his chance, rushing off to the morgue in time to pluck both of Einstein's eyes from their sockets, preserve them in formaldehyde, and place them in a safe deposit box, where they remain to this day. Harvey's actions may have been reprehensible, but his alleged intention was noble: to serve the interest of science. Over the years, he repeatedly made samples from his tissue collections available to researchers. He hoped right to the end that they would capture the essence of genius under their microscopes.

Since studies of Einstein's brain were virtually guaranteed to get publicity, it is no wonder that the experts jumped at the opportunity to report anything they found. The resultant studies claimed that the number of so-called glial cells was elevated, the inferior parietal lobe was larger than normal, and the Sylvian fissure was unusually shaped.

Were these findings the first steps toward understanding extraordinary creative power? Certainly not. Virtually all neuroanatomists have discounted these studies, calling them shoddy, unconvincing, and based on false assumptions. Of course, this brain did accomplish something colossal—but only in its interaction with many other brains. The world in which he lived was a crucial component of his brain's achievement. The researchers do not even know whether the deviations they measured in Einstein's nerve tissue—if they are of any significance whatsoever—are due to the fact that he continued to engage in intense intellectual activity to a ripe old age. How can they hope, then, to classify the unusual qualities they observed, which moreover apply equally to thousands and maybe even millions of other people?

In any case, they fail to shed light on Einstein's distinctiveness. What they do provide is further evidence that even after the close of a century defined by science, the misguided belief that qualities of the mind are reflected in the body has lost nothing of its power. And they demonstrate a longing for simple formulas that encapsulate the life and work of a mental giant of Einstein's stature. Ironically, the immortal Einstein's genius lay in generating simple formulas to explain the workings of the universe—but living beings cannot be reduced to such formulas.

Einstein was one of the most renowned people ever to walk the planet. Certainly no other scientist has come close to his degree of fame and mythic transfiguration. His seemingly paradoxical nature— bourgeois and bohemian, superman and scalawag—lent him an air of mystery. He could reconcile discrepant views of the world, but he was a walking contradiction. Einstein polarized his fellow man like no other. He was a friend to some, an enemy to others, narcissistic and slovenly, easygoing and rebellious, philanthropic and autistic, citizen of the world and hermit, a pacifist whose research was used for military ends.

On the one hand, he upheld the ideals of the French Revolution, advocating freedom and fraternity; on the other, he had a blind spot when it came to the female half of humanity. He carried moral authority, but he was rumored to have illegitimate children and syphilis. With his marked sense of justice, he had as much in common with a queen as with a vagrant, but the equality of the sexes was of no concern to him. Quite the opposite: he valued and used women as lovers, but never really accepted them as companions on an equal level (except perhaps as musicians), and could not tolerate displays of femininity. He failed miserably at marriage—twice.

Rarely has a single individual been so farsighted and myopic at the same time. He was one of the first to recognize the danger posed by the Nazis, the degree to which the Jews were being persecuted, and the threat to democracy in the United States by the American militarization after World War II, yet he never failed to startle his friends and colleagues with the extent of his political naïveté.

Discoveries that shook the world on the one hand, errors and miscalculations on the other. With his theory of relativity and his groundbreaking writings on quantum theory, he enhanced and transcended classical physics. But no sooner was he famous than he wielded his authority to impede further advances, and the younger generation regarded him as a stubborn mule who steered clear of progress.

Thanks to the power of his imagination, he could project his way into the essence of electrons just as well as into the destiny of distant stars. When it came to people who were close to him, however, especially his sons and the problems that bedeviled them, he had not a trace of empathy. He could be downright brutal, but he could show deep compassion for the poor, weak, and persecuted. He alternated between kind sage and incorrigible mule—an egocentric loner with a sense of responsibility for all of mankind.

Neither his brain tissue nor any other physical remains, such as his genes, reveal a thing about his extreme creative powers. The key to understanding Einstein lies not in biology, but in biography.

ONE

HIS SECOND BIRTH

THE FATEFUL YEAR 1919

When Albert Einstein woke up on November 7, 1919, a wintry gray Friday morning, in his apartment on Haberlandstrasse 5 in Berlin, his life had been transformed. The forty-year-old had no idea at this point what the next few weeks and months—and the rest of his life—had in store for him. His quest to "sneak a look at God's cards"[1] brought him about as close to the essence of nature as anyone had ever been. Even so, on the day of his "canonization" in the temple of science, he could scarcely envision the direction his life was about to take.

Until this point, Einstein had stayed out of the public eye. Now he learned firsthand that research and technology were not the only two forces that were shaping the twentieth century. The mass media had discovered him and made him the first global pop star of science. It is hard to imagine anyone more fully embodying the notion that fame feeds on itself. Today, Einstein's popular image—a craggy face encircled by a white mane, with a bulbous nose and a look of wide-eyed innocence—is better known than that of any other human being.

Fame and the mass media go hand in hand. *The Times* of London set off a chain reaction in the media that morning in November, and Einstein's fame was immediate. Newspapers and magazines were the voices of the epoch—the age of radio began about a decade later.

The British paper introduced to its readers "one of the most momentous, if not the most momentous, pronouncements of human thought."[2] This newspaper normally maintained a tone of genteel detachment and objectivity, but in this case the editors euphorically proclaimed a "revolution in science."

For the originator of the uproar far away in Berlin, the content of the report was no surprise. After all, the "revolution"—the general theory of relativity—had occurred four years earlier. And Einstein was already well aware that an astronomical measurement more than five months earlier had confirmed his "new theory of the universe."

Einstein had made several predictions as touchstones for the correctness of his model. One of them stated that large masses curve or bend space. If this curvature really existed, light would have to follow their forms exactly on its path through the universe. In the proximity of the sun, the closest massive body to the earth, it would have to be deflected by a tiny but quantifiable amount.

This amount can be calculated exactly using Einstein's formulas: 1.7 seconds of arc in geometric terms. In the cosmos, this is equivalent to a distance of the breadth of a match. The previous prediction on the basis of equations by Isaac Newton, the forerunner of modern physics, had predicted only half of this amount, and had yet to be tested. The moment was arriving to put Einstein's theory to the test. If his prophecy could be confirmed in practice, his model would triumph over Newton's, two centuries after the latter's death.

The requisite measurements are only possible every few years, when the moon covers the sun completely for a few minutes, from the vantage point of people on earth. Only then can stars in the vicinity of the sun even be detected, allowing a possible curvature of the light rays to be measured by the solar mass. Now the readers of *The Times* were learning that British researchers in the tropics had been successful in conducting this very test during a solar eclipse, on May 29, 1919.

Einstein had learned the results in the early part of the summer. On September 27, he wrote to his mother, who was bedridden with cancer in Switzerland, "Today some happy news. H. A. Lorentz telegraphed me that the English expeditions have locally verified the deflection of light by the Sun."[3] The formal announcement was held off, however, until a statement was read at a joint meeting of the Royal Society and the Royal Astronomical Society in London on November 6. This remarkable meeting altered the course of Einstein's life. The British mathematician and philosopher Alfred North Whitehead, who was present at the meeting, recorded his impressions:

> The whole atmosphere of tense interest was exactly that of the Greek drama: we were the chorus commenting on the decree of destiny as disclosed in the development of a supreme incident. There was dramatic quality in the very staging: the traditional ceremonial, and in the background the picture of Newton to re- mind us that the greatest of scientific generalizations was now, after more than two centuries, to receive its first modification. Nor was the personal interest wanting: a great adventure in thought had at last come safe to shore. . . . The laws of physics are the decrees of fate.[4]

At this moment, Albert Einstein was reborn as legend and myth, idol and icon of an entire era. The mortal Einstein had just passed his creative zenith, and the rather tragic second half of his life lay before him. An immortal of the same name then stepped onto the interna- tional stage—the Einstein that would be embedded in the conscious- ness of the twentieth century as the archetype of the adventurer of the mind whose philosophical quest embodied a conscience for mankind and put the principle of responsibility on a par with the standard of sci- ence and progress. His name became synonymous with genius even during his lifetime.

On November 10, *The New York Times* picked up the story with the headline, "Lights All Askew in the Heavens" and announced, "Einstein Theory Triumphs."[5] The paper reassured its readers that no one need bother trying to grasp the new theory. Only "twelve wise men" would be able to understand it. On November 11, another headline followed on the same topic, and for the rest of the year, additional stories on a nearly daily basis opened the eyes of the readership to the preposterous new world of relativity and its creator. These reports also played a major role in fostering Einstein's fame to the notoriously curious, sensation- seeking, and enthusiastic American public.

In Berlin, people were oblivious to these developments. They were burdened with other concerns one year after the end of the war. The majority of the population was hungry and cold. Early that month, the winter had set in prematurely, and the first snow had fallen. There was barely anything to eat and almost nothing to burn. The railroad had suspended passenger services for eleven days to transport as many potatoes and as much coal as possible to the city.

Nearly everything was in short supply. Even the small pleasures of life became big problems. *Der Abend* noted, "You are just as likely to win the lottery or be hit by lightning as you are to buy a bar of chocolate at a normal price."[6] Refugees from the East crowded into the congested city. Living quarters were hard to come by, and the homeless camped out in wind-protected corners. Sooner or later, the owners of large apartments were required to take in boarders—including the Einsteins, who had a seven-room apartment on Haberlandstrasse.

Einstein wrote to his mother in September 1919, "We have to relinquish a room (rent it out). Starting tomorrow, the elevator won't be operating anymore, so each exit will involve a climbing expedition, and in addition to that, much shivering lies ahead of us this winter."[7] The following March, he reported to his sons from his first marriage, Hans Albert and Eduard: "One week we had no lights, gas, occasionally even no water."[8]

Aside from such practical limitations, Einstein saw no reason to alter his daily routine on this morning in November. After waking up in his bedroom right next to the front entrance—the spartan furnishings were limited to a bed, a closet, a chest, a table, and a couple of chairs—he walked through the library and the living room to get to the bathroom at the other end of the apartment, which adjoined the bedroom of his cousin Elsa, whom he had recently married, just after divorcing his first wife. Afterward the family ate breakfast together. The Einstein household—which also included Elsa's two daughters, Ilse and Margot—did not have to endure hunger. The family was well supplied with what Einstein, a passionate eater, called "fodder," thanks to regular packages from Switzerland.

After breakfast, Einstein usually went upstairs to work at his desk in an attic room, where he spent most of his time. Two windows looked out over the roofs of Berlin. In one corner, next to his desk and a window, was his telescope, a basic model designed for amateurs. If he saw anything at all with it, it was more likely to be neighbors than stars. On the walls were pictures of Schopenhauer and of three great British physicists: James Clerk Maxwell, Michael Faraday, and, in a special spot, Newton.

Einstein retreated to his little empire for hours at a time. Sometimes, when he needed a change of pace, he went downstairs and improvised on the piano in the Biedermeier-style sitting room. Much to the chagrin of his neighbors, he played his violin, which had been his companion since childhood, only at night—in the tiled kitchen, because it echoed so nicely there.

The approaching storm of popularity had not reached him just yet. Letters addressed to "Professor Albert Einstein, Germany" would not get delivered to his house. The daily mail, which Otto, the doorman, would later bring up by the basketfuls, still fit in the mailbox. No statesman or queen was rushing to the telephone to congratulate him. The only telegram on record is from Hendrik Lorentz, a Dutch colleague he greatly admired. Lorentz told him that the results of the solar eclipse had been published in London.

While *The Times* was churning out yet another fact-filled story about the historic solar eclipse ("The Revolution in Science"), readers of the *Berliner Morgenpost* were being told to anticipate a far less consequential partial lunar eclipse the following night. Since the weather was overcast, there was little likelihood that onlookers would see much of this spectacle, but the newspaper told its readers exactly what to expect: "In Berlin, where the full moon will rise at 3:58 p.m., the moon will move into the shadow of the earth 2 minutes before midnight and will be located almost due South."[9]

For several centuries, astronomers have been able to predict exactly when solar and lunar eclipses would occur. Celestial phenomena of this sort have always fascinated people. Since antiquity (and probably far earlier), stargazers have explored the clockwork mechanism of the sky with increasing precision, first with the naked eye, and, after Galileo's time, with more and more sophisticated telescopes. At the beginning of the twentieth century, astronomic charts and maps became astonishingly precise. Anyone conversant in the laws of mechanics, as Newton formulated them more than 250 years ago, can pinpoint celestial events as precisely as most situations require. Their perfection is marred by trifling amounts—no more than small deviations after the decimal point—that would interest no one but pedantic specialists.

London was informing the world that an essentially unknown man named Albert Einstein in Berlin had challenged mankind's fabulous achievement of a perfect celestial formula with a model of the universe that was altogether new and, he predicted, more precise than Newton's. His utterly incomprehensible theory bore a suitably odd name: the general theory of relativity.

A man whose understanding of the motions of the stars and planets was no better than that of an amateur astronomer had assembled a strange system of formulas that described the cosmos better than any other scientist before him. He had no need to look through the eyepiece of a telescope—only to contemplate and calculate. Even though the old and the new systems deviated from each other by only infinitesimal amounts, their internal structure could not be more different. Newton's point of departure was perplexing long-distance effects, which he described in his equations, but could not explain; Einstein furnished a model to calculate celestial events and, in doing so, explained them.

Updates on the scope and consequences of these new developments were quick to reach the British and American victors in the Great War, but for Einstein's fellow countrymen, the magnitude of his success went unreported. Instead, the November 8 edition of the newspaper *Der Tag* carried a positive review of a book by someone named Johannes Schlaf, who actually advocated "helping the pre-Copernican worldview to triumph once again" and returning the earth to the center of the universe. There was not a single word about the boldest achievement of the twentieth century, which had enthralled British and American readers ever since its spectacular confirmation.

A precursor of today's mobile telephone, which had just been introduced by the "Society for Wireless Telegraphy," was creating a sensation among the Germans. The *Berliner Illustrirte Zeitung*, which, as always, had its finger on the pulse of the times, reported, "We will have to prepare ourselves for the fact that soon the telephone will be one of those things we will be carrying around with us all the time, like our watches, notebooks, handkerchiefs, and wallets."[10]

On November 15 there was finally an announcement about a devel-

opment in science that appealed to German pride: the Berlin researchers Max Planck and Fritz Haber were awarded 1918 Nobel
Prizes—Planck for physics, and Haber for chemistry—and Johannes
Stark received the physics prize for 1919. Each of these men came to
have an important role in Einstein's life, for better or for worse. Einstein himself would have to keep waiting for the telegram from Stockholm until November 1922, by which point he had been nominated for
the prize a total of ten times.

The country as a whole was hovering between collapse and a new
beginning. On the domestic front there was relative calm for the time
being. A planned general strike had just been called off. The failure to
strike dominated the headlines and was the talk of the town.

The young republic under President Ebert was negotiating with the
victorious powers about peace terms and reparations. Eleven days later,
General Field Marshal Paul von Hindenburg came up with one of the
dominant themes of the Weimar Republic, the "myth of the stab in the
back." This myth was one of the driving forces behind its eventual
failure.

On the same day, November 18, the *Vossische Zeitung* was the first
newspaper to provide a somewhat dull account of Einstein's breakthrough, which was derived from the reports in *The Times*. Additional
articles followed in other papers; their tone was generally matter-of-fact.
The British, by contrast, could not get enough of this story. Sir Arthur
Eddington, the scientific leader of the decisive solar-eclipse expedition,
wrote to Einstein on December 1, "All England has been talking about
your theory."[11] Paul Ehrenfest reported from Holland on November 24,
"All the newspapers are full of translations of agitated articles from *The
Times* about the solar eclipse and your theory."[12] In his reply, Einstein
commented on "cackling by the startled flock of newspaper geese."[13]

On December 14, however, the picture changed in Germany as
well. The front page of the *Berliner Illustrirte Zeitung* featured a photograph of a serious man with combed-back, dark hair and a thick mustache, his chin resting on the fingers of his half-open right hand, and
staring straight ahead. The caption read: "A New Celebrity in World
History."[14]

Up to that point, the public at large had taken very little notice of
the man pictured here, but before long nearly everybody had heard

about Einstein's achievements. A contemporary description reflects the sentiment of those days: "During this time no name was quoted so often as that of this man. Everything sank away in the face of this universal theme, which had taken possession of humanity. . . . In all nooks and corners social evenings of instruction sprang up, and wandering universities appeared with errant professors that led people out of the three-dimensional misery of everyday life into the more hospitable Elysian fields of four-dimensionality. . . . Relativity had become the sovereign password. . . . It was the first time for ages that a chord vibrated throughout the world. . . . The mere thought that a living Copernicus was moving in our midst elevated our feelings."[15]

All at once, Einstein's name was on everyone's lips. The uproar about his strange theory of relativity was remarkable in light of the fact that no one could understand it. Einstein attributed "the mass excitement about my theory" to the intriguing "mystery of its incomprehensibility. . . . I am certain that it is the mystery of *not* understanding that attracts people; it impresses them with the aura and magnetism of mystery."[16] He wrote to his friend Marcel Grossmann, "This world is a strange madhouse. Currently, every coachman and every waiter is debating whether relativity theory is correct."[17]

Until now, he could have kept thinking that the "relativity ruckus" would soon die down. "It was gracious destiny that I was allowed to witness this," he confided to Max Planck on October 23.[18] But in early 1920 he wrote to his friend Heinrich Zangger in Switzerland, "As for me, since the light deflection result became public, such a cult has been made out of me that I feel like a pagan idol. But this too, God willing, will pass."[19] He did not know at this time that even the gods were powerless in the face of his self-perpetuating fame.

His transfiguration drove him to distraction. He was besieged with nightmarish piles of unanswered mail. Following his lifelong habit of using poetry to express his emotions, he wrote a few lines of verse to vent his frustration about this epistolary excess:

> A thousand letters in the mail
> And every journal tells his tale
> What's he to do when in this mood?
> He sits and hopes for solitude.[20]

Very few were able to grasp his thoughts and fully appreciate the heroic fruits of his years of labor to create a new cosmic order. Einstein found himself turning to the "Old Man" (as he called the Creator) to discuss these ideas, ignoring the fact that Nietzsche had long ago declared God dead.

The worship as hero and saint with which Einstein was greeted after 1919, and which sometimes reached the point of hysteria, particularly on his travels around the globe, resulted in large part from his having razed the existing structure of physics with his monumental wrecking ball and established his new, and still valid, view of the world on the rubble. However, his lofty status stemmed from the effect he had on people at least as much as from his accomplishments.

Science had dealt mankind a triple blow. First Copernicus dethroned the earth, the crown of creation, from its place at the center of the universe, then Darwin undercut man's belief in divine creation, then Freud declared that people are ruled by their unconscious. Now Einstein emerged to offer consolation to the lowly human being, driven by instinct, derived from a primitive life-form, stumbling, lonely, on an insignificant little planet through the universe. He was living proof of man's enduring grandeur. By means of pure thinking, man's noblest art, he had succeeded in plumbing the depths of the universe.

Einstein entered the limelight—and drew out his enemies—for reasons that went well beyond this contribution to civilization. He wielded the authority he had attained as a physicist and prophet for humanitarian and political purposes. His quest and yearning for harmony and his crusade against any form of authority extended to humankind as a whole, and to the process of cultural progress. To a greater extent than any of his scholarly colleagues, Einstein combined his reputation as a scholar with a political agenda.

His inimitable, almost Chaplinesque appearance, coupled with his Groucho Marx–style spontaneous sense of humor, served him well in using the media for his own ends, just as the media used him for theirs. He was initially rather clumsy in handling the press, but he grew more and more skillful, all the while retaining a charming hint of awkwardness. His voice carried weight, his words made headlines, and his radio addresses were broadcast all across the country.

His poise in dealing with the press, radio, and film industry enabled him to create something that advertisers might now call a trademark. The Einstein brand blends the epitome of the absentminded professor and goofy bohemian, unconcerned with dress or speech codes, with the image of the clairvoyant analyst of our era and fearless fighter for freedom, human rights, disarmament, and world government.

When Einstein stuck out his tongue—at the world and the future— late in his life, he provided us with the image that signaled his complete transformation from a man to a metaphor. A breaker of taboos, part Galileo and part Gandhi, he succeeded in synthesizing artistic freedom with philosophical power. Einstein was a cross between Diogenes and Dali, as creative yet unassuming as can be.

However, now that the atomic bombings in Hiroshima and Nagasaki had cast a shadow on his star, the photograph also captured the wistful aspect of a man who was no longer able to align the playfulness of his naïveté with the gravity of his childlike nature.

The year 1919, with November 7 as its apex, divides the course of Einstein's life like a watershed. In the spring he divorced his first wife, Mileva, after years of discord, sealing his departure from his wild past and his youthful dream of a "vagrant life." A few weeks later, he married his cousin Elsa. The bohemian had returned to the bourgeois fold of his youth.

In late 1919, Einstein's mother, Pauline, who had had the greatest formative influence on the first half of his life, moved to her son's home on Haberlandstrasse. In the final phase of her terminal cancer, she wanted to die surrounded by family. She was able to experience the triumph of her son—"nourishment for Mama's already fitting motherly pride."[21] Now "Albertle," her little Albert, could begin his life as an adult.

TWO

HOW ALBERT BECAME EINSTEIN

THE PSYCHOLOGICAL MAKEUP OF A GENIUS

Before the undersigned appeared today, in person known, the merchant Hermann Einstein, residing in Ulm at Bahnhofstrasse 135 of Israelite religion, and reported that to Pauline Einstein née Koch, his wedded wife, of Israelite religion, who lives with him in Ulm in his residence on the fourteenth of March of the year one thousand eight hundred seventy-nine in the morning at half past eleven o'clock a child of male sex has been born, which was given the first name Albert.[1]

On a Friday shortly before the onset of spring, a cool wind was gusting through the Swabian city of Ulm. Under a clear blue sky, the temperature reached a high of seven degrees Celsius. In a well-heated corner house at the edge of the Old City, a newborn let out his first cries. No sooner had the young mother held her firstborn in her arms than his "extremely large and angular head"[2] gave her a terrible fright. At first she believed he was "deformed," at least according to the family legend that Maja, Albert's two-and-a-half-years-younger sister, recorded many years later. It would not be the last time little Albert would frighten his parents, who had married three years earlier in the Cannstatt synagogue near Stuttgart. His big head remained one of his most striking features throughout his life.

Statistically, he had a life expectancy of 35.6 years on the day of his birth. Infant mortality was still a major problem in those days. A doctor in Berlin named Robert Koch had just succeeded in growing bacteria in a pure culture, a first step in the eventual fight against these pathogens.

Improved hygiene was one of the reasons that the population in all of Europe was growing at a tremendous rate. Albert was born into a country with about 44 million inhabitants. When he left the country at the age of fifteen to move first to Italy and then to Switzerland, 52 million people lived in the German Empire; when he returned to Berlin in 1914, about 65 million people resided there. Even World War I left no more than a small dent in the rising curve. When Einstein had to turn his back on Germany for good in 1932 and immigrate to the United States, the population of the *Volk ohne Raum* had swelled to 66 million.

Entrepreneurial frenzy coupled with urban sprawl, radical shifts in art, redefinition of social norms, and rapid technological and scientific progress were the defining characteristics of Einstein's generation. The pace of man, merchandise, and media was accelerating, facilitated by telephones, pneumatic dispatch systems and transatlantic cables, highrises and elevators, electric streetcars, express trains, automobiles, and aircraft.

At the threshold of modernity, the kinetic energy of the newly evolving communication society was moving at a dizzying pace. The globalization brought about by new technology required uniform measurements of time. Clocks were synchronized, first in the cities, then beyond national borders, and eventually throughout the world. Right from Einstein's birth, he was exposed to the sounds of the revolution that would one day bear his name.

The house where he was born—which was destroyed in 1944—stood just a few feet away from the train station. The "lightning express" between Paris and Istanbul was now stopping in Ulm, and the city was connected to the world. The railway system had made significant inroads in overcoming Germany's proverbial provincialism. Eight years before Einstein drew his first breath, Bismarck had unified the German Empire under Wilhelm I. People who could not walk or bicycle (the new fashion of the time) to the train station took a carriage or hackney cab, still the only methods of individual transportation. At the time of Einstein's birth, the streets were redolent of horses, coal, and petroleum; hoofbeats and the racket of steam engines provided the auditory backdrop to life in the city.

On his fiftieth birthday he described the place in which he first saw the light of day: "For a place to be born in, the house is pleasant enough, because on that occasion one makes no great aesthetic demands; instead one begins life screaming at one's dear ones, without bothering too much about reasons and circumstances."[3]

Although he lived there for a mere fifteen months, the people of Ulm claim him as one of their own, even today. Who better to represent their motto *Ulmenses sunt mathematici*, with which they declare, full of ironic pride, that they are mathematicians, one and all? They attribute this remark to the mathematical wizard Johannes Faulhaber, who introduced the French scholar René Descartes to mathematics when he visited Ulm in 1620 and shortly thereafter apparently served as an assistant to the astronomer Johannes Kepler.

When he turned fifty, Einstein, in a letter of gratitude to the *Ulmer Abendpost*, dedicated a sentence to his birthplace: "One's place of birth attaches to one's life as something just as unique as one's origin from one's mother."[4] Although this did not necessarily reflect his own view, there was more than a grain of truth to the comment. Just like his parents and sister, he remained a Swabian for the rest of his life. When he lived in the United States, he nostalgically tinged his English with Swabian expressions, and he cherished the Swabian culinary skills of his second wife, Elsa. When his family moved to Munich in the summer of 1880, Einstein technically became a Bavarian, but he never defined himself as such.

Very few documents have been preserved from Einstein's childhood. The first photographs portray him as a cute, somewhat chubby boy gazing shyly into the camera. His paternal grandmother supposedly exclaimed, when she saw him for the first time, "Much too fat! Much too fat!"[5] This description is also secondhand, from his sister's recollections. According to her, when he first met his newborn sister, he asked where her wheels were, since his parents had promised him a new toy. That somewhat contradicts his own later depictions of a delayed speech development.

"It is true," he wrote in the year prior to his death, "that my parents were worried because I began to speak relatively late, so much so that they consulted a doctor. I can't say how old I was then, certainly not less than three."[6] First the shock of the shape of his head, then the discovery that he was a late bloomer—his mother's confidence was shaken to the core. Evidently little Albert displayed a pattern of behavior like that of some autistic children: he formed complete sentences in his head, tried them out in a muted voice, moving his lips while doing so, and only when everything fit together perfectly did he say them aloud with his child's voice.

His strange behavior lasted well into his early years in school. The housekeeper called him a "dope." A child like Albert, in the view of the German-born American psychoanalyst Erik Erikson, "today would be subjected to specialized examination and, perhaps, to treatment."[7] Fortunately, Einstein was spared that.

The boy appears to have been prone to violent temper tantrums, which faded away when he turned seven. If the little tot did not like something, his face would grow pale, his nose would turn white with fury, and he would let fly his attacks. One of his victims was a tutor who came to his home. According to his sister, Albert "grabbed at a chair and struck his teacher, who was so terrified that she ran away and was never seen again."[8] Maja had to endure his attacks as well. "Another time he threw a large bowling ball at his little sister's head, a third time he used a child's hoe to knock a hole in her head."[9]

During this phase, one experience in particular made a "deep and lasting impression" on Einstein: the day "my father showed me a compass."[10] He was surprised that its needle always pointed in the same direction without being touched. "Something deeply hidden had to be behind things."[11] The initiation of a genius? The "miracle" sheds light on the enigma of his uniqueness only up to a point. Nearly every child is amazed at the sight of a quivering compass needle or some other baffling physical phenomenon.

The other boys called him "Big Bore," because he did not want to join into their horseplay and roughhousing. When he did, it was only as referee. Because they considered him excessively devoted to truth and justice, they also nicknamed him "Straight Arrow." He remained devoted to fairness and compromise throughout his life.

In photographs he comes across as pensive, a bit pampered, stiff, and deadly serious—owing in part to the type of photographs taken at that time, when candid shots of people relaxing and laughing together were unknown. Nature, in the family's backyard or on family hikes in the foothills of the Alps, served the same function for him as did his music; he could unwind and think things over. Instead of playing outdoors, Albert preferred being alone, according to his sister, and he immersed himself for hours in activities that required patience and stamina. He owned an elaborate set of stone building blocks, which spurred him on to fashion intricate structures. He loved constructing houses of cards; by the age of nine, his steady hand allowed him to build them up to fourteen stories. He was fascinated by tricky fretwork. His first juvenile insights into the connections between heat and force came from playing with his little steam-driven engine, a present from Caesar Koch, his uncle from Brussels.

His mother, Pauline, is said to have remarked early on that "some day he may be a great professor."[12] Within the manageable confines of her two-child family, she provided him the optimal support to pursue this goal. His father, by contrast, was a cheerful, undemanding, and unambitious companion. "He was content to observe without wishing for more,"[13] Einstein wrote to his friend Heinrich Zangger in 1919.

Of all the rumors that have circulated about Einstein, one is especially persistent: that he performed poorly in school. That notion offers solace to parents whose children do not get the grades they would have hoped for, since they can reason, "If even Einstein . . ."—which in turn fuels the myth.

Albert's scholastic achievements showed great promise. He finished his first year of school at the top of his class, and he remained one of the best throughout high school. But in sports he was a washout. Physical exertion seems to have made him dizzy in no time at all. He flatly refused to participate in anything competitive, even a game of chess, but perhaps he just had an aversion to the organized drills in gym class.

Unlike most of his schoolmates, and schoolchildren today for that matter, the young Albert supplemented his education at school with his own self-styled curriculum at home and developed the skills he deemed important. He read and read and read. When he was hard at work, even the chaos of the family's constant chatter could not distract the little autodidact. "When I was young," he reflected later in life, "all I wanted and expected from life was to sit quietly in some corner doing my work without the public paying attention to me."[14]

He retained this trait for the rest of his life. "A separation seemed to take place between his mind and his body—not unlike the ecstasy of a saint,"[15] reported Antonina Vallentin, a girlfriend of his second wife, Elsa, who met him during the latter half of his life. "You could start talking noisily, or lapse into an even more embarrassing silence, with everybody staring at him, but he would neither see nor hear."[16] She noticed "that he was absorbed in his work, isolated as though on a desert island, and hardly noticed other people."[17] She also noted that no matter how wide open his eyes were, "they would be dark and lusterless as the eyes of a blind man."[18]

While the other children his age pursued adventures outdoors, he sought his "flow experience"[19] in his head. This term is used by psychologists to describe being "fully engaged with and absorbed by the object of [one's] attention."[20] He shared this type of fixation on "peak experiences"[21] with most other great minds. Einstein craved his rewards through the continued satisfaction of his unquenchable thirst for knowledge. He had to train the second-most-important part of his body—his backside, and its ability to sit still—at an early age as well. His very lively relatives watched an absentminded little Buddha sitting on the sofa and meditating about mathematical questions in a trance-like state.

Uncle Jakob Einstein, an engineer and a business partner of his father, taught him the fundamentals of mathematics while he was still in elementary school. Evidently his uncle had a rare ability to communicate the abstract world of diagrams and equations to a child in a playful manner. "Algebra," his uncle explained to him, "is the calculus of indolence. If you do not know a certain quantity, you call it x and treat it as if you do know it, then you put down the relationship given, and determine this x later."[22] When Jakob showed him the Pythagorean theorem, "the boy plunged into the task of solving the theorem for three weeks, using all his power of thought."[23] He came upon the correct proof on his own, using nothing but his intellect.

Had his creative fire already been sparked? Even though Einstein displayed talents that far surpassed those of the vast majority of his peers, who is to say when these talents crossed the threshold to the state of genius, which so very few attain? When and how is the course set that turns productivity into ingenuity? Are there developmental patterns from which principles can be derived for extraordinary creativity?

Psychobiographical research seeks out recurrent personality traits of this kind. Howard Gardner of Harvard University has compared Einstein with six other men who have been called geniuses of the twentieth century, including Picasso, Freud, and Gandhi. Gardner's investigations have led him to conclude that every creative breakthrough represents an intersection of childlike qualities and maturity. He goes so far as to contend that "the peculiar genius of the modern in this century has been its incorporation of the sensibility of the very young child."[24] In the spirit of Charles Baudelaire, who once called children the "painters of modern life," Einstein drew his revolutionary power as a mature scientist from the ongoing flow experience of his early years.

Gardner considers the fact that Einstein's parents left him alone with his dreams an essential element of his exceptional development, because solitude gives children "the opportunity to discover much about their world, and to do so in a comfortable, exploring way." In this way they accumulate "capital of creativity" on which they can draw for the rest of their lives.[25] Gardner therefore calls childhood a "powerful ally."[26]

Einstein's genius did not blossom overnight. As a rule, future geniuses go through a ten-year phase of ongoing practical and theoretical work—a kind of maturation process. Einstein required exactly one decade of mental toil, daily contemplation, and poring through books, often until the wee hours of the morning, to get from his high school diploma to his revolutionary discoveries in 1905, including the special theory of relativity. Mozart, Einstein's favorite composer, composed music for ten years before he was able to write the kind of music that made music history.

Researchers on creativity have also discovered that an overly high intelligence quotient is more of a hindrance than a help on the way to genius. Most people to whom the label "genius" applies are well above average, but the rare breakout genius with a super IQ (over 150) almost never turns out to be a creative genius.

Creative people are propelled by a high-octane motor: the sheer force of will. They feel the overwhelming need to be creative, and are distinguished by their determination and boundless perseverance. Einstein admitted to a "mulish stubbornness."[27] Wherever necessary, his maxim was to grit his teeth and bear it: "God created the donkey and gave him a thick hide."[28] This is why highly creative people tend to meet with disapproval, at least initially.

Einstein, an outsider, excelled only in physics and mathematics, but was otherwise a standard good student—except for the fact that he was too clever for school. "With Albert I long ago got used to finding not-so-good grades along with very good ones," his father said.[29] This did not stop the boy from picking fights with his teachers. He provoked them with rebellious contempt. When writing about his childhood, he blamed his recalcitrance on the barrackslike atmosphere of the Luitpold Gymnasium in Munich, which he attended starting in 1888. He compared the teachers to drill sergeants and criticized the "mindless and mechanical teaching methods, which, because of my poor memory for words, caused me great difficulties. . . . I would rather let all kinds of punishment descend upon me than learn to rattle something off by heart."[30]

In retrospect, he seems to have overstated his comparison between this high school and the cantonal school in Aarau, Switzerland, where in 1896 he completed his secondary education in the liberal spirit of the educational philosophy of Pestalozzi. The Luitpold was one of the more progressive educational institutions in Germany at that time. Although strict discipline was enforced, the atmosphere was not designed to break the spirit of the young people. When his teacher in Munich, Dr. Joseph Degenhart, advised him to leave the school, and Albert retorted that he had not done anything wrong, the teacher replied, "Your mere presence here undermines the class's respect for me."[31]

A desire to go it alone and great independence of thought distinguish people whose creative breakthroughs succeed. Their self-sufficiency is often misinterpreted as defiance, but it is more likely to express an unswerving attitude to rules and regulations—the typical tendency of young people not to accept opposing views when they regard their own as correct. Einstein remained true to himself in this regard throughout his life—and buttressed his outlook with an elegant saying: "Authority gone to one's head is the greatest enemy of truth."[32]

He rebelled against any kind of authoritarian structure: against rigid rules in school and at the university; against the dictates of bourgeois life; against conventions such as dress codes; against dogmatism in religion and physics; against militarism, nationalism, and government ideology; and against bosses and employers. His opposition to all forms of opportunism was one of the most remarkable of his personality traits.

"There was something enduringly childlike in Einstein's appearance and manner and in his veiled disregard of 'grown-up' standards," Gardner writes. "Even when quite old, he never lost the carefree manner of the child."[33]

Einstein himself tried to explain his secret with this argument when he had reached "an advanced age": "When I ask myself how it happened that I in particular discovered the relativity theory, it seems to lie in the following circumstance. The normal adult never bothers his head about spacetime problems. Everything there is to be thought about it, in his opinion, has already been done in early childhood. I, on the contrary, developed so slowly that I only began to wonder about space and time when I was already grown up. In consequence I probed deeper into the problem than an ordinary child would have done."[34]

It was Einstein who suggested that the Swiss child psychologist Jean Piaget ask children about their intuitive notions of speed, space, and time, and thereby provided the stimulus for one of Piaget's most fruitful areas of research. Unfortunately, Piaget had not had the opportunity to investigate the most interesting research subject of his era, young Albert. Had he done so, he would have been able to observe a remarkable phenomenon: Just as the tin drummer Oskar Matzerath in Günter Grass's novel *The Tin Drum* decides at the age of three to stop growing, the sculptor of the universe Einstein put a stop to a part of his maturation process at a crucial point in his development. He did not do so deliberately or consciously, but he retained something of his childlike nature for the rest of his life. He thus permanently mapped out a role for himself that would assume tragicomic dimensions. He remained, in Gardner's words, "the perpetual child." This term has also been applied to Mozart. Erik Erikson, who reached the same conclusion as Gardner, called Einstein "the victorious child."

While one part of his maturation lagged behind, all the other parts kept right on going. Einstein grew to be a man, a father, and a towering intellect. People regarded him as an "emblem . . . of the mature and reflective human being."[35] He retained his childlike essence, however. He kept retreating into it, and so erected his protective casing against the perils of life. From that point on, Albert Einstein was living with two personalities in the guise of a single individual.

This did not occur covertly; quite the contrary. The people around him regarded him as an utterly boyish personality who could enchant strangers but drive his friends to distraction. When they implored him to stop publication of a biography about him in 1921 because they considered its contents too personal, Hedwig, the wife of his friend and colleague Max Born, wrote him: "You do not understand; in such things you are a little child. You are loved, and you must obey: that is, judicious people (not your wife)."[36]

His best friend, Michele Besso, urged Einstein again and again to take care of his ailing son Eduard after the breakup of his family in 1914, but to no avail. In 1932, Besso wrote a letter to his friend, imagining how Einstein might react to these proddings: "What do you know, white-haired child, of the seeker's burden?"[37]

Innumerable reports attest to Einstein's playfulness, his wonderful rapport with children, his lighthearted wit, and his delight in practical jokes. His sense of fun comes through in his limericks and other witty poems:

> Duty in mind, pipe in hand
> That's how Captain Carefree stands
> Smiling wide, with eyes ablaze
> Nothing can escape his gaze.
> He looks out on the ship and sea
> His crew obeys him 1-2-3
> Calmly Carefree stands his ground
> And takes in everything around.[38]

His wife, Elsa, claimed that his personality had not changed since he played with her in the sandbox as a five-year-old. It is reported that she "fed" her husband, who was now over forty years old, because he forgot to eat when he was working, and that she paid him an allowance because he did not know how to deal with money. She called him a "silly, overly truthful and helpless child."[39] His sister, Maja, also "acted like a mother who comes to visit her son to see that everything is all right."

Even though his friends and family shook their heads at times when trying to keep him out of trouble, Einstein's childlike nature was charming, and it enhanced his charisma. The breathtaking ease with which he appeared to react to even his worst defeats, animosities, and crises also contributed to his superhuman aura. "Albert is a shy man," Elsa wrote to Antonina Vallentin.[40] Vallentin begged to differ: "Perhaps the word 'shy' is not the right word," she noted. "Albert Einstein knew nothing of these inner hesitations, of these nightmares that overcome the shy."[41]

Wherever he went, the same reports were sure to follow. Katia Mann, who was a neighbor of the Einsteins in Princeton when she spent two years in exile there with her husband, Thomas Mann, huffily remarked that he had "such big goggle eyes" and "something childlike in his nature."[42] Harry Graf Kessler's diaries recalled an evening "to which this really lovable, almost still childlike couple lent an air of naïveté."[43]

Superficial as these observations may seem, they reveal profound elements of Einstein's nature. Wisdom in the guise of a childlike nature is characteristic of all great clowns. He was not altogether naïve. "I always get by best with my naïveté, which is 20 percent deliberate, you know," he confessed.[44] Unlike his naïveté, which he even employed strategically to shield himself from others, his childlike nature was genuine, not feigned. This quality made him strong and impervious to dogmatism—there was nothing that could not be questioned. His childlike makeup offered him the guise of a jester's cap and lent him the aura of a pure, innocent child of God.

"He was a God, and he knew it,"[45] his friend and doctor Gustav Bucky remarked. This assessment is nothing short of astonishing. Even though in his later years Einstein kept coming back to the idea that he had been made into a "Jewish saint," deification, however figurative, was surely the last thing he sought. Still, he could not prevent other people from forming this impression.

The liberal journalist I. F. Stone, whom the physicist invited to his house in Princeton in the late 1940s, later recalled the visit: "It was like going to tea with god, not the terrible God of the Bible, but the little child's father in heaven, very kind and wise."[46] His secret was his closeness to the world of children. "Whenever Einstein speaks to a child one realizes what barriers exist in his relationship with adults. He is quick to reach an understanding with children and they have only to look at each other to become accomplices."[47]

By the same token, Einstein displayed traits of an adult when he was still a boy. The flip side of the perpetual child, as Einstein himself described it, is "a fairly precocious young man."[48] That is why the "second wonder of a totally different nature," namely his acquisition of a geometry book, surely merits more attention than the episode with the compass.[49]

In his autobiographical sketch, Einstein wrote that getting a "little book dealing with Euclidean plane geometry" when he was entering sixth grade was a true milestone in his life.[50] He not only studied its contents; he devoured it, and "at the age of 12–16 I familiarized myself with the elements of mathematics together with the principles of differential and integral calculus."[51] By the age of fifteen, he had mastered the entire mathematics curriculum at his school. He later recalled his first encounter with arithmetic as an almost religious experience, and remarked that its clear and logical structure was like music—especially the music of Mozart, whose Sonata for Piano and Violin in E Minor was his absolute favorite. His violin, on which he improvised, remained his true companion into old age.

These precocious achievements, which also included reading Kant's *Critique of Pure Reason* at the age of thirteen, would seem to indicate that he was a "child prodigy," which many consider him, or at least highly gifted. He considered himself "rather naïve and imperfect," as he told his uncle Caesar when he was sixteen years old,[52] but this is exactly where his ten-year period of maturation began. It bore incomparable fruit in 1905, his "miraculous year."

By 1894 the shy, somewhat delicate adolescent had been transformed into a heartthrob, with wavy black hair, sensual lips, and big dark eyes that were both pensive and provocative. His character was written all over his face.

Einstein's self-assurance, coupled with his extraordinary mathematical capabilities, saved him from languishing at the Luitpold Gymnasium. He devised a cunning plan to leave the school for good, to the undoubted delight of his teacher, Dr. Joseph Degenhart. First he secured a note from his mathematics teacher, Joseph Ducrue, stating that he had already reached the level of a graduating senior in that subject. Then he got a doctor who was friendly with the family to certify that he was suffering from "neurasthenic exhaustion," which had been *the* trendy illness in Germany since about 1880. Unbeknownst to his parents, he used this medical certificate to apply for permission to be excused from the school. It was granted.

His father, mother, and sister had been living in Milan for over a year while Albert stayed behind with relatives in Munich to finish his high school studies. His departure from the Bavarian capital, which has sometimes been regarded as heroic, was not rooted in his desire to be reunited with his family; he simply wanted to escape the claws of the school's authoritarianism. When the school dropout showed up unexpectedly at his parents' doorstep shortly before Christmas, 1894, he must have anticipated their astonishment. They could hardly believe that their dreamy little armchair scholar was standing there without a high school diploma. But Albert had yet another ace up his sleeve.

With the letter from his mathematics teacher, Einstein was able to enroll in the Polytechnic in Zurich, two years before he would be officially eligible to take the entrance examination. After a couple of weeks of dawdling in Italy and hiking in the Ligurian Mountains, the cramming began. Using his time-honored method of self-instruction, he plowed through Jules Violle's three-volume *Textbook of Physics*. As a warm-up exercise, he wrote his first scientific essay, "On the Investigation of the State of the Ether in the Magnetic Field," and sent it to his uncle Caesar in Brussels.

Einstein flunked the test, a failure that he considered "well deserved." The problem was not physics, but rather subjects like French and botany, which strained his "poor memory for words." However, the examining professor, Heinrich Friedrich Weber, was so impressed by Einstein's knowledge of physics that he bent the rules and invited him to attend his lectures for second-year students. The young man nonetheless chose a different path.

What started out as a minor calamity—Einstein was only sixteen and a half at the time—turned out to be a major stroke of luck. Fate handed him what may have been the crucial year for his metamorphosis from a precocious boy to a mature man. Not professionally—most likely he could have completed his study of physics with his eyes closed—but personally. In the little town of Aarau, just twenty kilometers from Zurich, he was able to complete his secondary studies at the cantonal school, which gave him an additional year of freewheeling high school life—one more year to figure out who he was, to experiment, to fall in love for the first time, and to get away from his parents.

The liberal school, with its less coercive climate of learning in its physics labs, and the tolerance for ideas and politics Albert enjoyed with his very liberal host family allowed him to draw a deep breath before adulthood forced him once and for all onto the treadmill of university studies, career, salary, and recognition. His schoolmate Hans Byland recalled that Albert "as a young man could not be fitted into any pattern."[53] He claimed that the "impudent Swabian" felt utterly "sure of himself" and displayed a "superior personality"; he also called Einstein a "laughing philosopher" whose "witty mockery pitilessly lashed any conceit or pose."[54]

Many others who formed close ties with him further down the line confirmed that the imperious Swabian could be downright unpleasant. His colleague David Reichinstein wrote, "Einstein can express a strong dislike, and can fly into a passion, becoming intolerant and even unjust."[55] His friend and doctor János Plesch warned, "It is difficult to make him your enemy, but once he has cast you out of his heart, you are done for as far as he is concerned."[56]

Albert lived at the "Rössligut" in Aargau, the home of Jost Winteler, a teacher (though not Einstein's teacher), his wife, Pauline, and their seven children. The Wintelers treated him like their own son. He was soon calling them "Papa" and "Mama." This familial bond remained intact throughout his life, and eventually he even became an official member of the family when his sister, Maja, married the Wintelers' son Paul. (His friend Michele Besso also married their daughter Anna.) Albert's brief dalliance with another daughter, Marie, came to an unhappy end for the young woman. When he went to Zurich to study there, he broke off their relationship while remaining close to the family. He later wrote to "Mama" from Zurich how pleased he was "to have a good chat with you, as if we were sitting together in the red room while the potatoes are getting brown with jealousy and the dear son and some other dear thing peep into the room."[57]

He could be as charming as a boy—and as hurtful. His peculiar combination of charm and superciliousness yielded a caustic arrogance that delighted him no end. "Long live impudence! It's my guardian angel in this world."[58]

But even if he had wanted to, he could not change the way he was. He had spent his whole life like a boy lost in a dream world, and came to terms with the reality of the insuperable gulf that separated his childlike frivolity from his exceptionally mature earnestness. His Faustian bargain was a "flight from the I and WE to the IT," namely the rationality of science.[59] He wanted to break free of what he called the "chains of the merely personal." Psychoanalyst Erik Erikson believes that Einstein was keenly "aware of some of the interpersonal conflicts that he thus learned to avoid and yet also to sublimate."[60]

His many friendships and get-togethers notwithstanding, Einstein paid the price for these conflicts with what would appear to be a profound, shocking loneliness to the end of his life. He considered his solitude both a burden and a boon. "I am a real 'lone wolf' who has never wholeheartedly belonged to the State, to my country, my circle of friends and not even to my family but who, despite all these bonds, has constantly experienced a feeling of strangeness and the need for solitude," Einstein wrote.[61] He defined himself by his tendency to embody extreme poles, and smugly delighted in his hermit's existence.

In a notable speech on the occasion of Max Planck's sixtieth birthday in 1918, Einstein categorized people like himself as "somewhat odd, uncommunicative, solitary fellows."[62] He mused about "what has brought them to the temple": "I believe with Schopenhauer that one of the strongest motives that leads men to art and science is escape from everyday life with its painful crudity and hopeless dreariness, from the fetters of one's own ever-shifting desires."[63]

All of us develop mechanisms to bear our biographical burdens. Einstein made "solitude" and "home" the bookends of his sense of self. He was, he noticed, happy "when I, with knitted brow, cultivate the golden scholarship in my room."[64] Sometimes it almost seemed as though he could not distinguish himself from his popular image, and thus he contributed to it. "Einstein the lonely genius" was partly a creation of his own making.

At the age of fifty-seven, he mused about the kind of "solitude that is painful in youth, but delicious in the years of maturity."[65] At that point he had already spent three years cut off from almost all contact with people who had been so valuable scientifically to him, on a distant continent, in the United States. He regarded himself as a "person with no roots anywhere . . . a stranger everywhere."[66] On one occasion he admitted how oppressive this burden could be: "Solitude can be tolerated only up to a certain limit, you know."[67]

He unflinchingly acknowledged a sense of detachment, even from those he was closest to, which came across as unapproachability. People who knew him well found that he was surrounded by an "impenetrable shell." "It was not easy for him," reports his assistant Leopold Infeld, "to emerge from his isolation and understand the way normal mortals speak and think."[68] Einstein's daughter-in-law, Frieda, said, "A kind of thin wall separated Einstein from his closest friends."[69] His fellow student Byland believed "he was one of those split personalities who know how to protect, with a prickly exterior, the delicate realm of their intense emotional life."[70] Deep in his shell, however, the perpetual child sought refuge in the cosmos.

Those closest to him were well aware of Einstein's inner conflict. "He needed to be loved himself," his son Hans Albert recalled. "But almost the instant you felt the contact, he would push you away. He would not let himself go. He would turn off emotion like a tap."[71] "This is simply done," his father said, "if you are indifferent enough to your dear fellow men."[72]

The young Max Brod provided remarkable insights into Einstein's inner life. The writer met the physicist in 1912 at a salon for Jewish intellectuals in Prague, which was run by the culturally active pharmacy owner Bertha Fanta. Brod's friend Franz Kafka occasionally accompanied him to these evenings, but Kafka did not record any impressions of the salon. Brod monitored Einstein with the keen eye of a writer who is adept at dissecting souls. He found Einstein's "strange, reserved way"[73] so fascinating that he modeled the character of Johannes Kepler in his novel *The Redemption of Tycho Brahe* on Einstein.

The novel describes this character's "pursuit of a single aim, which completely shut him off from the world without, rendering him inviolable, and at the same time utterly unreceptive to everything which did not directly concern his science."[74] Kepler, alias Einstein, responded to people "with the whole, pure, spotless surface of his soul, which he brandished before the world unswervingly, unhesitatingly, nay, with a certain rigor and ruthlessness,"[75] yet "the greater, more important part of his existence was enacted upon the stage of the unconscious and was in the truest sense of the term inaccessible to others, as to himself."[76] This effect was evidently unintended, however, since he was "in the strictest sense of the term incapable of calculation; he was not responsible for anything that he did."[77]

No contemporary description captured Einstein more perceptively. It was echoed by his later assistant Leopold Infeld: "In matters of logic and contemplation, Einstein understood everyone quite well, but he had a much harder time grasping emotional issues. It was difficult for him to imagine impulses and feelings unrelated to his own life."[78] In fact, Max Brod's remarks closely mirrored what Einstein said about himself. He revealed to Belgium's Queen Elisabeth, with whom he shared a close bond in his later years, "We see a man who succeeded in liberating himself, even though it had not been easy for him to be rid of the heavy burden of passions that Nature had dealt him for his passage through life."[79]

This was no mere self-stylization. A passage in one of his travel diaries is an incomparable act of self-analysis: "Hypersensitivity masquerading as indifference. Inhibited and alienated as a young man. Pane of glass between subject and other people. Unwarranted mistrust. Written word as substitute for life. Fits of asceticism."[80]

János Plesch even called him a "man without physical sensation": "He sleeps until he is awakened; he will go hungry until he is given something to eat—and eat until he is told to stop."[81] In another passage, Plesch notes, "Strangely, he laughs when others cry."[82]

Einstein certainly displayed characteristics of a "savant," as experts call highly talented people with tunnel vision. They can be intelligent individuals, cognitively gifted beyond all measure, even geniuses, who are distinguished only in the robotic, emotionless way they go through life. Naturally Einstein did not fit this description, but his troubled emotional life, most likely caused by hypersensitivity, may have been the price he had to pay for his often extreme rationality.

"Some supernatural, uncanny power must have accumulated in this being,"[83] Brod wrote about his Einstein-like Kepler. "Nothing, indeed, could turn him away from the sole aim of his being, in the service of which was accumulated all the immortal fire within his nature, all that was great and vital in his spirit."[84] He "expended his whole personality, heart and head alike, upon his scientific labors. Nothing was left over for human intercourse but a peevish insignificant little shadow of himself."[85] He also exhibited an "indifference toward all personal emotions"[86] and a "happy blindness for everything that diverted him from his scientific aims."[87]

Might people have privately been entertaining these thoughts about Einstein when he tactlessly interrupted conversations in a high-pitched voice to interject inappropriate jokes? Or when he absentmindedly scribbled his magic formulas in the notebooks he always carried with him at public gatherings? "But when he stops speaking," Antonina Vallentin observed, "the heavy door of the lost universe shuts behind him again."[88]

Researchers who attempt to use the paltry available data to make a posthumous diagnosis of Einstein's condition, particularly regarding his childlike development, arrive at similarly extreme conclusions. In the spring of 2003, Simon Baron-Cohen of Cambridge University, one of the leading researchers on autism, touched off a media sensation by claiming that Einstein may have been autistic.

Einstein's delayed speech, coupled with his early obsessions with scientific questions and his problems with social relationships, have led some researchers to conclude that he suffered from Asperger's syndrome, a form of autism that is generally not accompanied by learning disabilities.

Thomas Sowell of Stanford University has gone so far as to develop a special designation, "Einstein syndrome," for "intelligent children who are late in talking" and tend to be withdrawn. His goal is not to pathologize this behavior, but to encourage parents who have late bloomers to recognize that their offspring may not suffer from autism or some similar syndrome, but can develop brilliantly like the great physicist.

Sowell uses the term to describe a type of child who has several traits in common with the young Albert. Typically he finds that one or more of the child's close relatives are scientists, mathematicians, or engineers—in Albert's case, his uncle Jakob, as well as his father, Hermann, who earned a technical degree. Quite often a close relative plays an instrument—as did Einstein's mother, Pauline, who patiently arranged for him to take violin lessons. Often parents notice that a particular event suddenly triggers the onset of their child's speech. We do not know what that event may have been in Einstein's case. Nearly all of the children in question are male, strong-willed, and defiant, and are invariably classified as backward during their early years.

These characteristics are not always easy for others to endure. Max Brod took Einstein to task for his solipsistic remoteness. "As long as he was buried in some piece of work, he was utterly unconscious of himself and lived in the most perfect tranquillity."[89] Then again, this tranquillity was "almost superhuman . . . incomprehensible in its absence of emotion, like a breath from a distant region of ice."[90]

Even if we assume that Brod was using poetic license in these descriptions, no one else who knew Einstein—apart from the physicist himself—depicted his state of mind so bluntly. "This is simply how the Good Lord created his supposedly favorite creature," Einstein asserted defiantly.[91] "People are doomed by overpowering innate drives to make each other's lives a living hell."[92]

It is no wonder that he steadfastly rejected the notion of according psychoanalysis the status of a full-fledged science. Later in life, though, he came to regard Sigmund Freud and his findings in a more favorable light. In 1936 he wrote, "Since then, bits and pieces of my own experiences have caused me to accept Freud's beliefs, at least the major premises,"[93] and as a seventy-year-old, he conceded, "Dr. Freud was on the right track with his Oedipus complex, even though he may have greatly exaggerated it."[94]

Einstein adamantly opposed awarding a Nobel Prize to Freud, which exasperated the psychoanalyst no end. Even though Einstein accepted Freud's work, at least in part, he vehemently opposed Freud's way of puttering around with patients. "I am not in favor of psychoanalytic treatment," he wrote to his mentally ill son, Eduard, in 1932. "People can never break free of it."[95] He told a psychoanalyst who offered him therapy that he himself would "like very much to remain in the darkness of not having been analyzed."[96]

THREE

"A NEW ERA!"

FROM INDUSTRIALIST'S SON TO INVENTOR

It was a sight to make the guests forget the wintry weather. The spectacle began with "rockets roaring into the sky" and concluded with "a firecracker bringing to an end the extraordinarily successful firework display."[1] What followed on that historic evening for mankind in late February 1889 made the crowds on Pfarrstrasse in Schwabing, Bavaria, literally beam with excitement.

"All at once the festival site and the streets of Schwabing were aglow with the brightest arc lamps and electric light, which everyone greeted with a lively round of applause. Mr. Einstein, representing Einstein & Cie., which installed the street lighting, then presented the system to the city, expressing sincere gratitude for the exemplary commission," reported the *Schwabinger Gemeinde Zeitung*.[2]

Little Albert must have experienced this kind of highlight of his family life and moment of almost biblical profundity on several occasions. His father and uncle were bringing light to the people. Just a flick of a switch, and the harnessed fire of electric lamps chased away the dark, to the amazement and applause of the public. Light infused the crowd with a sense of well-being. Albert's father and uncle were the heroes of the evening—a boyhood fantasy come true.

Shortly before Albert's tenth birthday, Schwabing was one of the first communities in Bavaria to splurge on electrically generated light. The official opening turned into a grand party. In a procession of 150 cars, the crowd paraded through the newly illuminated streets to the Salvator Brewery, where the celebrations continued into the wee hours of the morning.

It was a thrilling moment for Hermann and Jakob Einstein as well. Their two hundred lightbulbs and ten glistening arc lamps, with a luminosity of a thousand "new candles," functioned without a hitch. Their Type G XIV two-shunt dynamos, powered by a 40-horsepower Deutz gas motor, ran flawlessly in the central station. The Einsteins had also installed electric lighting in the brewery and its tavern the previous year.

A mere ten years before the shindig in Schwabing, Thomas Alva Edison had developed electric lighting using a carbon filament. That invention, as old as Albert Einstein almost to the month, immediately began its triumphant advance. In 1885, companies in Germany manufactured about 15,000 electric lights; a mere six years later the number had risen to 2.3 million. The AEG (General Electric Company) in Berlin manufactured roughly half of them, now in mass production.

Like many others, the exuberant engineer Jakob Einstein wanted to share in this boom. The thirty-year-old was a master of persuasion. He was able to talk his brother Hermann, who was three years older, into making the momentous decision of closing down his thriving business in featherbeds in Ulm, which provided a secure and lucrative existence, and moving to Munich with his wife and son in the summer of 1880 to become Jakob's partner and commercial director of what was then a high-tech electrical engineering company. This was a major turning point in little Albert's life. Who knows whether he would have become the great Einstein if his father had opted for the security of his life in Ulm instead of embarking on this adventure?

In her memoirs of her youth, Albert's sister, Maja, assumed that her father's decision was motivated by his desire to ease his brother's financial straits. Without Hermann as a partner, Jakob would not have been able to launch his company. Hermann's indecisiveness made him a less-than-optimal entrepreneur, but Jakob was bursting with ambition. He did not want to limit himself to installing electric lighting. Maja reported, "His fertile and manifold ideas led him, among other things, to construct a dynamo of his own invention, which he wanted to produce on a large scale."[3] In the lighting system in Schwabing, Jakob would see an impressive realization of his plans.

Hermann's father-in-law, Julius Koch, funded a large part of the new company. In just a few years he had to write off his kindhearted contribution as a loss. For Albert, however, his father's fateful decision and his grandfather's imprudent investment marked a decisive step in his incomparable journey through life, in part because once the two families were living under one roof, at Müllerstrasse 3, Albert learned the fundamentals of mathematics at home from his uncle.

Very few theoretical physicists have spent their youth in as close proximity to technical equipment as Albert Einstein. Workshop, stockroom, and shop were located in the house in which the boy grew up. Before he could even speak in full sentences, his father and uncle were already introducing him to the magical world of their wares at the International Electrotechnical Exhibition in the Glass Palace in Munich. Ludwig II, the king of Bavaria, who was the patron of the show, asked his deputy, Duke Karl Theodor, to open the exhibition officially on September 16, 1882. As befitted the occasion, the opening ceremony took place in the evening, with artificial lighting, which was quite a novelty. Right from the start, the guests saw for themselves why the slogan of the exhibition was "More Light!"

The Einsteins introduced dynamos powered by twenty-two movable steam-driven engines in the nearby botanical garden, and a new type of telephone, "the Paterson System," as well as two telephone switchboards and eight microphones for broadcasting music. They transmitted a concert from "Kil's Coliseum" directly into phone booths on the exhibition grounds, which were constructed solely for this purpose and were shielded from outside noise. The booths were lit up by "Electric Lightbulbs Modeled on the Swan System," installed by Einstein & Cie. At home, Albert was able to hear "the steam whistle blow" in his father's factory at the start of each workday, according to the memoirs of a lathe operator, Aloys Höchtl, who had been employed by the company since 1886. He recalled, "The transmission, which stretched through the workshop for about 25 meters and was powered by a steam-driven machine, began to revolve."[4]

The boy saw and smelled the lubricating oil steam out of the overheated waves. He was able to touch insulators and switches with his own fingers. Anchors and windings—hands-on experience for the growing boy. Magnetic and electrical induction, important foundations of the special theory of relativity, were part of his everyday life. Anyone who has visited this kind of old production plant in which mechanics and electrodynamics work in combination (a standard subject in elementary physics), can appreciate the power of imagination that is sparked by familiarity with technology and its theoretical underpinnings.

Foreman Kornprobst must have spent a great deal of time showing the notoriously curious boy the interplay of thermally generated motion in steam-driven engines and mechanical gear transmission and engines generated by electrical power from the dynamos, and engineer Essberger must have explained the basics to him. In turn, the boy took many a problem that had been plaguing the experts for quite a while and solved it in the twinkling of an eye. The names of the two employees appear on the company's patent specifications. Between 1886 and 1894, the company acquired at least seven patents for arc lamps and gauges.

The extensive experience his male relatives had with inventions, technical innovations, and patent applications was part of Albert's daily life to the same extent that other boys were indulging in games and roughhousing. There was talk of lighting entire districts of the city, of volts, amperes, and ohms, and of using power sources to generate light. It seems somewhat logical, then, that Einstein earned his living as a patent official in Bern between 1902 and 1909—at the height of his creative years.

In December 1919, just after he had become world-famous virtually overnight, he recalled the patent office in a letter to his friend Michele Besso as the "worldly cloister where I concocted my finest ideas."[5] In the last year of his life, he looked back on his time as a "patent boy" and pointed out that "my work at the final formulation of technical patents was a true blessing for me."[6]

His early hands-on experience facilitated his ability to assess inventions in the often convoluted world of patent applications for machines and measuring tools. A born skeptic, he derived real pleasure from casting a critical glance at the papers of people who spent their time tinkering. Presumably he also had to test out proposals for nonsensical perpetual-motion machines. At the same time, he regarded this work as an ideal training ground for the brainteasers he loved, and he found he enjoyed it "very much, because it is uncommonly diversified and there is much thinking to be done."[7]

Einstein did not just evaluate patents, but also appeared in court as an expert witness, a function he continued to perform even when he had become a famous professor. During the time he worked at the patent office, he also came up with inventions of his own. In 1908 he published the basic idea for a device that could be used to measure the smallest electrical voltages. He had spent a year tinkering at the apparatus with Paul and Conrad Habicht, two brothers who were friends of his. At the University of Bern, he even had at his disposal "a small laboratory for electrostatic experiments that I have cobbled together with primitive components in order to work out the electrostatic method."[8]

The Habicht brothers ultimately constructed the "Electrostatic Potential Multiplicator according to A. Einstein," and Paul Habicht introduced the machine at the Physical Society in Berlin. As Einstein told his friend Besso, Paul had "enormous success. The *Maschinchen* [little machine] is head and shoulders above the filament electrometer, and its future is now secured."[9] Even though "the folks" in Berlin "were breathless,"[10] there was no prospect of a secure future or even of a steady income. The "little machine" was never marketed, and further technical developments soon rendered it obsolete.

Einstein never thought highly of his inventions; he was quite critical of the engineering profession, and he later made a point of advising his first son, Hans Albert, not to pursue this career. He told his first biographer, Alexander Moszkowski, in 1919, "My own parents originally wanted me to become a technical scientist, and I was expected to choose this profession to earn my livelihood. I was not, however, sympathetically inclined to it, for even at an early age these practical aims were to me, on the whole, indifferent and depressing. . . . Doubts, indeed, arose in me as to whether technical improvements and advances would actually contribute to the well-being of mankind."[11]

These doubts did not stop him from trying out his skills as an inventor time after time. In 1928 he asked Rudolph Goldschmidt, his colleague in Berlin and a professor of mechanical and electrical engineering, to collaborate with him. His request took the form of a pair of rhymed couplets:

> A little technology here and there
> Can interest thinkers everywhere
> And so I boldly think ahead:
> The two of us will lay an egg.[12]

The "egg" refers to a hearing aid that Einstein wanted to develop for a well-known singer who was hard of hearing. The hearing aid, which they developed and patented together, is designated as "ingenious" in the literature, but never resulted in a prototype for the hearing-impaired, not to mention an actual product. Goldschmidt's reply provides a glimpse of the fun the two professors had while working on this project:

> For us to lay an egg together
> Would be a difficult endeavor.
> The best solution, it is true,
> Would be, if it's all right with you,
> To lay till we can lay no more
> And have an omelet feast in store.[13]

Einstein enjoyed his greatest success as an inventor when he improved on a gyrocompass that could function purely mechanically, independent of magnetic fields. After working as an outside expert for Hermann Anschütz-Kämpfe, an industrialist in Kiel, to defend his patent in court in 1914, he helped Anschütz-Kämpfe to construct an improved version of the compass when the war ended. Anschütz-Kämpfe paid Einstein 20,000 marks for his assistance—in cash, so that he would avoid paying taxes. Einstein's contribution resulted in a patent for the company and ensured him a percentage of the income for the rest of his life.

One of the many paradoxes in Einstein's life is that in the 1930s, the navies of nearly every country (apart from the United States and Great Britain) used the gyrocompass he had helped construct for military purposes, specifically for orientation on the oceans, although he himself had been a pacifist for many years. Einstein appears to have had no qualms about constructing airplane wings right in the middle of World War I. His purely theoretical considerations laid the groundwork for the construction of an airfoil that experts later called the "cat's-back wing." As it turns out, the test flights were a disaster, and the test pilot was relieved "when I touched down on firm ground just before the fence at the end of the airfield at Adlershof after a pitiful straight flight."[14]

Einstein's attempts to help the American marines improve their torpedo techniques in World War II did not fare better. The officers quickly realized that this theoretician could not provide much help with their practical problems relating to the magnetic ignition of their torpedo heads. On September 1, 1943, Einstein finally admitted his limitation in a letter: "I don't believe that in this matter much can be achieved by mathematical calculation."[15] These failures were not the only reasons Einstein cannot be considered a great inventor. Noteworthy success kept eluding him. Still, he did acquire more than two dozen patents in his career, although they were always collaborative efforts. In 1936 he and his friend Gustav Bucky were assigned U.S. patent 2050562 for an automatic camera.

He acquired most of his patents with Leo Szilard, an exile from Hungary. In the late 1920s, the two of them developed a new type of pump for refrigeration machines. Since it was powered not mechanically, but by means of electromagnetic induction, it was designed to run refrigerators free of noise. "Things are going quite well with the icebox," Einstein reported to his son Hans Albert in 1928. "The AEG is very interested in it. We have come up with three different types."[16] Einstein and Szilard introduced their refrigerator using an evaporation system in 1928 at the Leipzig Fair. A container of methanol was placed underneath the "concrete refrigerator."

The two inventors were interested not only in the machinery, but also in such seemingly mundane problems as thermal insulation. "Szilard and I have applied for a patent for a very nice item," Einstein wrote to his son. "You use expensive pieces of cork. We use a few parallel paper-thin walls."[17] The fact that Einstein's inventions never reached the market seemed to echo the state of affairs of the Einstein family business, which would soon show signs of going under.

The International Electrotechnical Exhibition in Munich, where the Einstein brothers displayed their products for the first time, ended on October 15, 1882. It was a resounding success. Now people wanted the products in their homes as well. Sales soared. In 1884 the Einsteins got their first telephone line, number 722—an event that the city chronicle made a point of mentioning. After the exhibition, the number of subscribers nearly doubled, to 355.

With the increasing demand, Einstein & Cie. grew, but so did the competition. In 1885 the brothers acquired an estate at Rengerstrasse 14 (which was later renamed Adlzreiterstrasse 14) in the Sendling district. The property encompassed a factory building, storeroom, forge, shed, barn, and washhouse, as well as a garden with "English landscaping," as the children vividly recalled. The investments again far exceeded the liquidity of the business, however, and Julius Koch had to contribute another large sum of money. Shortly thereafter, when he moved to Munich to be with his daughter, son-in-law, and grandchildren, the space grew tight, and the brothers had an additional two-story apartment building constructed on the grounds.

Einstein grew up as a member of the upper middle class. Evidently, money was not a concern during his pre-college years. From 1882 to 1892 the staff of his father and uncle's company expanded from twenty-five to approximately a hundred workers. The business was turning a good profit, and breweries were particularly good customers. Georg Pschorr had a lighting system built first in his brewery and then in his tavern and private apartment. The Petuel brewery in Schwabing soon followed suit.

The hospital on the right bank of the Isar ordered its entire lighting system from the Einsteins. The export business, especially with Italy, did not fare badly, either. In the summer of 1887 the northern Italian city of Varese ordered electrical equipment and lighting from Einstein & Cie., and a similar commission seems to have come from Susa.

If any southern German city besides Ulm deserves the honor of being called an Einstein city, it is Munich, where Albert spent the critical years of his childhood and adolescence and acquired the tools for his extraordinary profession. This was also where his family's company became a vital part of the economic community. In 1884 it was awarded a commission from the city that was small but spectacular. In the fall, the *Centralblatt für Elektrotechnik* reported: "During the Oktoberfest the festival site was illuminated for the first time ever, with 12 arc lights from J. Einstein & Cie."[18]

The Oktoberfest edition of the local newspaper featured a touchingly poetic description of the new type of lighting: "The soft but intense gleam of the electric arc lamps casts an enchanted glow over the festival site with thousands of visitors and, in contrast to the red flicker of pitch lanterns and pale petroleum lights, shows off the unique charm that the silvery light of the moon produces when it is reflected in the green Isar River."[19]

Einstein & Cie. had come up with an additional innovation: supplying the electricity itself. The brothers laid a six-and-a-half-kilometer-long overhead cable from the factory to the festival site by way of Lindwurmstrasse. Visitors enjoyed the new illumination in what is now the standard manner: the lamps gave off light without creating any noise, while the steam-driven machines rumbled far in the distance.

In those days it was rare for technological innovations to catch on quickly, but electricity was a striking exception, at least in urban areas. In the space of just one decade, electricity went from obscure to ubiquitous—somewhat like the Internet more than a century later. Two developments went hand in hand to foster this development. The weak currents in telegraph and telephone systems enabled people to communicate at long range without a time lag for the first time since the beginning of the twentieth century. Suddenly, distance was inconsequential. Additionally, strong current from "power stations" resulted in new products, revolutionized production methods, and brought about a profound transformation of the workplace and everyday life.

The progress Einstein witnessed as a child had become more Janus-faced than ever. Mankind faced the prospect of enslavement by machines, as evidenced by assembly-line labor, yet technology also promised to liberate man from everyday drudgery. For the first time, there arose something resembling a leisure society, with mass mobility, large-scale sporting events, and public entertainment for everyone.

The remarkable developments in this field between the first International Electrotechnical Exhibition in Germany in 1882 at the Glass Palace in Munich and the following exhibition in 1891 in Frankfurt show how much it had advanced during the first twelve years of Albert's life. At the second show, the organizers tallied well over a million visitors to the fairgrounds. The exhibition fulfilled the promise of its slogan: "A new era!" Einstein & Cie. was one of the twenty-one companies listed in the exhibition brochure. The brothers from Munich were holding their own in the dynamo business. Their G75 machine provided light for the Pfungstätter Beer Hall, Café Milani, a maze, and a shooting gallery. If Hermann and Jakob had taken a closer look, however, they would have noticed that in this new era, their approach was missing the boat.

The "Frankfurt competition between the systems," which had been raging for years, was about to be resolved. Participants eagerly awaited the outcome of a demonstration to prove which of the two rivals—the widely employed direct-current power stations, which the Einsteins marketed, or the newer, technically more elaborate alternating-current power stations—would be better suited to supply power to the city. The correspondent of *The Times* waxed enthusiastic about alternating current, calling it "the most significant and important experiment in technical electricity . . . since this mysterious force of nature has become available to mankind."[20]

From that point on, alternating current was established as a viable alternative to direct current. Since each was superior in certain areas, most of the major producers began to offer both systems. Einstein & Cie., by contrast, continued to work exclusively with direct current, ignoring the fact that only alternating current could be transmitted over long distances without overly diminishing its effectiveness. That the Einsteins did not realize which way the tide was turning precipitated the imminent failure of their company.

The company's ensuing bankruptcy has usually been attributed to Hermann and Jakob's failure in 1892 to secure the major commission right in their hometown of Munich, where they had already installed lighting in the Schwabing section. After a tough battle, the contract to supply the rest of the city with electricity went to the competition. But that was not the sole reason for the company's demise. A price war forced the electrical industry to concentrate its resources, and only companies with sufficient capital on hand or an ample credit line at the bank could survive.

Nicolaus Hettler's dissertation on the company for the University of Stuttgart has established that "[t]he Einsteins evidently had neither."[21] Julius Koch's savings were insufficient to tide them over. In Hettler's view, the owners' personal failings were also to blame. Albert's father, Hermann, was incapable of making up his mind in critical business matters, and Uncle Jakob's confrontational manner made him unable to resist picking a fight with people who had a pivotal role in the survival of the company. He waged a feud with the head of the Technical Office in Munich, whose job it was to evaluate the products and services of the company—and who consequently subjected the brothers to close and quite critical scrutiny.

Good business prospects in Italy appear to have induced the brothers to close up shop in Munich and establish their base first in Milan and then in Pavia. In the space of only two years, however, they were forced to liquidate their business altogether. The Einsteins left their son Albert behind in Munich, intending for him to finish his secondary school and then study electrical engineering. He thwarted those plans, however, by leaving school and following his parents to Italy. They were quite horrified at his impetuous decision, since they were again facing bankruptcy. Prosperity and a secure middle-class standard of living were things of the past—for Albert as well. Still, the boy had gained valuable insights into the practical matters of technology and patents. No one could take from him what was now in his mind.

FOUR

OF DWARFS AND GIANTS

A BRIEF HISTORY OF SCIENCE,

ACCORDING TO EINSTEIN

In your schooldays most of you who read this book made acquaintance with the noble building of Euclid's geometry, and you remember—perhaps with more respect than love—the magnificent structure, on the lofty staircase of which you were chased about for uncounted hours by conscientious teachers.[1]

Einstein was thirty-eight years old, and at the height of his creative powers. It was the middle of World War I, and his productivity was unprecedented. Still, he decided that the time had come to compose an eighty-page "generally comprehensible" essay for a broader audience of "dear readers" on the special and general theories of relativity.

He took pleasure in formulating the essay in the same manner in which he had acquired his understanding of science as a child. "Have you, my friendly reader, ever thought about how the world would have appeared to us humans if we had been created without eyes?"[2] This passage comes from Aaron Bernstein's twenty-volume *Popular Books on Natural Science*, which the young Albert was given between his tenth and twelfth birthdays. Bernstein's compendium was "a work that I read with breathless attention."[3] Books like these shaped the young bookworm's outlook as he perused them on the couch in his parents' home in Munich. The Bernstein edition was a standard item on the bookshelves of the enlightened bourgeoisie. Einstein's own wise statements a quarter of a century later sound like the enchanted echo of his past readings as a boy—as though he were trying to pass along to his readers the gift he had had the good fortune to receive at the beginning of his journey through life.

These were the roots of his later thought experiments, his mental journeys into the microcosm and macrocosm. Here he got to know the pleasure, the excitement, and the satisfaction that come with a total submersion in scientific material. "Out yonder there was this huge world," the perpetual child Einstein wrote in his autobiographical notes of his constant amazement during his youth, "which exists independently of us human beings and which stands before us like a great, eternal riddle."[4]

According to his sister Maja's memoirs, the medical student Max Talmud provided him access to the reading material he sought. Once a week the impoverished future doctor took his seat at the Einsteins' dining room table in Munich and ate his fill in accordance with an old Jewish tradition. Albert was not content simply to curl up in his easy chair to read these books; he insisted that Talmud discuss the contents with him in detail.

Talmud also introduced Albert to Kant's *Critique of Pure Reason* and Ludwig Büchner's international bestseller *Force and Matter*. The latter's older brother Georg had become world-famous with his dramas *Danton's Death* and *Woyzeck*. Ludwig's *Force and Matter*, which was subtitled *Foundations of a Natural World Order*, represented a radical attempt to "recast the previous theological-philosophical worldview."[5]

Talmud and Albert also read and discussed Alexander von Humboldt's *Cosmos*, a fascinating five-volume *Attempt at a Description of the Physical World*, as the subtitle indicated. In the mid-nineteenth century in Germany, this book was read more than any other apart from the Bible.

Einstein never forgot what he learned in these books. Albert, like adolescents in every era, threw himself wholeheartedly into his quest to find meaning in his world and develop his character accordingly. But unlike other adolescents, who commonly orient themselves to friends or cliques, the young Einstein focused on the printed word, which opened up new worlds to him.

The books he read conveyed a sense of what it was like for Galileo to look through a telescope, for Newton to describe the orbit of the moon, and for many other researchers in centuries past to explore and conquer stars and atoms, light and electricity, space and time. Just like the mathematics his uncle Jakob taught him when he was in elementary school, just like the little geometry book he later pronounced "holy," and just like his experience with technical equipment, inventions, patents, and constructions, these vivid descriptions opened his eyes to the world of natural scientists and discoverers.

"Keep in mind, my dear reader," Bernstein wrote, "that anyone who lives in great times like these and cannot imagine how discoveries of this kind are made does not deserve to take part in this era."[6]

Albert took Bernstein's guidance to heart. Those weighty tomes enabled him to conjure up his own images of the world; they gave him a well-rounded education in the best sense of the term. If there is a lesson to be learned by today's parents from the development of the young Einstein, it is above all a realization of what he gained from his comprehensive exposure to age-appropriate works about the adventure of science. Every child has a sense of wonder; it is just a matter of cultivating it early and with the proper means.

His books provided him with a kind of role model, and made him aware of what he wanted to make of his life. On the playground of his fantasy he was able to envision himself as one of the greatest discoverers in the West's "triumphant progression of the natural sciences," long before he had truly joined their ranks.[7]

Albert enthusiastically explored questions of space, time, energy, and matter. He grew familiar with scientific concepts, and read about speculations and axioms. He worked on hypotheses, theories, and laws. He regarded experiments, prognoses, and proofs as elements of the most magnificent game of all time. Bernstein's books also taught him quite a bit about very practical matters, such as the interpretation of weather maps, the function of lightning rods, and the "dangers of brandy." Conceivably his lifelong aversion to alcohol stemmed from these readings. In any case, sheer concentration gave him the high he needed.

Büchner sang the praises of the Enlightenment battle against superstition and magic: "How quickly the power of the spirits and gods fade away at the hands of science!"[8] What well-informed boy could resist identifying with the knights of reason as the saviors of mankind?

In Büchner he also discovered a political appeal sure to strike the fancy of a growing boy: "Thoughtful and freedom-loving people will embrace the notion that the world as such is not a monarchy, but a republic governed by eternal and unalterable laws."[9] This idea might have failed politically, but stood a chance of success in the realm of science: "The truth has an attractive charm," wrote Büchner.[10] Freedom, enlightenment, eternal laws, and the heroes of truth were the essential signposts that guided Einstein along his extraordinary path in life from his early childhood.

His self-styled curriculum enabled him to see beyond the limits of the individual disciplines, and gave him the tools to achieve his theoretical breakthroughs within just a few years. By the age of sixteen he knew more about the essence of science than many seasoned professors. He still lacked insights into the current problems in physics; they would not appear on his personal curriculum until the following decade. What he did know, however, was that natural science "strove for a generalization and simplification of concepts."[11] Humboldt left a lasting imprint of the "highest, rarely attained goal" on the young Einstein's mind.

"The highest and ultimate aim of *our science* will always be," Aaron Bernstein announced at the beginning of his twenty volumes, "to adopt the most straightforward possible approach for all things, to trace back all facts to *one* explanation."[12] In view of the forces of attraction in nature, the author wondered, for instance, "whether they all originate from a natural force that we know nothing about, of which gravitation is just one manifestation."[13] This question would become Einstein's veritable idée fixe during the second half of his life. Bernstein was addressing the eternal dream of the world formula.

Ludwig Büchner, convinced that there was an "ultimate purpose of existence," provided Albert's lifelong mantra: "Simplicity is the hallmark of truth."[14] Einstein would later be the one to unite "force and matter" (energy and matter) in a succinct formula, the profoundest simplification in the world: $E=mc^2$.

Einstein internalized the long-established fundamental principles of scientific thinking at an early age. Any major unifying theory begins with a process of observation, collection, and organization. Natural scientists take apart the phenomena of the world and subdivide them using increasingly specialized traits and distinctions. Their craft is defined primarily by their ability to distinguish on the basis of sensory observation—fixed stars from planets, circular paths from ellipses, horizontal from vertical force, iron from bronze, protein from cell sap, or simply oil from water.

After sorting comes organization of sensory impressions; after gathering comes systematization. Man's response to the diversity of nature is abstraction in classification, which may hold the secret to unifying principles, for plants and animals, crystals and stones, the sun, moon, and stars.

Time was one of the first things man classified at some point in the dim and distant past. Because calendars could be used to plan ahead, they facilitated settled cultures with agriculture and livestock breeding. They originated hand in hand with the development of astronomy. As the mother of all disciplines, astronomy gave birth to mathematics and geometry; moreover, in the words of the British historian of science John Desmond Bernal, astronomy was "the grindstone on which all instruments of science were sharpened."[15]

The young Einstein was amazed to read about the incredible precision possible in celestial observation. The ability to ascertain "the number of stars that is clearly discernable to the naked eye" ("4,022 over the horizon of Berlin alone," according to Humboldt)[16] is unfathomable to the typical stargazer. Registering even the tiniest changes and shifts between the points of light within the cosmic constellations without optical aids requires a high degree of skill, but constructing a system to describe and even predict the orbits of all the planets would appear to entail a superhuman effort.

Still, people have been doing so since antiquity. Until the sixteenth century, the Ptolemaic system was considered accurate. This system locates the planets in the correct order on their orbits, with one slight flaw: it has the sun, rather than the earth, orbiting between Venus and Mars. And the earth stands in the center, instead of the sun. Nicolaus Copernicus finally corrected this misconception and moved the sun to its proper place at the center of the system. He was one of the first heroes to emerge from Einstein's readings as a boy; twenty years later, Max Planck hailed Einstein as the "new Copernicus."

The Polish astronomer's *De revolutionibus orbium coelestium*, a mathematical model of the universe, was not published until 1543, the year of his death. Einstein learned that aesthetic and philosophical arguments can also provide support for a theory. Nature does not produce anything superfluous, Copernicus argued, which was one of the reasons he decided against the Ptolemaic system, with its enormous number of makeshift constructions, in favor of a "simpler" solution—an earth that revolves on its own axis and in an orbit around the sun.

Of course, his system also featured "epicycles," complex orbits in which the centers of small circles move around the circumference of larger circles. However, that is not the crucial point. The Copernican revolution was rooted not in minor mathematical or technical details, but in a major advance: With his heliocentric system, Copernicus had created nothing short of a new view of life. Man in his universe moved from the center to the periphery, but space essentially opened out into infinity.

It took quite a while for Copernicus's new way of looking at the universe to gain acceptance. One of his advocates, the Italian Giordano Bruno, was even convicted of heresy by the Roman Inquisition and burned at the stake in 1600. The same year saw the publication of *De Magnete*, a book by William Gilbert, the personal physician to Queen Elizabeth I. Gilbert's book introduced the effect of geomagnetism to a wide readership half a century after the discovery of the compass principle. Magnetism also provided what seemed to be a plausible answer to the question of what—if not God—held the planets in their orbits. If all celestial bodies were magnetic, couldn't this best-known form of attraction be the force that held them on what time-honored notions of cosmic impeccability considered their perfectly circular tracks?

Johannes Kepler, who carried on the work of Copernicus, did away with the concept of orbits. This first great Protestant scientist, who was the son of impoverished parents, worked briefly as an assistant to the astronomer Tycho Brahe in Prague. Kepler not only relied solely on Brahe's exact measurements of the celestial bodies, which far exceeded everything that had been measured up to that point, but also used Brahe's data to question the prevailing notion of planetary orbits in his mystical, almost fanatic wish to penetrate the mysteries of the cosmos. After endless tinkering—for the orbit of Mars alone he needed six years to fill up nine hundred pages with calculations—he discovered that the orbits were actually ellipses.

Kepler's story showed the young Einstein that persistence and a belief in the right solution could pay off. The astronomer's work also offered him a glimpse into one of the most astonishing phenomena in science, one that would continue to fascinate him: the mathematicization of the world. How can mathematical formulas and geometric figures, which are essentially constructs of the human mind, give us such a clear picture of the world, as though the book of nature were written in our own language?

In the third century BCE, Archimedes expanded the method of determining pi to five places after the decimal point. Two thousand years later, Newton revolutionized physics by developing infinitesimal calculation from these beginnings. Archimedes also derived formulas to calculate the surface areas and volumes of complex geometric figures such as cylinders and cones. Soon after, in Alexandria, Apollonius of Perga investigated the mathematics of "conic sections," and introduced the terms *hyperbola*, *parabola*, and *ellipse*. Kepler, who had recognized that circular orbits were incorrect, used Apollonius's tools to show that ellipses fit the orbits of the planets perfectly.

However, Kepler's curves were not altogether correct, either. Soon, fixed elliptical orbits no longer sufficed to provide a mathematical representation of astronomers' increasingly exact measurements. It took three hundred years for Albert Einstein's general theory of relativity to provide the necessary tools to map out the far more complex orbits.

Galileo Galilei of Pisa, a contemporary of Kepler, was an equally important influence on the young Albert. Galileo's systematic studies of free fall made him one of the fathers of experimental physics. By linking controlled experiments with precise mathematical formulations, and by attempting to explain how nature worked, he overturned current concepts and founded modern natural science. Legend has it that he used the Leaning Tower of his hometown, Pisa, to show that heavy objects did not fall faster than light ones. According to his famous laws of falling bodies, feather and stone fall to earth at the same speed in the absence of restricting factors like air resistance.

In 1633, Galileo was summoned to Rome to face the papal Inquisition and disavow the Copernican view of the world as the only true one. The Church kept the trial secret for fear of generating publicity. The accused was given a rather nominal prison sentence. That he murmured the phrase *Eppur si muove* (and yet it does move) to characterize the earth while facing the Inquisition is the stuff of legend. He was able to serve out his sentence in the palace of a friend. There he completed his works about relations between bodies at rest ("statics") and in motion ("dynamics"). He thus became one of the founders of the art of engineering based on scientific principles.

Galileo had many reasons to remain true to his belief, which was based in large part on an astronomical discovery he had made in 1609. In Holland, he had heard, an optician had made an important chance discovery while constructing a pair of eyeglasses. The man was struck by the fact that by combining lenses, faraway objects could be made to appear closer. Galileo immediately built himself a primitive "telescope." With this new invention, he was able to observe the nocturnal sky in hitherto unseen detail.

With the advent of the telescope, and of the microscope shortly thereafter, a previously invisible world opened up to natural scientists. These achievements brought to light a virtually unprecedented principle of scientific progress: New tools make possible new knowledge, which in turn forms the basis for new types of tools. By the end of the eighteenth century, science had profited far more from industry than the other way around.

The approach of sorting out and distinguishing previously unknown details and particulars entered a new phase with the enhanced vision afforded by telescopes and microscopes. Each improved version of optical equipment revealed new levels of the magical world, both small and large, to natural scientists and to the young Einstein, who was glued to his science books at home on his sofa. Microscopes allowed for a view of the fine structures of life, such as tissues, cells, and their individual components. Telescopes brought astronomers beyond the visible solar system, and eventually even the Milky Way, and offered them access to increasingly distant worlds.

With his first telescope, Galileo discovered that moons revolve around the planet Jupiter. Looking at this microcosm of the Copernican system, in which once again the earth is not in the center, convinced him once and for all of the correctness of his view of life.

Galileo did not make his most significant discovery far out in the solar system, but back home on earth, in his experiments with projectiles and balls on inclined planes. He astutely recognized that the "parabolic path" of a projectile, for instance a cannonball, has two components: the horizontal shot and the vertical fall. And he made an additional groundbreaking discovery. His observations of swinging pendulums showed him most convincingly that the natural motions of bodies are uniform and follow a straight line, as long as no forces act on them—a fundamental insight that contradicted the Aristotelian view of life, also held by the Church. No force is necessary to sustain motion. This insight formed the foundation for the coming revolution of Isaac Newton.

Galileo formulated a principle that came to be called the "Galilean principle of relativity" many years after his death. If the conductor in a train that is traveling at 100 kilometers an hour walks in the direction of travel at 5 kilometers an hour, he is moving 105 kilometers an hour relative to the platform. If he goes against the direction of travel, he moves at only 95 kilometers an hour from the perspective of the observer standing outside. Within his "frame of reference," namely the train, however, he is still walking at 5 kilometers per hour, regardless of the direction he takes. From his perspective, in turn, the observer on the platform races by at 95, then at 105 kilometers per hour.

In classical mechanics, this method of calculation is absolutely sufficient. However, it contradicts unimaginably quicker velocities of light and electrons, which were unknown at that time. Two hundred sixty-three years after Galileo's death, Einstein's theory of relativity would revolutionize these concepts.

It must have been exciting for the high school student to be a passenger on a scientific journey through time. It must have fascinated him to see how each observation builds upon another, idea upon idea, discovery upon discovery. And it must have been exhilarating to read about the founder of mathematical mechanics, and how he was able to unite apparently divergent physical systems.

There are intriguing parallels between Newton and Einstein, the two central figures of early modern and modern physics. Like Einstein, Newton is believed by some to have been afflicted with Asperger's syndrome. Both of them wanted to merge celestial and earthly events and to understand the work of the Creator. Newton said, "He endures forever, and is everywhere present; and, by existing always and everywhere, he constitutes duration and space."[17] Einstein mused shortly before his death, "In the beginning (if there was such a thing), God created Newton's laws of motion together with the necessary masses and forces. This is all."[18]

The special theory of relativity of 1905 used one of Newton's ironclad principles as its basis. "The motions of bodies included in a given space," Newton wrote in his magnum opus, *Philosophiae naturalis principia mathematica* (*Mathematical Principles of Natural Philosophy*), "are the same among themselves, whether that space is at rest or moves uniformly forward in a straight line."[19]

At the time of his death, Einstein had attained the status of a saint, just as Newton had when he died in 1727. On Newton's gravestone in Westminster Abbey, the Latin inscription reads: "Here lies what was mortal of Isaac Newton." Einstein's family scattered his mortal remains (with the exception of his brain and eyes) to the winds—as befitted a life that was not tied to one place.

Newton was Einstein's foremost role model. It was Newton's view of life that he would revise, Newton's mechanics that he would revolutionize, and Newton whom he would reestablish. Near the end of his life, he humbly declared, "Newton, forgive me. You found the only way that, in your day, was at all possible for a man of the highest powers of intellect and creativity."[20]

One of Newton's lasting contributions was to bring the celestial clockwork down to earth. His formulas, which drew on the work of Kepler and Galileo, showed that terrestrial events could theoretically be calculated into all of eternity. As long as the positions of the constellations and their velocities are known, even their future constellations can be predicted.

Virtually prophetic powers have accrued to science—a circumstance that would influence Einstein's development in no small measure. Newton and his contemporaries predicted that in 250 years, on May 29, 1919, the sun would disappear behind the moon for a few minutes in the late afternoon in the tropics. The total eclipse of the sun, by means of which Einstein would finally supersede the reverently worshipped Newton, was written up in astronomy books during Newton's lifetime. Einstein later paid tribute to this achievement in a poem:

> Look to the Heavens, and learn from them
> How one should really honor the Master.
> The stars in their courses extol Newton's laws
> In silence eternal.[21]

The resounding success of Newton's new physics was based on a mathematical method he created. Every student learns infinitesimal calculus in the context of differential and integral calculus—in the style that the mathematician and philosopher Gottfried Wilhelm Leibniz invented at roughly the same time as Newton. Both of their accomplishments marked the pinnacle of a development that goes back almost to the beginnings of advanced human civilization, to the Babylonians.

Newton's indisputably unique achievement was his new method of solving a multitude of physical problems. His three laws of motion made it possible to describe virtually any mechanical process in precise terms. Entire disciplines, such as hydrodynamics and aerodynamics, draw on his studies. Aerodynamics found its first application in Einstein's era; it was used for the construction of airplanes.

Mass has a double role in Newton's laws. As inertia, it describes a body's resistance to acceleration. The greater the inertial mass of a carriage, for example, the more horsepower it takes to get it moving, to accelerate it. On the other hand, gravitational mass provides the gravitational force with which one body attracts others. Even Newton's proverbial falling apple exerts a force on the earth. Newton explained Galileo's law of falling bodies, according to which all bodies, independent of their mass, fall to the ground equally quickly, showing that inertial and gravitational mass have the same value. With an increase in inertia, the gravitation increases to the same degree. However, he was not able to explain why.

Newton earned his immortality by establishing laws that unite the mechanics of the heavens and the earth in one system. The force that makes the apple fall to the earth is the same as the one that keeps the moon on its track around the earth and the earth on its track around the sun: gravity. Newton was perplexed as to what this force of attraction might be. It would be up to Einstein to provide an answer, in his general theory of relativity.

Why is it that the moon does not fall onto the earth, which attracts it? What keeps the celestial bodies in their orbits? In Newton's world there is an exact equilibrium between gravity on the one hand and centrifugal force resulting from the motion of the world on the other. If one were to stand atop the highest mountain in the world and fire a cannonball from an extremely powerful cannon in such a way that its centrifugal force exactly equaled its gravitational force, in principle it could keep flying around the earth indefinitely like the moon (or a satellite). Einstein had been inspired by the pertinent chapter in Aaron Bernstein's book; this chapter bore the title "The Path of the Moon Compared with the Path of a Cannonball."

Nothing of this sort occurred in Galileo's world. The cannonball always landed on the ground after the same period of time, regardless of the velocity with which it left the cannon and whether it was propelled forward or fell straight downward. Only the distance of its flight is dependent on the velocity. Newton's laws, which make allowances for all forces, including gravitational and centrifugal force, describe physical reality far more precisely.

The question remains: What moves toward what? Does the cannonball or moon move relative to the earth or the earth relative to the moon? At this point Newton introduced the concept of absoluteness, which other thinkers like Leibniz disparaged; Einstein was later able to classify it correctly. According to Newton, space and time are absolute. Even if nothing else existed, space and time would be there.

Absolute space forms a kind of receptacle for events on earth. Everything moves in this space, not just the earth and the moon. Newton's deep religiosity and reverence for the Creator made him regard this space as altogether at rest, and intrinsically constant. The physicist thus created a privileged frame of reference, with the result that Galileo's relativity was not entirely valid: if a train passes by a platform, the observers and travelers are unable to say that the one is moving relative to the other, and vice versa. Instead, the one is at rest, and the other is moving—in relation to absolute space.

Newton also dedicated his second absolute to the omnipotence of the Creator in his *Principia*: "Absolute, true, and mathematical time, of itself and from its own nature, flows equably without relation to anything external." Accordingly, there is only one clock for the entire universe. The idea of an absolute time that moves to the steady beat of this clock corresponds perfectly to our everyday experience—but not to reality, as it has been understood since Einstein. His theory of relativity would revolutionize the seemingly incontrovertible notion of absolute time.

Although Newtonian mechanics fails to account for some things, it provides a nearly perfect description of physical reality. Automobiles, trains, ships, airplanes, and spacecraft adhere to his equations. Newtonian mechanics proved persuasive from the start mainly because of its enormous power of prognostication.

Aaron Bernstein's *Popular Books* devoted many pages to a celestial phenomenon that has continued to frighten and terrify man: the comet, a cosmic projectile with a radiant tail, which appears suddenly, lingers on the horizon for a while, and then disappears again in the expanses of space. Views about the comet provide a perfect illustration of the transition from the Middle Ages, which inclined toward magical explanations, to the modern Age of Reason. Until Newton's day, comets were considered a mystical medium to predict destinies, and were often regarded as evil omens.

Bernstein invited his readers to join him on a "fantastic journey through the universe": "Are we traveling by sea? On horseback? By train? Not at all! We are traveling with the aid of an electric telegraphic device!" This must have enthralled the young Einstein the way other children feel their first time on skis. Into the expanses of space with the speed of electrical current! Albert read up on how the astronomers gradually learned to predict the appearance of comets. Once their unexpected appearance and disappearance could be prophesied, however, they lost their eeriness.

Eight years after Newton's death, physics celebrated its first great prophetic triumph. Using Newton's equations, the English natural scientist Edmund Halley predicted the precise time that a comet (which later bore his name) would return. Sure enough, it appeared in the sky in the fall of 1675, right on schedule.

Suddenly researchers were able to travel through time by means of mathematics. They could gaze into the future. The oracle had turned rational. Fate adhered to man's seemingly magical formulas. God's clockwork mechanism appeared perfectly clear. Einstein would later show that the Creator had constructed the universe in a manner that was far more complex than Halley and his contemporaries were able to infer from Newton's work.

The young Albert Einstein found an even more incredible story of astronomical clairvoyance in Volume 16 of Bernstein's *Popular Books*. He read there that "in 1846 a natural scientist in Paris named Leverrier, without looking at the sky, without observations, figured out purely by means of calculation, that 600 million miles away from us there had to be a planet; that this planet made its orbit around the sun in 60,238 days and 11 hours; that it was 24½ times as heavy as our earth, and at a certain time at a certain place in the sky would be visible if our telescopes were only good enough to let us see it. Isn't that astonishing?"[22]

On September 23, 1846, Johann Galle, an astronomer at the well-equipped observatory in Berlin, received a letter from Leverrier requesting that he "lie in wait for the new planet at the precisely designated spot in the sky." That very evening, Galle pointed his telescope in the indicated direction. He did indeed locate the heavenly body, which was later given the name Neptune. It was an incredible triumph of the mathematical physics that Copernicus, Kepler, Galileo, and Newton had founded.

At the close of this chapter, Bernstein reassured his readers, "I am not aiming to make an astronomer out of you, but I hope that I will be able to explain the miracle of this discovery to you."[23] Then he explained how Leverrier, in complete accordance with Newton, deduced the existence of Neptune on the basis of the other planetary orbits. "That is why you should honor this science! Honor the men who cultivate it! And honor the human mind, which is more perceptive than the human eye!"[24]

Bernstein did not make an astronomer out of Einstein, but his "human mind, which is more perceptive than the human eye" was already hard at work. He himself would later make predictions on the basis of his own formulas that not even the Newtonian system yielded. Above all, his theory would provide an explanation for a peculiarity in the solar system that was like a last little blemish in Newton's perfect celestial mechanics once Neptune had been discovered: the perihelion of the planet Mercury—the curve that its position closest to the sun describes in the course of centuries—deviates per year ever so slightly from the predicted orbit.

Just as in the case of Neptune, the astronomers of that era had long suspected that an additional planet that had yet to be discovered was causing the perturbation. They even gave it a name: Vulcan. Vulcan was never found, however, and could not be found, because it did not exist. Newton's equations had furnished an incorrect prognosis. They could not explain the behavior of Mercury. Time and again, researchers attempted to do away with this inconsistency without requiring an additional planet. Einstein's general theory of relativity provided the first explanation that got to the bottom of the peculiarity of the Mercury perihelion.

The young Albert must have been even more amazed to read about how astronomy traces light. This phenomenon would fascinate him throughout his life. His 1905 essay on the "light quantum hypothesis" was a groundbreaking study of quantum theory that elucidated the double nature of light. Aaron Bernstein had marveled over "light, the messenger that brings us knowledge of what goes on far away,"[25] and emphasized that "this law of light . . . is valid throughout the universe [and] applies to any kind of light, be it far or near, large or small."[26] Albert's readings described the quest for the speed of light like a detective story.

Ole Rømer of Denmark had the first spectacular success in 1676, a mere seventy years after the telescope was invented. He made his discovery while studying the same Jupiter moons that once assured Galileo that he was right to support the Copernican view of the universe. The satellites of the largest planets darken just like the earth's moon when they enter the shadow of Jupiter. Rømer observed Io in particular, the innermost of the four moons of Jupiter that were known at the time. He was struck by the fact that the calculations indicated that lunar eclipses on Jupiter occurred with regular delays, the difference amounting to twenty-two minutes.

These delays, he correctly surmised, resulted from the position of the planet (and its moons) in relation to the earth and the sun. If Jupiter is on the same side of the sun as the earth, the light (and hence the "news" of the eclipse) arrives earlier; if it is on the far side of the sun, it occurs later. In other words, light has a longer distance to travel, which takes time. The speed of light can be measured on the basis of distance and time.

Rømer's measurements were initially off by five minutes. The speed of light, calculated from his data, was 210,000 kilometers per second—which was considered an enormously rapid tempo at that time. It was also an enormous concept to grasp. Many of his contemporaries found it difficult to accept the fact that light takes time at all, that it covers ground and moves. Subsequent calculations brought Rømer closer to the actual number, which today is set at 299,792 kilometers per second.

A great number of startling similarities to Einstein's later ideas can be found in Bernstein's books. One example is the way in which Bernstein explained the phenomenon of "aberration" of light by the movement of the earth: "Let us imagine that someone standing at the side of a train fires a pistol at one of the cars moving quickly past him."[27] The bullet pierces the first wall, flies across the train, and pierces the opposite wall as well. Since the train was in motion during this time, "A person sitting in the car . . . seeing the two holes will assume that the shot must have been fired at the train at an angle."[28]

For astronomers, this means that a beam of light from a celestial body entering a telescope describes a slightly angled path within the tube of the telescope—and even in the eye of the observer—owing to the movement of the earth. "In the small amount of time that the light takes to go through the telescope, the earth (and the telescope) moves a little bit farther in its orbit," Bernstein explained. "Thus, if you want to see a fixed star, you have to tilt the telescope a few degrees."[29] Repositioning the telescope compensates for the aberration of the light, somewhat like a pedestrian holding an umbrella sideways to avoid being pelted with rain.

The astonishing discovery drew on observations by the British astronomer James Bradley in 1725. Bradley had intended to measure the distance of fixed stars from the earth, but the quality of the optical instruments was not equal to the task. Bernstein wrote, "As so often occurs on the way to an important goal, Bradley did not find the truth he was seeking, but a different and equally important truth."[30] The crucial point for Bernstein was Bradley's realization that the "aberration" is the same for all stars, regardless of how far away they are. He concluded correctly that all light moves at the same speed, and this became one of the two basic principles in Einstein's special theory of relativity: the speed of light is constant.

But what is the actual nature of light? According to Newton, it consists of tiny particles, so-called corpuscles, which whiz through space like projectiles. His great rival, the Dutchman Christian Huygens, proposed a wave theory of light. In 1801 the English doctor, Egyptologist, and physicist Thomas Young sided with Huygens. He used a characteristic of light known as interference, in which two beams of light can either reinforce each other or cancel each other out, like two waves in the water.

Because Newton was regarded as virtually infallible at that time, Young's idea did not take hold at first. However, experiment after experiment, especially those in the laboratory of the French scientist Augustin Fresnel, that studied the patterns of shadows confirmed the wave theory of light. Within a quarter of a century it had gained complete acceptance. Now it needed to be determined what medium was carrying the light waves. If it was not water, as it is for ocean waves, or air, as in sound, what was vibrating? Physicists of the time, who firmly believed in mechanics, came upon an idea that would determine their thinking for a century to come. Since in their view there had to be a mechanically ductile something that transported the light waves, they postulated a hypothetical medium: ether. Even though no one was able to weigh, see, or in any other way measure this ether, it seemed ideally suited to explain the wave characteristic of light. Bernstein went on the assumption "that now astronomy has allowed us to eliminate any doubt as to the existence of an ether."[31] This was an error, which Einstein would correct once and for all.

Albert soon realized that Bernstein's books encompassed something resembling his research agenda for his later years. To a degree that was virtually unparalleled, he would in theory travel to both extreme ends of the universe, into the microscopic world of the tiniest particles and into the macroscopic farthest galaxies. Bernstein said, almost prophetically, "Just as light-time is now a scientific yardstick for what was previously immeasurably large, someday the length of light waves will most likely be used to measure the hitherto invisible and inestimable smallness of the atom."[32] Like the traveler in Jonathan Swift's famous novel, Einstein journeyed to the giants and the dwarfs. Unlike Gulliver, however, he traveled to both simultaneously. When he arrived, he would try as never before to unite the lands of Lilliput and Brobdingnag in a single big nation.

Albert's readings opened the doors to nearly everything that natural science had established up to his adolescence. Chemistry had made tremendous progress in the nineteenth century—especially since the Frenchman Antoine Laurent Lavoisier had brought order to the chaos of the chemical elements and the Briton John Dalton had introduced atomic theory into chemistry. Physics still needed a full century to accept the existence of atoms, which the Greek Democritus had assumed so far back in the past. Einstein's study of "A New Determination of Molecular Dimensions," written in 1905, his "annus mirabilis," would make a vital contribution to this eventual acceptance.

Albert found everything in his books, the answers he would need to rebuild physics. The chapters about magnetism and electricity, the "twin sister of light," opened up the world of his father's factory to him from the perspective of discoveries and theories. Bernstein continuously pointed out that words like *matter, primary matter,* and *electrical fluid* were only names for things whose true essence was yet unknown. Einstein would later write: "Physical concepts are free creations of the human mind."[33] Even so, Bernstein felt confident "that the forces of nature we observe are only the diverse expressions of a single force of nature."[34]

Well into the nineteenth century, electricity and magnetism were considered two distinct entities, although natural phenomena such as lightning led scientists to believe that there was a connection between them. They magnetized iron and redirected compass needles. In 1820 the Danish physicist Hans Christian Ørsted discovered that electric current flowing through a wire had the same effect on compasses. Eleven years later the brilliant English experimental physicist Michael Faraday figured out that the effect also functioned the other way around. If a magnet moves along an electric conductor, it induces electric voltage. This "magnetic induction" still forms the basis of the generation of current in dynamos and generators today. At that time it unleashed a wave of additional discoveries, including the electrotechnical revolution that was the basis of Albert's father and uncle's business.

Faraday's achievement centered on the relationship between the two phenomena, which were united in the new science of electromagnetism. His studies demonstrated that the two were connected not statically, but dynamically. His "law of induction" made it possible not only to generate electricity, but also to utilize it to run machines. Ever the pragmatist, Faraday spoke of lines of force that in combination yield a "field" in regard to magnetic effects—for instance, on iron filings on a piece of paper over a magnet.

The concept of the field furnished natural science with a brand-new model, which Einstein pondered for the rest of his life. Faraday's concept of a field, as produced, for example, by a magnet, fills space, and exerts a force on a body, such as a piece of iron.

Just as Newton in his day embedded Galileo's ideas in a theoretical work of the highest mathematical elegance, another Briton gave Faraday's intuitive ideas an elegant mathematical structure. In the 1860s, the Scot James Clerk Maxwell created a system of unprecedented complexity in physics: his field equations. One aspect of Maxwell's work in particular captured Albert's fancy: the synthesis of two different domains—optics (the theory of light) and electromagnetism—into a single theory. Maxwell realized that light, too, consisted of electromagnetic waves.

The young Albert must have been intrigued to read about a term Aaron Bernstein coined to compare receiving a letter in the mail with transmitting information from distant parts of the galaxy. Using the analogy of the postal service, he called the transmission of light from far away the "*Postenlauf* of light." Bernstein concluded, "In this sense we always see only the past and never the present."[35] The time it takes for light to reach the earth varies according to the distance it has to travel. It takes eight minutes from the sun, up to fifty-two minutes from the planet Jupiter, and more than two hours from Uranus. The fixed stars reveal completely new dimensions. The light of the closest one, Alpha

Centauri, is not visible to the human eye until three years after it has been emitted; the second closest, Vega, in the constellation Lyra, takes a full twelve years and one month.

Alexander von Humboldt wrote, "Cosmic occurrences of this sort are like voices that reach us from the past. It has been justifiably claimed that when we look through our big telescopes, we are also exploring space and time [because] the sight of the star-studded sky offers a non-simultaneous array of images."[36] The young Einstein had his life-long topics handed to him on a platter: that space and time are tightly interwoven, that light and time have some special connection, that the simultaneity of events depends upon the standpoint of the onlooker. Felix Eberty, an author in Berlin, used these interconnections as a basis for his book *Stars and the Earth, or, Thoughts upon Space, Time, and Eternity*. This work was first published anonymously in Germany in 1846, and was widely read throughout Europe and the United States. When Eberty's book was republished in 1923, this time disclosing the author's name, Einstein contributed the preface.

This remarkable work reads in part like science fiction, although it is based on solid scientific facts. The author turns the tables—or the light beam—and has observers look down on the earth from stars located at a variety of distances out in space. In this way he imagines a kind of time machine.

Eberty wrote, "In the immeasurably great number of fixed stars which are scattered about in the universe . . . reckoning backwards any given number of years, doubtless a star could be found which sees the past epochs of our earth as if existing now."[37] The "omniscience of God with relation to past events"[38] suddenly takes on a somewhat natural explanation. "If we imagine the eye of God present at every point of space, the whole course of the history of the world appears to Him immediately and at once."[39]

He conjures up a "picture . . . embracing space and time together, and representing both so entirely and at once that we are no longer able to separate or distinguish the extension of space from that of time."[40] Some of his lines read as though he himself had traveled into the future and taken a look at the world according to Albert Einstein: "Thus we have here the extension of Time, which corresponds with that of Space, brought so near to our sensible perception, that time and space cannot be considered as at all different from one another. For those things which are consecutive one to the other in point of time lie next to one another in space."[41]

This idea was crucial for the special theory of relativity: Time travels with light. Or the other way around: when time is at the same level as light, it is in effect frozen. Eberty gave free rein to his imagination with this idea, in comments such as this: "Not only upon the floor of the chamber is the blood-spot of murder indelibly fixed, but the deed glances further and further into the spacious heaven."[42] If a photograph of the earth on which a clock can be made out were sent into space, the image of the time on the clock would reach the faraway observers together with the light, as though the clock were at a standstill during the journey. "Thus, that record which spreads itself out further and further in the universe by the vibration of the light, really and actually exists,"[43] the reason being that "the events, not only of *our* world, but of *all* worlds, are at present expanded in space as the greatest and truest History of the Universe."[44]

The world of science fiction now appears in a very different light. "Clearly," the time traveler in H. G. Wells's novel *The Time Machine* (1895) says, "any real body must have extension in *four* directions: it must have Length, Breadth, Thickness, and—Duration."[45] The idea is original, but hardly new. In 1907 the mathematician Hermann Minkowski would give it the finishing touches, using Einstein's special theory of relativity to weld together the four dimensions to make spacetime.

Eberty had stated in 1846, "Time is only the rhythm of the world's history."[46] Earlier in his book, he wrote, "The duration of time is, therefore, unnecessary for the occurrence of events. Beginning and end may coalesce, and still enclose everything intermediate."[47] His "microscope for time" is like a cosmic cinema, contracting the film of celestial events in slow motion or expanding it in time-lapse photography—fifty years before pictures were able to move. The author even played with contractions of space and accelerations of time. What would happen if "the year passes by in six months"?[48] The answer must have thrown his contemporaries at first, since they were unaccustomed to theoretical musings of this kind: "We should be cognizant of no change . . . as we can determine the lapse of any period of time only by comparison, or by measuring it with some other period."[49]

The same applies to space. Eberty pondered whether we would notice any change if we all shrank in size. "We might, like Gulliver's Lilliputians, fairly consider ourselves perfectly grown men."[50] Einstein would later supply the scientific foundation for these kinds of fantasies, as though he had written for an imaginary traveler who moves at virtually the speed of light without noticing it.

Karl Clausberg, a professor of art history at the University of Lüneburg, regards this realm of the "fictional and imaginary" as the "breeding ground for a new kind of scientific fantasy, which came to be held in the highest regard in the twentieth century: thought experiments." Albert Einstein was the undisputed master of the art.

Einstein mulled over the questions "What would it be like to chase after a light beam? And to ride on one?" He claims to have posited these hypothetical considerations at the age of sixteen, but his autobiographical notes fail to mention that there is a source for this highly inventive and seemingly original idea. The question can be found almost verbatim in Bernstein's writings. The special theory of relativity is essentially an answer to this question.

FIVE

THE BURDEN OF INHERITANCE

EINSTEIN DETECTIVES IN ACTION

To one degree or another, everyone takes secrets to the grave. The legacy of normal mortals fades in the course of generations, but Einstein, who was already a legend in his lifetime, was denied—or spared—this merciful oblivion.

"With me," he once said, "every peep becomes a trumpet solo."[1] Once he had "drawn his last scholarly breath" (to use his own phrase), the brief autopsy procedure was followed by a prolonged process of picking apart the genius on the dissection table of scholarly research.

Jerusalem, Hebrew University: an austere library building. Far away from the circulation desk and the hustle and bustle of students—the Einstein Archives, the Holy Grail of Einstein research. Approximately 80,000 pieces of paper, the majority of which are documents he penned: letters, memos, notes, diaries, sketches, drafts, partial calculations, manuscripts, any bit of scribbling that did not disappear in the maelstrom of history. There are also passports, documents, and photographs. All this is stored to resist the ravages of time in gray cardboard acid-free folders in acid-free cartons made out of gray cardboard as well, side by side in shelves that go up to the ceiling. Long corridors of pure Einstein.

Very few people are granted access to these gems. Authorized experts are permitted to have a look at the originals only under the steely gaze of an archivist in a windowless adjoining room. The items you can hold in your hands here, most of them yellowed pieces of paper, would fetch many millions on the auction market. However, even the experts are barred access to many documents. Strict trustees and concerned heirs have made sure that bank books and divorce papers as well as love letters and other personal correspondence are kept under lock and key.

State guests in Israel invariably ask to visit the archives. They are shown the physicist's meticulously ironed evening wear, his little library, including a book by Bertrand Russell in which Einstein wrote "incredibly naïve," a letter to Bertolt Brecht about his play *Galileo*, a letter to a little girl whom he assures that he has "no hair on [my] hands, as is so often the case with ugly men,"[2] and a 1912 manuscript, carefully preserved in a large flat box, with a now-famous formula on page 38: E equals m times c^2. Ultimately, then, Einstein, who felt worshipped as "the Jewish saint," left far more relics for posterity than he could have imagined.

Pasadena, the California Institute of Technology: a two-floor imitation half-timbered house with a manicured lawn. At the entrance is a blue sign with white lettering to announce the center of Einstein research. The sign reads EINSTEIN PAPERS PROJECT, and is followed by the physicist's initials. Rows of green hanging files, labeled index tabs, pale yellow cardboard cases, fireproof steel cabinets, photocopies of the originals from Jerusalem, adding up, once again, to roughly 80,000 documents—a kind of parallel universe to the Einstein Archives in Israel, securely housed in the basement of the building. The people who work here, most of them historians and physicists, are publishing Einstein's complete works as a set of heavy tomes with black covers and a somewhat old-fashioned layout. This unparalleled undertaking is likely to take forty or even fifty years to complete. No Einstein researcher or biographer could get around working with these volumes. Since their first publication more than twenty-five years ago, *The Collected Papers of Albert Einstein* have served as the richest source of what we know about the physicist. Part of the printed material comes from his estate. The other part, however—and this is what makes the people here especially proud—turned up in the course of their own investigations. They scatter in every direction, and follow up on every lead in archives and private collections around the world. And when these leads turn up previously unknown documents, "a historian's heart leaps for joy."

These are the words of A. J. Kox, a professor from Amsterdam, who comes here on a regular basis from the Netherlands for several months at a time. While poring through sources, Kox found a reference to a letter Einstein wrote to Adriaan D. Fokker, a colleague in Holland, which had previously gone unnoticed. Kox went downstairs to the inner sanctum, the archives in the basement. The electronic data bank had entries for seven letters from Einstein to Fokker, of which one was undated. Location: drawer 73, document 264. That had to be it. Holding the sought-after paper in his hands, Kox let out a cheer. "I knew it!" The letter was from 1919, written in Lucerne, Switzerland, and dated July 30. It was not a sensational find, but it provided an additional little piece of the immense Einstein puzzle. The letter reveals why Einstein did not visit Fokker when he was ill in Arosa, Switzerland. "I have to spend the couple of days I can spare here with my mother, who is a patient at the clinic."[3] Her cancer had entered its terminal phase. The letter is a classic example of Einstein's determination to dodge the "merely personal." "Life doesn't make it easy for anyone. But it is lucky when we are able to emerge from our own uncomfortable confines to some extent and focus on objective matters that are beyond the reach of the wretchedness of life."[4]

This document, however, also reveals something about "the German political mentality"[5] in this turbulent era, and about Einstein's view of the Germans: "These people have no clear notion that they have become the blind instruments of an arrogant and unscrupulous minority, which is why their professed indignation about a 'dictated peace' is not a hollow phrase or sham, but reflects their actual experience." At the beginning of the letter, he interjected this comment about science: "It does seem as though the masses of the world are virtually at rest in a statistical equilibrium in the proper frame of reference."[6]

Einstein's letter, encompassing reflections on his dying mother, the torment of personal experience, the question of guilt in World War I, and the physical masses of the universe, provides unique insight into his disposition in the months preceding his entrance into the hallowed halls of physics and before the eruption of the first serious political hostilities and the death of his mother.

Even with 80,000 documents at our disposal, it is difficult to form a balanced assessment. Most of these documents are *from* Einstein, but far fewer are *to* him, and there is much too little *about* him, particularly concerning the first half of his life, the period prior to 1919. A great deal of what was written *about* him after that date also suffers from distortion stemming from blind reverence (and not infrequently from hatred). The Einstein myth is bigger than Einstein himself. The legend distorts our view of the man. Moreover, with Einstein's sudden burst of fame in 1919, the *Collected Papers* project hit a brick wall. The sheer volume of material could no longer be dealt with in the previous manner. A torrent of mail and publications, as well as countless articles in newspapers and magazines, make the watershed in Einstein's life quite apparent. When Volume Nine was about to go to press, the team began to realize what an explosive situation it was facing. In view of this situation, Diana Kormos-Buchwald, the director of the project, convened an editorial meeting. Buchwald and Kox gathered around the conference table with an international group of scholars, which included a German, an Iranian, an Israeli, a Hungarian, and several Americans.

The director had no need to go into great detail about the situation. As the first person to occupy this post without long years of expertise in Einstein research, she has been subject to intense scrutiny in the scholarly community. The group concurred that the new edition would not cover the last two months of Einstein's life and would end on April 30, 1920. Would the publisher support that? Buchwald went back to her office and placed a call. The reply took quite a while. The door stayed closed for a long time, while Buchwald sat tight and smoked, which was not exactly a daily occurrence on a California campus. Finally the door opened, and smoke signals emerged. The publisher had agreed. The edition could go into the final revision. The team has learned to work in smaller increments. Volume Ten spans a mere few months, from May to December, 1920. If this tempo is retained in preparing future volumes, the project may entail hiring on several more generations of researchers.

Professor Buchwald invited her team to lunch at the Athenaeum, the venerable Caltech guesthouse situated only a few hundred yards from the team's half-timbered cottage. The group happily took advantage of the opportunity to have a look at the so-called Einstein suite on the second floor. The double room in which the heroic subject of their investigations always spent the night when he came to Pasadena was now unoccupied. Still, he appeared in their path as they made their way upstairs. An alcove on the stairwell holds a bust of Einstein by the sculptor Helaine Blum, showing him as the world often imagines him: pensive, earnest, and impenetrable.

The windows of the suite look out onto the parklike campus with its palms, orange groves, fountains, and columned walkways. Einstein liked this Californian Arcadia under an ever-blue sky. In 1931 he noted in his travel diary: "Pasadena is a giant garden with a rectangular grid of streets and villas in little gardens with palm trees, small-leaved oaks, pepper trees."[7] He added playfully, "On behalf of the latter all dishes are over-peppered."

The Einstein team posed for a group photo in the wood-paneled living room of the suite. The only one missing from the picture was Professor Robert Schulmann, the longtime director of the project, and in a way its soul for nearly a quarter of a century. Quick, clever, creative, and, when need be, aggressive—these are the characteristics that assured him his place in the world of Einstein. He is the truffle pig among the detectives. The academic world owes crucial insights, some of which are quite sensational, to his nose for news and his charm and tenacity. He lives five hours away from Pasadena by airplane, and a mouse click away in cyberspace.

In Bethesda, Maryland, a roomy apartment serves as the base of operations for the master detective on Einstein's trail. The Schulmanns' dinner table is redolent of Europe, complete with sauerkraut, meatballs, and rye bread. Judit, his wife, is from Hungary, and her cooking reflects her heritage. Exiles never seem to shed their heritage altogether.

Schulmann, who was born in the Philippines, speaks a soft Bavarian German with a hint of a Viennese accent. His parents grew up in Bavaria. He learned his clear, streetwise English in Los Angeles, which is where his family ended up when he was four years old. He became a professor of history, and in 1981 he joined the Einstein Project, six years before the first volume of the *Collected Papers* was published. His contributions were key to its sensational success, in large part because the documents he uncovered revealed quite a few dark sides and abysses in Einstein's life, which Einstein would have preferred to keep concealed for all time.

The trustees Einstein named, Helen Dukas and Otto Nathan, had pictured a very different outcome. Dukas came into Einstein's life on Friday, April 13, 1928. Einstein was in bed recovering from a serious illness and sought an assistant who could aid him during his convalescence. Dukas was unemployed at this time, but she would never be so again until the time of her death. Her future was decided on that fateful Friday the thirteenth; her destiny was Albert Einstein. Her new employer liked to introduce her as follows, "This is Miss Dukas, my faithful helper. Without her nobody would know that I was still alive, because she writes all my letters for me."[8]

Roger Highfield and Paul Carter describe Dukas as a "tall, slender, and outwardly austere young woman, whose diffidence concealed a tough and sometimes sardonic personality."[9] She became Einstein's secretary, and later also his housekeeper and cook. She accompanied his family to the United States, and after the death in 1936 of Einstein's second wife, Elsa, she took over Elsa's function as a Cerberus to shield him, the man of mythic dimensions, from his "natural enemies"— namely journalists. When she passed away at the age of eighty-seven, on February 10, 1982, her comrade-in-arms Otto Nathan announced to her "enemies" from the press: "Einstein died a second time when she died."[10]

Nathan was the second trustee. In the Weimar period he had been a prominent economist and government adviser, and then had fled from the Nazis. Having met the Einsteins shortly after their emigration, he became a family friend in Princeton, and their adviser in all financial matters. Also, he became a close ally of Helene Dukas, who was now known as Helen.

Schulmann said, "Nathan worshipped the ground Einstein walked on. And he was a real terrier when it came to protecting Einstein." Since Einstein was intent on casting out everything "merely personal" from his image for posterity, it was only natural for him to entrust his complete estate to Dukas and Nathan in his 1950 will. Schulmann feels that these two "regarded Einstein as God's gift to the world."

Nathan assumed his new role just a few hours after Einstein's death. He was on hand when Thomas Harvey opened the dead man's body and removed the brain. He let Harvey take possession of this organ, and stayed in contact with the pathologist for the rest of his life to ascertain whether research on Einstein's brain would yield successful results. There is no doubt as to how he pictured this success. "Nathan envisioned researchers presenting Einstein's brain as the super brain of a genius," Schulmann said.

Helen Dukas continued living in Einstein's house on Mercer Street in Princeton until her death. As the secretary of the immortal man, she kept on replying to letters to her professor after 1955. She also continued to receive indirect remuneration from him: Not only did he leave his personal possessions and an additional $20,000 to her, but he also stipulated that as long as she was alive, she was entitled to the income from his writings.

Their loyalty paid off. Nathan and Dukas made life difficult for anyone who tried to gain access to the approximately 42,000 items in the archives at that time. But even they understood that the floodgates would not hold indefinitely. Hence it was not surprising that important papers vanished the way incriminating material from secret service archives tends to vanish after a change of regime. It is uncertain how many documents were removed from Einstein's estate after his death. There is no doubt, however, that documents casting Einstein in an unfavorable light, at least in the opinion of his trustees, were eliminated. His will stipulated that his works—texts, documents, ideas, ideals, and view of the world—were to be his legacy. He would have been disheartened to see how his image was presented to the world. Even if the executors sincerely believed that they were acting in accordance with his wishes, the picture of Einstein they tried to paint for posterity bore little resemblance to the actual man.

More often than not, their approach came back to haunt them. The more unattainable the sources, the more dogged the investigators, the more spotty the information, the more pressing their questions. The greater the secrecy, the more suspicions and conjectures escalated.

It must have been in the 1950s that Gerald Holton, a young physicist at Harvard University, first came to visit Helen Dukas in Princeton. Initially he planned only to take a look at a couple of sources in preparation for a symposium. At this time the estate was split in half: many of the personal letters were still in Einstein's house on Mercer Street, while the other material, which was primarily scientific and political in nature, was housed at the Institute for Advanced Study, where Einstein had spent the last twenty-two years of his professional life. Holton found the elderly woman he had come to see on the campus of the Institute, in a large, walk-in vault in the basement of the main building. He recalled, "I was not prepared for what I found there."[11] Helen Dukas was sitting all by herself in the glow of a single lamp, surrounded by an array of tall filing cabinets, engrossed in her work. "The whole scene reminded me of Juliet in the crypt."[12]

The physicist discovered a hopeless jumble of papers in the cabinets. Helen Dukas was clearly in over her head with the task of transforming the estate that had been entrusted to her into an orderly archive. Now, with her permission, Holton and a few students took over, and the floodgates opened. The stern literary executor began handing over the writings that had been stashed away in the private residence to the Institute as well. How many documents never made their way there is one of the imponderables of Einstein research.

Finally, in 1971, the keepers of the Grail negotiated a contract with Princeton University Press that is still the basis for researchers today, but the ambitious project started up at a snail's pace. In due course, a suitable director was found; John Stachel of Boston University was appointed in 1976. Stachel, a theorist of relativity, moved to Princeton in 1977, but soon realized that Dukas would not readily cede control over her cache. "Access to the archives was very much on a personal basis, depending on how your relationship with her worked."[13]

No scientist can work like that. Stachel insisted on complete editorial freedom, but Otto Nathan refused to go along with Stachel's demand and would not sign a contract to this effect. A court of arbitration decided in favor of Nathan's adversaries. He did not give up, however, and fought the proceedings through every possible court. It took him until 1980 to exhaust all of his juridical options. The project could now begin.

Nathan had his back to the wall. In the fall of 1981, Robert Schulmann signed on in Princeton. The old economist now faced his most dogged opponent with this young historian, although the two had several points in common. Both had strong ties to their German roots. Both took a very dim view of the American government, each in his way. Both understood nothing whatsoever of physics. And both were obsessed with Einstein, but with opposite agendas: the one was shielding what the other was seeking.

Driven by inquisitiveness and a desire to set the historical record straight, Schulmann struck out in a new direction. He did not limit his search for clues to Einstein's estate. Instead, he set out for Europe, scouted obscure archives, crawled around in attics, sifted through forgotten papers, called on contemporary witnesses, got acquainted with descendants of Einstein's friends, conducted interviews, posed questions, and gradually acquired a knowledge base that kept yielding new leads.

In November 1985, in a suburb of Zurich, at an evening gathering at the home of the physicist Res Jost, the subject of Otto Nathan came up, and the discussion turned to one of his earliest deeds to protect the deceased genius. The host had spent nearly a decade working next door to Einstein in Princeton, and had collected money in Europe for the *Collected Papers* project. The meandering story that Schulmann would gradually piece together from Jost's account began during Einstein's lifetime and ended with a sensation.

Mileva, Einstein's first wife, died alone on August 4, 1948, in a clinic in Zurich, after a long period of suffering. As is so often the case in these instances, her daughter-in-law was saddled with the task of settling her estate. The daughter-in-law was Frieda, the wife of Einstein's son Hans Albert. When she entered Mileva's apartment on Huttenstrasse in Zurich, she had no idea what she was about to stumble upon. Her first discovery was a bundle of cash adding up to 85,000 Swiss francs, stashed under a mattress. Mileva had evidently saved this money for her son Eduard's medical care. Eduard was living out the sad life of a schizophrenic in a nearby mental hospital, in the care of ineffectual mental health professionals.

Frieda then discovered an impressive bundle of letters. Her mother-in-law had saved her correspondence with Albert, whom she had once loved so dearly, from the first years of their relationship. Frieda took the bundle and brought it to Berkeley, where she lived with her husband, who was a professor of hydraulics at the University of California. Nothing happened for nearly ten years, although the trustees of the estate appear to have been aware of the existence of the letters without knowing their contents. Right after Einstein's death, Otto Nathan attempted to find out what was in them from Einstein's son, Hans Albert, but to no avail. In early 1957, Nathan insisted on getting copies of the letters for the archives. Hans Albert stonewalled, and for good reason. His wife, Frieda, had decided to write a book portraying the human side of her father-in-law and to publish it with excerpts of the explosive letters. She made arrangements with Origo Verlag, a publisher in Zurich. The proceeds would go to caring for her ailing brother-in-law Eduard in the mental hospital.

Now Nathan and Dukas pulled out all the stops. They brought the matter before a Swiss court in 1958. An ugly lawsuit ensued. Hans Albert and Frieda's lawyers contended that the documents contained information about the family and could therefore be published by the family. The opposing side contended that the letters were part of Einstein's literary estate. Ultimately the judges sided with the trustees' argument. Frieda died in October 1958, before she had a chance to appeal the verdict. From that point on, the trail of the letters turned cold. Today, Einstein experts still have no idea what they contained, or even whether they still exist at all.

Twenty-seven years after Frieda's death, when Robert Schulmann was eating dinner at the Josts' home in Zurich, he sensed that "Einstein was in the air." The conversation turned to the mysterious correspondence, and Hilda Jost, the lady of the house, almost casually remarked that about six years earlier, in 1979, when the whole world was celebrating Einstein's hundredth birthday with festivities, conferences, and exhibitions, she had run into the daughter-in-law of Hans Albert and Frieda, who told her that the letters were still in the family's possession and were exquisite.

All kinds of ideas began swirling through Schulmann's head. Virtually nothing was known about the period of these letters, 1897 to 1903, during which Einstein experienced both personal and scientific major changes and achieved breakthroughs. Schulmann told the group how excited he was—how fervently he desired to get hold of these manuscripts, and what they would mean for Einstein research. However, he had not compiled enough evidence at this point.

As it happened, he was invited to the Josts' home for dinner once again just a few days later. This time Gina Zangger, the daughter of Einstein's longtime friend Heinrich Zangger, had been invited, too. Schulmann vividly recalled "a tough old lady who would sometimes dribble things out."[14] Gina, who had often supplied him with information in the past, needed to get something off her chest that night. "The psychology of these things," said the detective, "is that they want to tell you, but they don't want to tell you."[15]

Stories of this sort—involving relatives and their relatives, friends, and acquaintances—often entail major detours before getting to the point; a description may inch along like a flame along a fuse, and eventually there is a big explosion.

It turns out that in the 1950s, Gina Zangger had gone to high school at the same Swiss boarding school as Evelyn Einstein, the adopted daughter of Einstein's son, Hans Albert, and his wife, Frieda. It happened that Gina was a good friend of the boarding school director's wife, who in turn confided to Gina that Frieda and her husband had agreed to raise Evelyn only as a favor to Einstein. The old lady would not divulge any further details, and did not respond to Schulmann's intense probing. However, even in the absence of additional information, it did not take him long to put two and two together. The intimation could only mean that Evelyn was Einstein's biological daughter.

Schulmann later asked his host Jost in private what he thought of the whole story. The physicist confirmed that he had had wind of this rumor as well. Without revealing his source, he stated that Evelyn was the product of Einstein's affair with a New York dancer in 1940. Schulmann could hardly contain his excitement, not only at the prospect of having come across an illegitimate child of the physicist, but at the sudden realization that Evelyn Einstein in Berkeley might know the fate of the missing letters.

In January 1986, Schulmann called up Einstein's adopted granddaughter. They arranged to meet. He is one of those astonishing people to whom nearly anyone would feel moved to confide the story of his life on the spot, a cross between a perpetually friendly father confessor and an avuncular detective. Evelyn told him about her upbringing "as an unloved child in a dysfunctional family" and how her relatives gave her the cold shoulder after the death of her adoptive father, Hans Albert, in the summer of 1973.

When her money ran out at one point during that period, she called up Otto Nathan and asked him for financial assistance. After all, her adoptive grandfather's estate was bringing in millions in reprint rights and photo and advertising rights. Nathan refused her request. Schulmann described the problems with the *Collected Papers* to her, explaining that the trustees were to blame for the fact that not a single volume had been published in the thirty-one years since Einstein's death. Schulmann did not dare to broach the question of a possible paternity at this first meeting. He waited until a later visit that year to confide his secret to her. Evelyn was shocked. The longer she contemplated the idea, however, the more it began to make sense. Many questions to which she had been unable to find answers seemed to clear up in an instant. Little by little, she even warmed up to the idea of being Einstein's biological daughter.

She was not the first to be pegged by the rumor mill as an illegitimate child of the physicist. In the summer of 1993 the Czech physicist Ludek Zakel told *The New York Times* that he was born as Einstein's illegitimate son in Prague. In one instance, Einstein even hired a detective agency to look into a story of this sort, but the investigator came up empty-handed. Evidently the question of whether he might have laid "eggs on the side" was of great interest to his acquaintances as well. In 1932 he sent one of his poems on this subject to his friend János Plesch, a doctor:

> All my friends are hoaxing me
> —Help me stop the family!
> Reality's enough for me,
> I've borne it long and faithfully.
> And yet it would be nice to find
> That I'd possessed the strength of mind
> To lay eggs on the side—so long
> As others didn't take it wrong.
> A. *Einstein, Stepfather*[16]

Six years passed before there was further progress in the case of Evelyn. Robert Schulmann had let all speculation rest for the time being, saying, "As a historian, I need hard facts." Suddenly an opportunity arose to get precisely these "hard facts," complete with the seal of approval of exact science. The geneticist Charles Boyd at Rutgers University in New Brunswick, New Jersey, was planning a genetic evaluation of vascular diseases, and contemplated using the Einstein family for his study. This family had had a striking frequency of vascular illnesses that were fatal or contributed to the cause of death. Boyd's colleagues urged him to contact Schulmann, who would be able to recite the medical history of the Einsteins off the top of his head, and might possibly know where an important tissue sample for the genetic examination could be found: Einstein's brain. Schulmann was unable to help him, since he did not know where the brain was. But he pointed out that two of Einstein's grandchildren, Bernhard Caesar, the biological son of Hans Albert, and Evelyn, the adopted daughter, were still alive.

The moment Schulmann mentioned Einstein's brain and his adopted granddaughter Evelyn in the same breath, he saw his chance: Could the tissue be useful for something after all—for a paternity test? He said to Boyd that the surest way to find the organ would be to take into custody the man who had absconded with it after Einstein's death: Dr. Thomas Harvey.

Boyd managed to track down the pathologist in Lawrence, Kansas, in 1993. As a token of his generosity, Harvey gave him a chunk of brain tissue: the piece marked 47, from the rear area of the right frontal lobe. Evelyn Einstein, curious about her genetic heritage, had a dermatologist take a tissue sample from her and pass it along to Boyd.

The geneticist got down to work. Piece number 47 of Einstein's brain landed in a blender. From the mush Boyd extracted the genetic material and tinted it with a special dye. But the outcome of the examination was disappointing. The material was so denatured, and in such minute pieces, that a genetic analysis was out of the question. Thus, after more than seven years, this trail of Schulmann's investigations, which had been sparked by a rumor during a dinner in a suburb of Zurich, reached a dead end. Einstein's adopted granddaughter and presumptive daughter was living in a pitiable condition in an apartment near Berkeley. She was quite ill—liver disorder, lupus, cancer—and was barely able to get around without her electric wheelchair. The burden of the great name had become an even greater liability for her with the issue of the unresolved paternity. She no longer wished to discuss it: "How would knowing it help me?" In her family it was "an open secret," but the family had turned its back on her, and she did not offer any reasons why.

Schulmann threw in the towel. He did not believe any answer to the question of paternity would be found in letters that were lying sealed in the Einstein Archives in Jerusalem. Then he turned to the other rumor concerning the letters, a rumor on which Schulmann was far more fixated than he was on any possible corrective to Einstein's family tree. At the very first meeting with Evelyn, when the two were just beginning to build a mutual trust, he had asked her whether she knew anything about the whereabouts of the letters. No, the granddaughter replied dejectedly. What she did possess, however, was a manuscript in the handwriting of her adoptive mother, Frieda—the introduction to the book whose publication had been averted by the intervention of Otto Nathan in 1958.

Schulmann made it clear to Evelyn that he absolutely had to see it. She promised to have a copy sent to him. Even before he had left the city, he got a telephone call he would never forget. When Evelyn, back in her apartment, pulled the manuscript out of its plastic folder, she saw several pieces of paper behind it that she had never noticed before. And as far as she could tell, they were copies of letters. Schulmann dropped everything and set out for her apartment.

The moment he held the pages in his hands, this connoisseur was certain what Evelyn had unwittingly had in her possession all these years. He was looking at copies of letters that Einstein had written to his son Hans Albert between 1914 and 1955. Suddenly, however, his eyes lit on something far more startling: copies of the long-sought love letters between Albert and Mileva.

In a flash, Robert Schulmann figured out how to get at the originals as well. The staff of the Einstein Papers Project and the Hebrew University soon entered into negotiations with the trustees of Hans Albert's family. Otto Nathan had no say in this matter. On the evening of April 18, 1986, the thirty-first anniversary of Einstein's death, the goal was finally reached: The family lawyer personally removed the original letters from a vault and made two photocopies of every sheet of paper in the bank—one for the Hebrew University, and one for Schulmann's research team. The originals remained in the possession of the family. In 1986 they were auctioned at Christie's in Manhattan for $442,500.

In Boston, where the Einstein Papers Project had moved in 1984, several hectic months ensued. The correspondence was incorporated into the first volume at virtually the last possible moment, although its publication was held up again as a result. Hundreds of details needed to be clarified. Again Schulmann and his colleagues traveled to Europe to research details on particular events in the correspondence in libraries and newspaper archives. The ratio between the letters Einstein sent and those he received was quite lopsided. Only eleven of the fifty-four letters were in Mileva's handwriting, and the rest were in his. She had saved and preserved; he had lost and thrown away—often after using the reverse sides as scratch paper for calculations or notes.

In the course of his life, Einstein moved more than twenty-five times. In Switzerland alone, from 1895 to 1911, he changed addresses more than fifteen times. In Bern, Albert and Mileva called six addresses home, which kept their household possessions to a minimum. Even quite a few letters from the world-famous Max Planck to the then-unknown patent clerk fell victim to a constant need to declutter.

Otto Nathan died at the age of eighty-seven on January 27, 1987. He did not live long enough to witness the publication of the first volume of Albert Einstein's *Collected Papers*, just a few months later. He had held up its publication by many years. Now, thanks to the delay, Volume One contained the key to understanding Einstein's personality that the executor had wanted to keep concealed at all costs.

Suddenly the world got to know another Einstein, genius and lover, in his magnificence and his cruelty. And a secret was revealed that Albert and Mileva had both kept even after their separation and marital strife, and that each of them had taken to the grave.

SIX

"ELSA OR ILSE"

THE PHYSICIST AND THE WOMEN

He was a man of extremes, as the women in Einstein's life were well aware. At times he drew them in, and at others he pushed them away. Sometimes he burned with desire—"I miss the two little arms and the glowing little girl full of tenderness and kisses"[1]—and his wishes were fulfilled: "How delightful it was the last time, when I was allowed to press your dear little person to me the way nature created it."[2] He enjoyed wisecracking—"The only thing she is missing is someone to dominate her"[3]—and putting his foot down about any plans a woman might have to change him: "If I were the way you want me to be, I wouldn't be Albert Einstein."[4]

Einstein had what amounted to a split personality, even in his early years. On the one hand, he retained a childlike quality and innocence of thought—quite possibly a crucial prerequisite for his colossal scientific achievements. On the other, he was a young man with a healthy libido: ardent, dreamy, lascivious, playful, curious, and insatiable, a lover in the making. His love letters use the language of a young man-about-town in the first flush of his success in the arena of budding eroticism.

> Many, many thanks, sweetheart, for your charming little letter, which made me endlessly happy. It is so wonderful to be able to press to one's heart such a bit of paper which two so dear little eyes have lovingly beheld and on which the dainty little hands have charmingly glided back and forth. I was now made to realize to the fullest extent, my little angel, the meaning of home-

sickness and pining. But love brings much happiness—much more so than pining brings pain. . . . You mean more to my soul than the whole world did before.[5]

Now it's again your special turn, love. Know that you are kissed, hugged, and loved, the way your faithfulness deserves. I think about it so often each day. Dear Miezchen is now working hard again, but in the evening I think: now she thinks of me with love, and kisses her pillow in her bed. I know how it's done![6]

I now have someone of whom I can think with unalloyed pleasure and for whom I can live. Had I not already sensed it, your letter, which waited for me here, would have said it to me. We will have each other, which until now we have missed so terribly, and we will give each other the gift of equilibrium and a happy view on the world.[7]

Interestingly, these excerpts are addressed to three different women. The first letter, written in 1896, went to Marie Winteler, who was apparently Einstein's first flame. The third was sent in 1913, to his cousin and future second wife, Elsa Löwenthal. Only the second letter, dated 1901, was written to Mileva Marić, his lover during his student days, then his first wife and the mother of his children.

His early correspondence with Mileva shows the world the other Einstein, for better and for worse. Once the letters had been published, his love life sparked the imagination of the public even more than did his scientific achievements and political engagement. The dramatic structure of the revelations fuels our curiosity. We are riveted by his relationship with his cousin Elsa, which was published in Volumes Five and Eight of the *Collected Papers*. Readers react as they invariably do when the personal lives of prominent people enter the limelight: they relish indignation. They may protest that their hero is being demeaned, but they still demand to know the juicy details.

Einstein's correspondence with women invariably reveals that he regarded them almost as playthings. He simply had to have them, yet the moment he did, the curve of desire turned downward. He was evidently most tender from afar, in his own quirky manner: "I cannot wait to have you again, my all, my little beast, my street urchin, my little brat," he wrote to Mileva in September 1900.[8]

Much to the chagrin of Einstein researchers, there is precious little documentation of his affair with Marie Winteler. Later in life she stated that the two of them had shared a purely platonic love. Perhaps that was one of the reasons that the relationship remained just a passing incident in his life. Einstein was turning eighteen and saw that the time had come to sow his wild oats. He sought and found sex without marriage, or at least before marriage, with Mileva, a progressive woman, and later with Elsa. In the phase of early flirtation with Elsa, he boasted, "Let me categorically assure you that I consider myself a full-fledged male. Perhaps I will sometime have the opportunity to prove it to you."[9]

Marriage often signals the moment when the pinnacle of passion has peaked, and Einstein and his two wives were no exception. Einstein expert Robert Schulmann believes that in the case of Elsa, "it is even uncertain whether Einstein slept with her at all after 1920."[10] However, he most certainly found outlets for his sexuality in quite a few extramarital affairs after the major turning point in his life in 1919. The first indications of his sometimes ruthless behavior to his partners were evident in his relationship with Marie Winteler.

Albert and Marie met at her parents' house in Aarau, Switzerland. Jost and Pauline Winteler had accepted the schoolboy into their home like a son while he was finishing up his secondary-school graduation examinations in the cantonal school. He repaid their kindness by breaking their daughter's heart. For Marie, Albert was what she could never be for him: the love of a lifetime.

"You know quite well all that dwells and lives in my heart for you and you alone," wrote the young woman to the man she adored, in one of two extant letters, when the matter was evidently already finished for him, "and I could never describe, because there are no words for it, how blissful I feel ever since the dear soul of yours has come to inhabit my soul; all I can say is that I love you for all eternity, sweetheart, and may God preserve and protect you."[11]

The affair with Marie allowed Einstein to test out his erotic feelings. Once his "Schatzerl" had fulfilled this purpose, he dropped her without letting her know. She was permitted to keep doing his laundry—

which probably kept her hope alive. She was not allowed to visit him in Zurich, which was only twenty kilometers from Aarau.

He had now begun his study of physics at the university in Zurich, and had met Mileva Marić. This classmate, the only woman in the group, had everything that Marie could never offer him. As a free spirit she shared his youthful urge for a "vagrant life" in the blissful simplicity of the bohemian student world. She was more or less his intellectual equal. And like him, she was captivated by the foundations of physics. "Once I am back in Zurich," he wrote her in 1899, "we will start immediately with Helmholtz's electromagnetic theory of light."[12]

The first daughter of a well-to-do couple, Milos and Marija Marić, Mileva was born in late 1875, in the province of Vojvodina, in what is now Serbia. This territory still belonged to the Austro-Hungarian Empire at that time, and her father had learned his German while serving in the military. Mileva was brought up bilingual, and attended a German-language university. Her devotion to learning and outstanding achievement in school earned her the nickname "Saint." She was one of the first girls in the empire to obtain special permission to attend a traditionally all-male college preparatory secondary school. She passed her final examinations in 1894 with top grades in mathematics and physics. Her outstanding school record enabled her to continue her studies in Switzerland, which was liberal in matters of educating women. She completed her degree at the School for Young Ladies in Zurich, which entitled her to pursue a university education.

In the winter semester of 1896, she was the only woman to take up a study of physics in Section VI of the Eidgenössische Technische Hochschule (Swiss Federal Technical University, known by its initials, ETH), where heartthrob Albert Einstein was also enrolled. The two soon struck up a friendship. A self-assured letter that Mileva sent him in October 1897 from Heidelberg, where she was spending a semester as an auditor, provides insight into her personality:

> I do not believe that the structure of the human brain is to be blamed for the fact that man cannot grasp infinity; he certainly could do so if in his young days, when he is learning to perceive, the little man wouldn't be so cruelly confined to the earth, or

even to a nest between four walls, but would instead be allowed to walk out a little into the universe. Man is very capable of imagining infinite happiness, and he should be able to grasp the infinity of space.[13]

This first extant letter of their correspondence also indicates their sense of being kindred spirits. "How lucky I am to have found in you a creature who is my equal, who is as strong and independent as I am myself!"[14] he exulted. With her independence and what Robert Schulmann called her "healthy dose of cheekiness," she conquered the heart of the oddball who focused his longings on her.

"My only hope is you, my dear faithful soul," he wrote to her in the late summer of 1900. "Without the thought of you I would not want to live any longer in this sorry human crowd. But possessing you makes me proud & your love makes me happy. I will be doubly happy when I can press you to my heart again and see your loving eyes, which shine for me alone, and kiss your dear mouth, which trembles in bliss for me alone."[15] A year later he said, "I always find that I am in the best company when I am alone, except when I am with you."[16]

While their love was beginning to blossom, Einstein had his eye on another lover, one whom he would never rebuff: his science. He decided at this point that Mileva was his only suitable partner in love. In a remarkably clear instance of self-analysis, Einstein explained the change taking place within him to his jilted lover Marie's mother, whom he lovingly called "Mama":

It fills me with a peculiar kind of satisfaction that now I myself have to taste some of the pain that I brought upon the dear girl through my thoughtlessness and ignorance of her delicate nature. Strenuous intellectual work and looking at God's Nature are the reconciling, fortifying, yet relentlessly strict angels that shall lead me through all of life's troubles. If only I were able to give some of this to the good child! And yet, what a peculiar way this is to weather the storms of life—in many a lucid moment I appear to myself as an ostrich who buries his head in the desert sand so as not to perceive the danger. One creates a little world for oneself, as lamentably insignificant as it may be in comparison with the perpetually changing size of real existence.[17]

Marie probably meant a great deal to him at that time, as he did to her—primarily as a young woman who attracted and probably also aroused him when he pictured how her "dainty little hands glided sweetly" on the stationery. A noteworthy passage in an early letter to Mileva provides an indication of the emotions Marie had stirred in his heart: "You don't have to be afraid at all that I'll be going to Aarau very often now. Because the critical little daughter is coming home, with whom I fell so terribly in love 4 years ago. To be sure, I feel otherwise very safe in my high castle peace of mind. But if I saw the girl again a few times, I would surely get crazy again, I am aware of that & fear it like fire."[18]

His remark was apparently intended to comfort Mileva, but considering her sensitive nature, which coupled anxiety with jealousy, these words must have set off alarms. She was surely also startled by Einstein's brutal frankness, which she was hearing for the first time, but not for the last.

Nearly a year later, when the two of them were enjoying a greater degree of intimacy, and the polite *Sie* form of address had given way to the familiar *Du* and they called each other by the nicknames "Doxerl" or "Miezchen" (for Mileva) and "Johonesl" or "Johannzel" (for Albert), Mileva had an even ruder awakening. Einstein, who was vacationing in Melchtal with his sister and mother, did not mince words when describing to Mileva a "scene" his mother had made on her account.

We come home, I into Mama's room (just the two of us). First I have to tell her about the examination, then she asks me quite innocently: "So, what will become of Doxerl?" "My wife," I reply, equally innocently, but prepared for a real "scene," which ensued right away. Mama threw herself on the bed, buried her head in the pillow, and cried like a child. After she had recovered from the initial shock, she immediately switched to a desperate offensive. "You are ruining your future and blocking your path through life." "That woman cannot gain entrance to a decent family." "If she gets a child, you'll be in a pretty mess."[19]

Why did he go into such detail? Did he want to show her how gracious he was to want to marry her despite his mother's adamant opposition? Did he wish to torment her? Or did his merciless bluntness simply reflect his childlike nature, and he failed to realize how hurtful he was being? Einstein's stern mother had been so happy about his love for Marie that she had hinted at a maternal blessing for the union in a series of letters to Marie's mother. Now she did everything in her power to oppose his new relationship with Mileva: "This Miss Marić is causing me the bitterest hours of my life," she wrote to Pauline Winteler. "If it were in my power, I would make every possible effort to banish her from our horizon; I really dislike her."[20]

Albert conveyed his mother's feelings of aversion to his fiancée nearly verbatim. In relating "the scene" to Mileva, he emphasized his refutation of "the suspicion that we had been living in sin with all my might," but he did not spare her the harsh words of his "Mama's" additional diatribes, notably these stinging statements: "She is a book like you—but you ought to have a wife" and "When you're thirty, she'll be an old hag."[21]

Hermann and Pauline Einstein now added prophecy to their rebukes, and a parental curse began to hover over this blossoming relationship. Albert told Mileva,

My parents are very distressed about my love for you, Mama often cries bitterly and I am not given a single undisturbed moment here. My parents mourn for me almost as if I had died. Again and again they wail to me that I brought disaster upon myself by my promise to you, that they believe that you are not healthy. . . . oh Doxerl, this is enough to drive one crazy! You wouldn't believe how I suffer when I see how both of them love me and are so disconsolate as if I had committed the greatest crime and not done what my heart and my conscience irresistibly prompted me to do.[22]

He was suffering, and, under the guise of his parents' concern for him, reproached her for her physical disability, which had caused her so much trouble well before his harsh words. She was not ill, but she had had a limp since childhood, apparently as the result of tuberculosis of the bones. A friend once talked to Einstein about Mileva's limp and asked him how he could stand being with such an ugly person. He replied, "Why not? She has a lovely voice."[23]

Albert the vagabond had found a gypsy woman in Mileva, an exotic creature with a southern complexion—he called her "my little Negro girl" at one point[24]—and Slavic intensity. She was both a physicist and a "wild witch,"[25] a reticent, fiery lover whom, "in the evening and at night" he could "kiss . . . and squeeze . . . to my heart's desire."[26] Any criticism of Mileva set him off. His mother would have been better off trying to thwart their relationship by approval than by opposition. As it was, however, Mileva became a pawn in his lifelong struggle against any form of authority. She served as an unwitting means for him to break away from his domineering "Mama" by offering him the degree of motherliness he needed to satisfy his childlike nature. She kept house for him, although, compared with her bourgeois Swiss neighbors, she fell far short of the perfect housewife. She cooked and laundered for him, and brought as much order as she could to his perpetual clutter. He noted with satisfaction, "A good housewife is someone who finds a middle ground between a filthy swine and a neatnik."[27]

When all is said and done, she assumed a role in his life that few others achieved, at least in the early phase of their relationship. He let down his guard with this "only sweet little woman"[28] and made her his confidante. "Apart from you, all the people look so alien to me as if they were separated from me by an invisible wall."[29] Mileva was his only true companion. No one after her would understand even his darkest sides the way she did. In all his other relationships, including his later romantic liaisons and his second marriage to Elsa, he never again achieved the degree of intimacy, respect, and common ground that he experienced, at least for a while, with his "Miezchen."

The literature on Einstein has generally painted a very different picture of Mileva. We read about a difficult, ill-humored, melancholy, taciturn, depressive, pathologically jealous partner who spoiled the life of her witty, cheery, carefree, good-hearted contemporary. Quite possibly, however, it was just the other way around, and he was largely responsible for the fact that she gradually wasted away, lost her mettle, and became "gloomy, laconic, and distrustful."[30] In the early letters there are indications that he was anything but the affable, endearing man at her side. In September 1900 he wrote, "Now that I think of you, I just believe that I do not want to make you angry & tease you ever again, but want to be always like an angel! Oh, lovely illusion! But you love me, don't you, even if I am again the old scoundrel full of whims and devilries, and as moody as always!"[31]

Two years later he told her, "You can't imagine how tenderly I think of you whenever we're not together, even though I'm always such a mean fellow when I'm with you."[32] On another occasion, he begged her forgiveness: "It was only because of nervousness that I was so awful to you."[33]

Mileva knew what she was getting herself into with him. In any case, he harbored no illusions about his own behavior. When Julia Niggli, an old friend he had met in Aarau, asked him for advice regarding her relationship with an older man in 1899, he replied to her, "I know this sort of animal personally, from my own experience as I am one of them myself. Not too much should be expected from them, this I know quite exactly. Today we are sullen, tomorrow high-spirited, after tomorrow cold, then again irritated and half-sick of life—and so it goes—but I have almost forgotten the unfaithfulness & ingratitude & selfishness."[34]

Mileva cannot have failed to notice the uncompromising nature of his priorities. His other love, physics, was at the top of the hierarchy, and she felt increasingly excluded from that field. She wrote to her friend Helene Savić in 1900, "It is better this way for his career and I can't stand in its way; I love him too much for that, but only I know how much I suffer because of it."[35]

Soon there would be another cause for suffering. The love letters that Robert Schulmann dug up in 1986 reveal that Mileva got pregnant in about May 1901, traveled to Serbia to her parents in the late summer, gave birth to a daughter named Lieserl that winter, and returned to Switzerland without the illegitimate girl a year later. Einstein probably never got to see his daughter. The fate of the child remains a mystery.

Once the sensational news circulated, speculations ran wild. The American author Michele Zackheim examined markings that Mileva had made in a book about *The Sexual Question* and in a brochure about *Alcohol Poisoning and Degeneration* and surmised that Lieserl was handicapped. Zackheim further concluded that Einstein (who was a lifelong teetotaler) may have suffered from syphilis, citing as evidence the ongoing speculation that he regularly sought out prostitutes with Mileva's full knowledge. Hadn't Einstein's doctor in Berlin, János Plesch, announced after Einstein's death that the rupture of his enlarged abdominal aorta was a delayed effect of untreated syphilis? That is how it always works with Einstein: speculations evolve into anecdotes that are then proliferated in book-length studies like Michele Zackheim's *Einstein's Daughter*.[36]

Even so, it is remarkable that Einstein never saw his first child and did not visit his future wife after the birth. It would not have been impossible to do so. A train ride from Switzerland to Serbia took less than a day, and he had ample time to travel. Of course it would be too simple to place the blame for this failure on Einstein's contempt for women, nourished by his reading of Schopenhauer. He once disparagingly remarked to the wife of one of his colleagues, "Where you females are concerned, your production center is not situated in the brain."[37] On the subject of "women's suffrage" he wrote to his second son, Eduard, in 1928, "Only the manly ones fight for that."[38] Still, how might he have reacted if his firstborn had been a boy?

The fate of Lieserl piques our curiosity, and researchers have tried time and again to fill in the picture. Did Albert and Mileva give the child away? If so, there ought to be adoption papers, but no information has come to light. Did the child die of scarlet fever? If so, there ought to be a grave somewhere in Mileva's homeland. Robert Schulmann was one of the researchers who tried to track down every available clue, but Lieserl's fate remains elusive to this day.

Underlying all these efforts is the attempt to understand why Albert and Mileva's marriage later collapsed. The literature contains a single, indirect reference to the drama with Lieserl, which remained puzzling up to the time the love letters were discovered. It is found in the first major biography following Einstein's death. This book was published in 1962. Trustee Helen Dukas provided the author, Peter Michelmore, with only the sources that showed Einstein in a favorable light. He knew just as little about the broken marriage as about Einstein's existing relationship with his cousin Elsa.

Michelmore had the opportunity, however, to converse at length with Hans Albert, the couple's first son, who was born in 1904 and died in 1973. Hans Albert evidently made reference to a dark side of his parents' relationship, specifically to an "incident" that occurred before his birth. "Friends had noticed a change in Mileva's attitude and thought the romance might be doomed," noted the biographer. "Something had happened between the two, but Mileva would only say that it was 'intensely personal.' Whatever it was, she brooded about it and Albert seemed to be in some way responsible."[39]

What could he have been "responsible" for? Had he forced her to give up their child in 1902, or had he made her choose between Lieserl and himself? In any case, the modest beginnings of his professional career were at stake. An illegitimate child would probably have precluded him from getting his job at the patent office in Bern. Michelmore commented, "Friends encouraged Mileva to talk about her problem and to get it out in the open. She insisted that it was too personal and kept it a secret all her life."[40]

Why, then, did she save the letters, even the few in which Lieserl was mentioned? Was she resigned to the possibility that they would be found after her death and even published—or was she perhaps hoping that would be the case? If so, what was she hoping to achieve? A posthumous confession, or belated revenge against her famous ex-husband? In any case, she could hardly doubt that the explosive contents, if they were to be made known, would portray him, rather than her, in a bad light. He would be assigned the role of the perpetrator and she of the victim.

When Mileva agreed to marry Albert on January 6, 1903, after a five-year relationship, the fire had probably already gone out of their relationship as far as he was concerned. She, however, had maneuvered herself into a dependency on him that he wanted no part of. Shortly after the wedding, she wrote to her friend Helene, "I feel, if this is at all possible, even closer to my dear sweetheart than I did in Zurich. He is my only friend and companion, and I am happiest when he is with me."[41]

It did not take long for Albert to break the promise of their love and give up the private dream world they shared. He was still consoling her and writing, "I long after you every day, but I do not act that way because it's not 'manly.' "[42] However, the more time he spent with friends, fellow students, and colleagues, all of them men, the more he neglected his wife. When she gave birth to their son Hans Albert in 1904, her life suffered a major blow despite her delight in her son. She had been one of the few women to carve out a career in the natural sciences, but then, under the burden of a concealed pregnancy, she experienced the end of her self-liberation like so many before her and since: in the classic role of wife and mother, whose husband continued on with his life. "The pearls are given to the one, the other gets the case," she wrote resignedly to Helene. "I am very starved for love and would be so overjoyed to hear a yes that I almost believe wicked science is guilty, and I gladly accept the laughter over it."[43]

Her friend received this letter in 1909. "Wicked science," from which Mileva now felt excluded, had by then almost completely detached Albert from his wife. He now enjoyed a certain degree of renown, at least among experts. In the same year, 1909, he left the patent office and became a professor at the University of Zurich—the first step in his delayed academic career.

In this period there is a report of an episode in which Mileva was evidently unable to rein in her jealousy. Einstein had written a few tender but innocuous lines in the personal poetry album of a woman named Anna Schmid ten years earlier during a vacation with his mother and sister, and now this woman sent him congratulations on his appointment after reading about it in the local newspaper. Anna Schmid saved Einstein's poetry for the rest of her life.

> You girl small and fine
> What should I inscribe for you here?
> I could think of many a thing
> Including also a kiss
> On the tiny little mouth.
>
> If you're angry about it
> Do not start to cry
> The best punishment is—
> To give me one too.

This little greeting is
In remembrance of your rascally little friend.[44]

Einstein replied quite cordially to the woman, who was now married, and invited her to Zurich. Her response to this letter evidently wound up in Mileva's hands, and she suspected an affair. Incensed, she wrote to Anna's husband to protest this correspondence. Einstein had no choice but to intervene. He apologized to the man for his wife's conduct, explaining that it was "excusable only on account of extreme jealousy . . . to behave . . . the way she did." He promised that "I shall not do anything ever again that might disturb your happiness anew."[45] He could not bring himself to forgive Mileva for a long time. Even five months later he wrote to his friend Besso, "Mental balance lost because of M. not regained."[46]

Mileva was suffering the fate of innumerable other women. With her marriage already on the rocks, she discovered she was expecting another child. Their son Eduard was born in 1910. While Eduard was still a baby, Albert's career caused Mileva still more grief. Against her will, he accepted an appointment at the University of Prague. She lived in seclusion there, had no friends, and, in the sixteen months of her husband's professional intermezzo, never settled in, while he continued to distance himself from her. He left her alone for weeks at a time, traveling to lectures and conferences. In October 1911 he received a poignant letter from her. "I would have loved only too well to have listened a little, and to have seen all those fine people," she wrote. "It's been an eternity since we have seen each other; will you still recognize me?"[47]

Their son Hans Albert would later recall that around the time of his eighth birthday, in May 1912, all the signs of a crisis in his parents' marriage were evident. That was no coincidence. At this very time, Einstein had revived an old contact from his childhood—with his cousin Elsa, who was divorced and living in Berlin with her two daughters Margot and Ilse. The beginnings of this liaison remained a mystery well into the 1990s. That is a tribute, in the words of the authors Roger Highfield and Paul Carter, "both to his skill at covering his tracks and to the devotion he inspired in the people around him."[48] Again there were letters, this time from Albert to Elsa, which were preserved and gave posterity a belated record of the affair. Einstein's son-in-law Dimitri Marianoff, who would marry Einstein's stepdaughter Margot and live with her in the household on Haberlandstrasse beginning in 1930, evidently knew of them. In his biography of his father-in-law he wrote, "His letters to her, if ever published, will take their place among the love letters of the world."[49] That is certainly an exaggeration. But when the *Collected Papers* published the correspondence in 1993, there was ample confirmation that Einstein remained true to his womanizing pattern of behavior.

"I have to have someone to love, otherwise life is miserable," he wrote to Elsa on April 30, 1912. "And this someone is you; you cannot do anything about it, since I'm not asking you for permission. I am the absolute ruler in the netherworld of my imagination, or at least that is what I choose to think."[50]

He backed down once again and declared an end to their secret correspondence, but around the time of his thirty-fourth birthday, his passion flared up again. "What I wouldn't give to be able to spend a few days with you, but without . . . my cross!" he said in one of the extant letters.[51] In the meantime, his "cross," the once-beloved Mileva, was getting her hopes up once again: After the brief interval in Prague, the Einstein family was again in Zurich, to her great delight. Albert had been appointed Professor at the ETH, which was the former Polytechnic. Might it not be possible for them to revive their good old days together?

The return to the city of their young love was not enough to rescue their marriage. "He works on his problems tirelessly; you might say that he lives for them alone,"[52] a disappointed Mileva wrote to her friend Helene. Passages in a family friend's diary even suggest that Einstein was beating his wife. Reports from his older son, Hans Albert, attest to the fact that he was certainly capable of using physical force. Supposedly the divorce papers—which are under lock and key in Jerusalem—also make mention of violence in their marriage.

Albert promised his cousin Elsa—"itinerant folk that we both are"[53]—a freewheeling student lifestyle, just as he once had Mileva: "How nice it would be if one of these days we could share in managing a small bohemian household."[54] The opportunity would soon present itself. Einstein got a lucrative offer from the Prussian Academy in Berlin. The fact that Elsa lived there, and he could look forward to "our rendezvous in your room,"[55] made this decision quite a bit easier.

"I am living a very secluded and yet not lonely life, thanks to the loving care of my cousin, who has drawn me to Berlin in the first place, of course," he later confessed to his friend Heinrich Zangger.[56] He was already calling Elsa's daughters "the little stepchildren."[57] He was clearly contemplating the end of his marriage to Mileva. For her, the news that they would be moving again came as an unpleasant shock. She had just settled her family in Zurich, which she loved dearly, and now she had to face the prospect of leaving yet again. On top of that, she would have to live near his relatives in Berlin, who hated her so much. Einstein described his wife's wretched quandary to his new love with an almost sadistic relish:

> My wife whines incessantly to me about Berlin and her fear of the relatives. She feels persecuted and is afraid that the end of March will see her last peaceful days. Well, there is some truth in this. My mother is otherwise good-natured, but she is a really fiendish mother-in-law. When she stays with us the air is full of dynamite.[58]

Perhaps Elsa was pleased by these lines, or perhaps she felt sympathy for Mileva. Albert had just told her recently, "I shudder at the thought of seeing her and *you* together. She will writhe like a worm if she sees you even from afar!"[59] Oughtn't Elsa to have seen that as a warning of what she was letting herself in for? For now, her lover wrote, "I treat my wife as an employee whom I cannot fire. I have my own bedroom and avoid being alone with her."[60] Just ten years later, Elsa would have to endure a similar arrangement.

Did Mileva know that her marriage was already over by this point? Shortly before they moved to Germany, Einstein reported to Elsa that his wife "sniffed out some kind of danger"[61] in her. He acknowledged Mileva's desperation almost perfidiously: "Until now she practically never had any dealings with anyone else but me, poor devil."[62] In April 1914, Mileva and their two sons arrived in Berlin. Albert drew up a set of "conditions" she would need to adhere to if they were to live together. If she had had a shred of self-respect by this point, she could not have accepted them:

A. You make sure
   1. that my clothes and laundry are kept in good order and repair
   2. that I receive my three meals regularly *in my room*
   3. that my bedroom and office are always kept neat, in particular, that the desk is available *to me alone*
B. You renounce all personal relations with me as far as maintaining them is not absolutely required for social reasons. Specifically, you do without
   1. my sitting at home with you
   2. my going out or traveling together with you
C. In your relations with me you commit yourself explicitly to adhering to the following points:
   1. You are neither to expect intimacy from me nor to reproach me in any way.
   2. You must desist immediately from addressing me, if I request it.
   3. You must leave my bedroom or office immediately without protest if I so request.
D. You commit yourself not to disparage me either in word or in deed in front of my children.[63]

Mileva was so desperate that she acquiesced at first. However, he made matters still worse by telling her that he was interested only in their sons. Companionship between the two of them was out of the question, and if she did not stick to the businesslike conditions, he would split up with her on the spot. Eventually he suggested a formal separation and offered to let her keep the children and to send her 5,600 marks a year—nearly half of his income. Mileva had no choice. She threw in the towel. On July 29, 1914, she left—forever. Albert brought her to the Anhalt train station and kissed the boys good-bye. As they rode off, he cried like a baby. The next day he had their furniture packed and shipped to her. Nearly five years would pass until their divorce was finalized.

After Mileva's departure, he enjoyed, as he wrote to his friend Michele Besso, "the extremely agreeable, really fine relationship with my cousin, the permanent nature of which is guaranteed by a renunciation of marriage."[64] As a matter of fact, he had already promised Elsa this very thing—marriage. "We men are deplorable, dependent creatures, this I gladly admit to anyone," he confessed to his friend Besso in 1916. "But compared to these women, every one of us is a king; for he stands more or less on his own two feet, not constantly waiting for something outside of himself to cling on to. They, however, always wait for someone to come along to use them as he sees fit. If this does not happen, they simply fall to pieces."[65]

Albert regarded Elsa as the opposite of Mileva; he saw Elsa as a replica of his mother and a suitable companion—but not as a partner for a genuine marriage. The two of them had played together as children in Munich and shared the kind of lifelong intimacy that arises only in a sandbox. She spoke the same dialect as he, understood his ribald Swabian humor, had the same taste in food, and was corpulent like his mother. She had none of the frailty and exoticism of her predecessor, nor did she have any scientific ambitions. She was an open, strongwilled woman who loved being surrounded by people she could cook for and provide the form of comfort that is known as German *Gemütlichkeit*.

In her knowing simplicity, Elsa left her Albert to his science and did not even attempt to penetrate the mysteries of his thought. "Although she may get on my nerves at times and is no mental brainstorm," Einstein later conceded to his son Hans Albert, "she is exceptionally kindhearted."[66]

When she met him, she was a divorced mother and a professionally trained actress who was well provided for, but earning some extra money with voice instruction. Occasionally she would delight a goodsized audience with her art at public literary readings. She enjoyed some degree of repute in the better circles of Berlin and later introduced her husband to these groups. She went to great lengths to make him into the gentleman at her side, but achieved only moderate success.

As the years went by, the two of them came to resemble each other more and more. In some photographs they look like brother and sister.

"She had allowed herself to age prematurely, either through laziness or resignation," her friend Antonina Vallentin wrote later. "Her face had grown heavy, her hair gray before its time. She who reproached her husband for the carelessness of his appearance was equally open to reproach herself."[67] At some point even her signature began to resemble his. She stopped making a plain letter E in "Elsa" and added the same squiggle in the upper portion that Albert used to make the E in "Einstein."

Although she neglected her own appearance, as though vying with him for the more disheveled look, her vanity prevented her from hiding her beautiful blue eyes behind glasses. She was so nearsighted, however, that at a banquet she mistook the centerpiece for the main course, and heaped flowers onto her dinner plate. Although she could barely see anything six inches in front of her face, she was the one who cut Albert's hair, since he refused to go to the barber, and was therefore most likely the originator of his signature hairdo.

Being able to stand at the side of her famous husband in the public eye enabled this gregarious woman to tolerate the bitter reality of everyday life. "It fell to my mother," her daughter Margot recalled, "to look after Albert for all kinds of things—from his clandestine smoking, which he was not allowed to do, to his eating and sailing. If I may say so, he had stayed a child. I recall, for instance, that my mother often said during lunch, 'Albert, eat; don't dream!' "[68]

She took care of him, fed him, and shielded him from intrusive journalists and petitioners. Einstein preferred running around in the same old clothes day after day, so she dressed him and made sure that he was properly attired for public appearances. "She retired into his immense shadow, perfectly at ease in this shelter," said Antonina Vallentin. "Her sympathy tried to soften the effects of Einstein's intransigence."[69]

To a certain extent, she fulfilled the very "conditions" he had imposed on Mileva. Although she traveled with him, went out with him, and "sat with him at home,"—activities prohibited in the conditions he'd spelled out for his first wife—he had his own bedroom at the opposite end of the apartment. She was allowed to enter his attic study only with his permission. Once, when she was bringing a guest upstairs, she inquired after his health and asked whether his trip had been pleasant.

Her husband shouted at her, "You're disturbing me. You have no idea how you disturb me." When she included him in the plural pronoun *we*, as spouses generally do, he jumped out of his skin with anger. "Speak of you or me, but never of us."[70]

The "new Copernicus," as Max Planck once called Einstein, could be quite nasty. To add insult to injury, his personal habits did little to enhance their life together. "My husband's snoring is unbelievably loud," Elsa confided to a friend. "You cannot sleep next to him."[71] He did not wash and shave properly, and he had sweaty feet. Once, at the beginning of their relationship, when Elsa complained timidly about his poor hygiene, he wrote her that if he was not to her liking the way he was, she should go out and look for "a friend who is more palatable to female tastes."[72] He closed the letter as follows: "So, a foul profanity and a hand kiss from a hygienic distance from your really filthy Albert." Perhaps the idea of separate bedrooms was not so distasteful to her after all.

Shortly before that, he had taken devilish pleasure in asserting his independence: "Official report: The hairbrush is being regularly applied, other cleansing also relatively regular. Otherwise conduct so-so. Toothbrush retired off again for purely scientific considerations having to do with the dangerousness of hog bristles: Hog bristles bore through diamonds; how then should my teeth be able to withstand them?"[73] By the time they got married in 1919, shortly after his divorce from Mileva, the passion had died down—at least on his part. Elsa, whose dowry came to 100,000 marks, had to bear witness to his quest for the joys of love outside the confines of their household.

Opinions differ sharply as to Einstein's taste in women. János Plesch reported, "In the choice of his love-partners, he was not too discriminating, but was more drawn to the robust child of nature than to the subtle society woman."[74] Plesch's son put it even more bluntly: "Einstein loved women, and the commoner and sweatier and smellier they were, the better he liked them."[75] Son-in-law Dimitri Marianoff said, "I always felt that Einstein was attracted toward the ugliness of women only because of his large compassion."[76] Einstein's housekeeper Herta Waldow had a different recollection: "He liked beautiful women, and they in turn adored him."[77]

Some of his affairs during his years in Berlin suggest that Einstein was certainly not averse to "subtle society women." At any rate, he arranged to be picked up at home by rich, elegant women in their chauffeured limousines and be invited to concerts and the theater. Elsa, who did not know how to handle money, had to provide him with at least as much pocket money as he needed to be able to pay for his own cloakroom ticket.

He often met up with Estella Katzenellenbogen, a fashionable woman who owned a chain of flower shops in Berlin. He regularly spent the night with Toni Mendel, a well-to-do widow, in her villa at Wannsee. Herta Waldow reports that Toni indulged Elsa's legendary love of sweets by bringing her chocolates. The key source for details on this relationship has been lost, because Albert stipulated that her heirs were to destroy all of his letters to her after her death.

The correspondence with Betty Neumann, who served briefly as his secretary, is kept under lock and key in the Einstein Archives in Jerusalem. The historian Fritz Stern once got his hands on "the primitively wrapped (but closed) file" by accident. "Before I might have been tempted to tamper, the file was quickly snatched away from me."[78] According to Stern, Einstein's affair with Betty Neumann lasted more than a decade.

Once a week, Margarete Lebach came to Einstein's summerhouse in Caputh. She, too, attempted to appease the aggrieved wife with sweets, and brought Elsa home-baked vanilla pastries. Herta Waldow recalled, "The Austrian woman was younger than Frau Professor, and was very attractive, lively, and liked to laugh a lot just like the Professor."[79] The housekeeper reported that on the days of Lebach's visits, Elsa "left the field clear, so to speak": "When she came, Frau Professor would always go into Berlin to do some errands or other business. She always went off into the city early in the morning and only came back late in the evening."[80] Unconfirmed reports claim that Lebach also gave birth to an illegitimate child with Einstein.

When Elsa finally vented her anger to her daughters, Margot and Ilse told her that she had to make a choice between separating from "Father Albert" and accepting his affairs. Their mother decided tearfully to continue running her errands on the days Lebach visited. "Her love was indivisible," said Konrad Wachsmann, the architect of the summerhouse in Caputh and a friend of the family. "She simply did not understand that her husband was occasionally interested in other women."[81]

"He acted on women as a magnet acts on iron filings," reported Wachsmann. "But he also enjoyed being in the company of women and was captivated by everything feminine. In any case, the trouble between Albert and Elsa Einstein began with Elsa's fit of jealousy. . . . Einstein watched her behavior for a few days and eventually got angry, because he regarded behavior of this kind as childish. This was usually the point at which both of them discussed the issue of divorce."[82]

Marriage was "the unsuccessful attempt to make something lasting out of an incident," Einstein once said.[83] It was "slavery in a cultural garment."[84] When he was older, he decided that "Marriage was surely invented by an unimaginative pig."[85] Shortly before his death, he acknowledged that it was "an undertaking at which I failed pretty miserably—twice."[86]

When people examine the way Einstein dealt with women, they sometimes measure his character by a moral yardstick that would never be applied to comparable shining lights in literature, music, art, or politics. For Kennedy, Picasso, Mozart, or Brecht, amorality served more to enhance than to hurt their reputations. By contrast, Einstein seemed better suited to the image of a childish sage devoid of sexuality than that of a licentious skirt-chaser. As the Gandhi of natural science, or the Moses of modernity, he was thought to combine the purity of the prophet with the innocence of the pacifist. He himself reacted to this sort of moral imperative with pragmatic humor: "To a pure soul, everything is pure; to a pig, everything is piggish."[87]

Even before Elsa married him, she saw how cruelly her "old man" could treat her when he subjected her to one of the most humiliating ways in which a man could treat a woman. In May 1918 he presented her and her daughter Ilse with a horrendous choice.

This story did not come to light until a few years ago, and then only because the recipient of its intimate contents did not comply with the urgent wish of the writer—"Please destroy this letter immediately after reading it!"[88] The estate of Georg Nicolai, a professor of medicine, was found to contain a letter from Ilse Einstein, who evidently had a relationship with the local celebrity and bon vivant Nicolai.

Yesterday, the question was suddenly raised about whether A. wished to marry Mama or me. . . . Albert himself is refusing to take any decision; he is prepared to marry either me or Mama. I know that A. loves me very much, perhaps more than any other man ever will, he also told me so himself yesterday. . . . I have never wished or felt the least desire to be close to him physically. This is otherwise in his case—recently at least.—He himself even admitted to me how difficult it is for him to keep himself in check. . . . Help me![89]

There is no record of Nicolai's reply. Ilse, in any case, took herself out of the running. Her mother agreed to marry Albert a year later, despite this incident. He did not make any headway with Elsa's frail daughter, who was always dressed in the latest fashion and spent hours on end in beauty salons. However, he dropped hints to the effect that his fire had not been extinguished. In August 1919 he told Ilse she could become his secretary—but only if she gave up her part-time job as a laboratory assistant at the university: "Reason: preservation and possibly enhancement of your maidenly charms."[90] And in 1920, when he was preparing to leave on a trip to Norway, he wrote to his friend and colleague Fritz Haber, "I would take with me only one of the women, either Elsa or Ilse. The latter is more suitable because she is healthier and more practical."[91]

SEVEN

THE MIRACULOUS PATH
TO THE MIRACULOUS YEAR
EINSTEIN'S ANGELS

Did she or didn't she? Few other questions have brought Einstein research into the headlines as frequently as this one: Did Mileva lend her beloved Albert a helping hand and lead him along the path to success? Did she make significant contributions to his epoch-making studies in 1905? Might she have been the one to furnish the ideas that later made him the star of physics? Was she in fact the "mother of the theory of relativity," as the magazine *Emma* called her in an article that caused a sensation in October 1983? Did she merit a loftier status in his biography than that of a physically flawed impermanent partner with marriage certificate and children, whom he had to sacrifice at the holy altar of science? Or would he have achieved the same goal without her assistance, as a genius sui generis?

The *Emma* article is essentially based on a single source, a biography of Mileva by the Serbian writer Desanka Trbuhović-Gjurić. Describing the special theory of relativity, which was published in 1905, Trbuhovic-Gjurić commented: "We cannot help but be proud that our great Serbian woman Mileva Marić was part of its genesis and revision."[1]

Einstein reportedly told Mileva's father, "I have Mileva to thank for everything I have created and achieved. She is my brilliant inspiration, my guardian angel against transgressions in life and even more in science. Without her I would never have begun or completed my work."[2] In an oft-quoted letter from Einstein to Mileva on March 27, 1901, he declared to her, "How happy and proud I will be when the two of us together will have brought our work on the relative motion to a victorious conclusion!"[3]

This story was reported throughout the world, and scholars of Einstein set out to find evidence of Mileva's participation in Einstein's scientific discoveries. An investigation by John Stachel, the first director of the Einstein Papers Project, "leads to the conclusion that [Mileva] played a small but significant role in his early work."[4]

What might this role have been? His letters to her used "I" and "my" far more frequently than "we" or "our" when he was discussing science, but in late April 1901 he wrote, "I am quite curious whether our conservative molecular forces will hold good for gases as well."[5] Shortly after that, when describing a discussion he had had with Professor Gustav Weber of the Winterthur Technical College, he said, "I gave him our paper. If only we soon have the good fortune to continue pursuing this lovely path together."[6]

Quite possibly the end of their collaborative "lovely path" had already been reached by this point. Mileva's pregnancy marked the turning point in her scientific career. She may have pictured herself and Albert as the illustrious research couple, similar to the married Curies. Now she was becoming the mother not of the theory of relativity, but of Einstein's flesh-and-blood offspring.

What had preceded this turn of events? As any creative individual knows, inspiration is sparked not only by suggestions, ideas, and other concrete input, but also indirectly, when someone points up flaws and inconsistencies, for instance, or asks for clarifications, or simply engages in free association while brainstorming at the kitchen table. Mileva was Einstein's most important companion on his way to the top. Even though she was able to accompany him with the full complement of her talents only in the early stages, she may have helped show him the way to the celestial heights of his accomplishments. Einstein made no mention of her participation when later recounting the genesis of the theory of relativity, which should come as no surprise, since he had long since parted ways with Mileva by that time. Although he kept producing new versions of his theory to a ripe old age, Mileva's name never came up. Of course, hers was not the only name to be omitted; other key associates also went unmentioned.

As a man of the nineteenth century, inspired by the Romantic cult of genius of his childhood readings, which depicted the discoverer as a solitary hero, Einstein never saw himself as part of a team. He was happy to accept help from those who contributed their efforts, and he often felt indebted to them for life, yet he did not acknowledge them in his opus—as though these individuals had simply appeared before the genius in response to a law of nature.

The image of the solitary genius is nothing but a relic of idealized notions of the seemingly superhuman abilities of an individual, but it has little bearing on reality in complex systems of human interaction. Since each of our lives develops from a combination of gains and losses, the concept of genius really ought to be diverted from the individual to the system, according to psychologist Ursula Staudinger, who directs the Jacobs Center for Lifelong Learning and Development at the International University in Bremen. An individual like Einstein accordingly embodies a kind of creative melting pot in which the diverse elements combine to form an altogether new quality. "Genius" is ultimately a matter of using unerring instinct to recognize, absorb, and weld together to form a greater whole, integrating all the settings, ideas, and component parts encountered along the way. It begins as a raw diamond that many help to cut.

Viewed in this way, the question of Mileva's contribution to Albert's work forms only one facet of a far more comprehensive complex of the Einstein system when we investigate who contributed what, and where these contributions occurred. Before, during, and after her role in his life, a series of other individuals kept appearing like angels at exactly the right time and supporting his work. As much as Einstein treasured his solitude, it seems improbable that he would have become an object of international acclaim without the input of these people.

In addition, he had the good fortune to develop his talents in progressive cosmopolitan milieus. When the "prodigy"—he was only seventeen and a half at the time—enrolled at the Polytechnic in Zurich in October 1896, he was coming to the largest city in Switzerland: a banking metropolis, an international financial center, the backbone of Swiss industrialization. Engineers and natural scientists were in high demand. When Einstein arrived, there were approximately one thousand students—no more than the number of pupils in his high school in Munich. Five students enrolled in Section VI A of the training program for teachers of mathematics and science: Mileva, the only woman in the section, Jakob Ehrat, Louis Kollros, Marcel Grossmann, and Einstein. Grossmann became his first true friend and one of the most helpful "angels" in his life. "This Einstein," Marcel is said to have told his parents quite soon after the first time he had met Einstein, "will one day be a very great man."[7]

Albert lived in student housing in the middle-class Hottingen section of Zurich, just a short walk from the "Poly." He resided at three different addresses, as a tenant and in guesthouses. His wealthy relatives in Italy contributed one hundred francs a month for his living expenses. His father's company had gone bankrupt once again, and his parents were temporarily in a tight spot. "What depresses me most is, of course, the misfortune of my poor parents," he wrote to his sister Maja. "I am nothing but a burden to my family. . . . It would indeed be better if I were not alive at all."[8] However, since he was, in his own description, "a cheerful fellow and barring an upset stomach or something of that kind [had] no talent whatsoever for melancholic moods,"[9] his student years were relatively carefree, despite his budgetary restraints. He played music (his violin was always on hand), hiked, spent nights in smoky bars, and joined in the usual hijinks. Einstein was a typical student, apart from the fact that he never touched alcohol.

Just as Max Talmud had eaten at the Einsteins' home during his student years, the young Einstein was able to enjoy a weekly dinner at the table of the Fleischmann family, business partners of his relatives in Genoa. Acquaintances of his parents arranged for him to receive a warm welcome from the Stern family. Alfred Stern, a history professor who was well known far beyond the borders of Zurich, took Einstein under his wing "in fatherly friendliness." Einstein later expressed his gratitude in a letter to Stern: "more than once I went to your place in a sad or bitter mood and always regained there my joyfulness and inner equilibrium."[10]

Einstein was evidently bitterly disillusioned with his university studies. "For people of my kind of pensive interest, university studies are not necessarily beneficial," he wrote shortly before his death. "So I gradually learned to make my peace with a somewhat bad conscience." He soon resigned himself to being "a mediocre student."[11] Just as in high school, he was exasperated by the quality of the instruction and the course offerings. "It is, in fact, nothing short of a miracle," he mused in retrospect, "that the modern methods of instruction have not yet entirely strangled the holy curiosity of inquiry."[12]

He did not make matters easy on himself. He let the golden opportunity of studying with Professors Hermann Minkowski and Adolf Hurwitz, two of the preeminent professors of mathematics in all of Europe, slip through his fingers. "I saw that mathematics was split up into numerous specialties, each of which could easily absorb the short lifetime granted to us," he wrote much later in life. "Consequently I saw myself in the position of Buridan's ass, unable to decide upon any specific bundle of hay."[13] His carelessness would come back to haunt him. The very same Minkowski, who often found fault with Einstein's flippant attitude toward his studies, would turn out to be one of the younger scientist's angels. In 1908 he provided a mathematical formulation of the special theory of relativity, which went a long way toward smoothing the path to the general theory of relativity.

Marcel Grossmann was the first to toss his friend Albert an anchor. With final examinations in mathematics just around the corner, Einstein realized that he was ill prepared. Grossmann, who was one year older, lent him his own notebooks with meticulous course notes so that Einstein could cram. Einstein and Grossmann also prepared for their intermediate examinations together. Einstein earned the top grade, ahead of his friend. This same friend later helped him out of a jam when he was working on the general theory of relativity. Since the "Physics Lab for Beginners" was not to Einstein's liking, he often cut class, and was given a written reprimand; on one occasion he even got the lowest possible grade, but he was pleased to report to his girlfriend that physics professor Heinrich Friedrich Weber, who had broken the rules after Einstein failed the entrance exam by inviting him to his classes at the Polytechnic, displayed "great mastery" as a lecturer.[14]

His notes for Weber's introductory lecture series are among the very few surviving documents from this period. The marginalia reveal Albert's budding interest in forces between molecules. In Weber's well-equipped laboratory, he was "diligently and delightedly" given the opportunity for "hands-on experience," particularly in electrical experience,"[15] which tied in quite nicely with the practical knowledge he had amassed at his father's and uncle's factory.

Once he discovered that Weber was excluding the latest developments in the subject that was closest to Einstein's heart, the student began to " 'cut' quite often and study the masters of theoretical physics with a holy zeal at home."[16] Forced once again into learning the material on his own, he took a unique approach to the study of modern physics. Mileva was fully engaged in at least this aspect of his intellectual adventure, and physics offered them a wonderful smorgasbord of educational opportunities.

Einstein's readings included James Clerk Maxwell's and Heinrich Hertz's studies of electromagnetism, as well as newer inquiries by Ludwig Boltzmann and Hermann von Helmholtz. Boltzmann had derived the basic principles of the theory of heat from the statistical theory of mechanics, and Helmholtz provided a new underpinning for these principles with his fundamental laws of thermodynamics.

The latter's irrefutable proposition about the conservation of energy stood before him like a prototype. It stated "that all forces in nature that have hitherto been examined separately . . . can be measured in the same units, namely in units of energy; their sum total in the universe neither increases nor decreases."[17] Echoes of Einstein's boyhood readings are impossible to miss.

At the final examination in the spring of 1900, Albert placed fourth among the five candidates in Section VI A. On a grading scale of 1 to 6, with 6 as the best grade, he achieved an average of 4.91 and passed, but Mileva failed with a score of 4.0. This is where their professional paths diverged. Albert went his way with the innocent insouciance of a willful child. Mileva had to repeat the examination, and was held back. Einstein completed his official studies fairly effortlessly, in contrast to his private studies, which he continued to pursue at home in his easy chair or smoking a pipe in the coffeehouse. He also read epistemological treatises, focusing especially on the writings of the Austrian Ernst Mach. Yet another angel in his life had brought these books to his attention at just the right time—an engineer named Michele Besso, who was six years his senior.

Albert saw a fitting intellectual and personal counterpart to Marcel, a disciplined, intensely focused, outstanding mathematician, in Michele, who was an unsystematic, spontaneous thinker with flashes of brilliance. Marcel Grossmann, with whom he went "once a week like clockwork to Café Metropol am Limmatquai [was] not the kind of vagabond and eccentric I was."[18] By contrast, he regarded his friend Besso as a "Schlemihl," a cross between a walking disaster and a good-for-nothing. "He is an awful weakling without a spark of healthy humaneness, who cannot rouse himself to any action in life or scientific creation," Albert confided to Mileva, "but he has an extraordinarily fine mind, whose workings, though disorderly, I watch with great delight."[19]

Grossmann and Besso were a study in contrasts in their physical appearance as well. Grossmann had an elongated Nordic face with severe features, a broad forehead, and a long nose, while Besso was a southern type with dark curls and a full beard. Einstein's look was a little of each. During his student years his appearance matured from that of a juvenile fellow with a shy gaze and a hint of downy hair around his lips into that of a self-confident young man with strong features and the beginnings of the mustache that came to typify him as he stared jauntily into the camera.

Ernst Mach's ideas, which Einstein discussed at length with Besso, guided his philosophical thinking in the very phase in which he worked on the conceptual bases of science. The Viennese physicist and philosopher advocated "economic" scientific models that leave no room for unverified hypotheses and makeshift solutions. According to Mach, physical concepts such as speed, force, and energy need to be clearly linked with sensory knowledge. Einstein based his later epistemological credo on this principle: "I do not consider it right to conceal the logical independence of a fundamental concept from the sense experiment," he said. "The connection is not comparable to that of soup to beef, but more to that of cloakroom ticket to the overcoat."[20]

Mach contended that notions of absolute space and absolute time, which had gone virtually undisputed since Newton, were nothing but assertions lacking any basis in experience. No one has ever seen or heard the tick of a clock that sets the pace for terrestrial events—because it does not exist. The ominous ether, a makeshift solution for understanding light, is equally incapable of being observed or measured.

Mach also adamantly opposed the notion of atoms, the existence of which can be inferred only indirectly. No one had ever established any direct proof of their existence. On this point, however, Einstein held a different view. He was fascinated by ongoing discoveries of new types of rays, especially when they pertained to electricity, with which he had become so familiar in his childhood. Cathode rays, for instance, are created when electrical current makes an arc between two poles, a cathode and an anode.

In 1895 the Frenchman Jean-Baptiste Perrin was able to prove that cathode rays are made up of particles. In 1897 the Briton Joseph John Thomson described the tiny, negatively charged particles as "electrons," and in doing so unveiled the first building blocks of atoms—although they were not recognized as such. While experimenting with cathode rays in 1895, the German Wilhelm Conrad Röntgen happened to observe the energy-rich rays that later bore his name. These rays could even penetrate solids. In 1901 he received the first Nobel Prize.

The Frenchman Antoine-Henri Becquerel discovered the natural "uranium rays" in 1896, and his compatriots Marie and Pierre Curie became aware of the more general phenomenon of radioactivity two years later. In 1897 the Briton Ernest Rutherford provided proof that there were at least two kinds of rays, which he called alpha rays and beta rays.

Three years later, it was again Becquerel who established that beta rays were actually electrons, and the tiny charge-carriers turned out to be components of atoms, which emitted them. In light of these and similar discoveries, Einstein became an atomist early on. Not only did he develop the theory of relativity in 1905, his "miraculous year," but he also laid the foundations for an atomistic theory of matter in three lesser-known but equally significant studies.

The path to success was fairly rocky at first for the freshly minted graduate. Quite possibly the hurdles he faced were ultimately a blessing in disguise; life was giving him the space and time he needed to develop, although he had to cope with strained finances and professional insecurity. Instead of landing right on the career track, Einstein was forced to seek a roundabout route of entry to his chosen profession. Mileva was his mainstay during this period, although his parents were bitterly opposed to her presence in his life. Besso accepted a position as a technical expert for an electrotechnical firm in 1900, and he and his wife, Anna, left for Milan. In 1901, Grossmann became a teacher at the Thurgauer Cantonal School in Frauenfeld.

When Einstein returned from a visit to Milan, he learned that he was the only graduate in his department not to get an assistantship. His failure to secure a position was most likely the consequence of his lackadaisical and rebellious attitude while a student. When a university professor bristles every time his student violates basic etiquette by addressing him as "Herr Weber" instead of "Herr Professor," he is less likely to promote that student's job prospects.

Undeterred, Einstein pursued another route of entry to the scientific establishment: he published articles in scholarly journals. His first two publications reveal how uncertain he felt feeling his way through this new terrain, but at the same time they radiate a casual self-confidence. After all, he now had something to show for his efforts—although, as he later admitted, these were nothing more than "my two worthless first papers."[21]

"You can imagine how proud I am of my darling," Mileva wrote to her friend Helene. Mileva's ambition was starting to be wrapped up in Albert's. "This is not just an everyday paper, but a very significant one; it deals with the theory of liquids. We also sent a private copy to Boltzmann, and would like to know what he thinks about it. Let's hope he is going to write to us."[22]

There is no record of any reply from Ludwig Boltzmann. However, the very fact that the brand-new graduate and his girlfriend sent this work to one of the most important physicists shows that Einstein had a healthy dose of self-esteem. His cheery confidence in his own abilities again surfaced when he contemplated sending a private letter to the renowned Giessen physicist Paul Drude "to draw attention to his mistakes."[23]

Two months later, after following through on this idea, he received a reply from Drude that infuriated him. Einstein considered Drude's letter "such irrefutable evidence of its writer's wretchedness that no comment by me is necessary. From now on I'll not turn any longer to these kinds of people but rather will attack them mercilessly via journals, as they deserve."[24]

Authority and hierarchy were one and the same to Einstein when it came to truth and accuracy. He enjoyed quoting the German poet Ludwig Uhland's declaration that "the brave Swabian is not afraid."[25]

After a falling-out with Professor Weber during his first attempt at completing his doctorate, he approached Weber's colleague Alfred Kleiner. He appears, however, to have been unsuccessful with Kleiner, too, as indicated by a receipt for 230 francs Einstein received after withdrawing his thesis. "To think of all the obstacles that these old philistines put in the way of a person who is not of thcir ilk, it's really ghastly!" he fumed in a letter to his girlfriend. "This sort instinctively considers every intelligent young person as a danger to his frail dignity."[26]

He gave up his plan to complete his dissertation for the time being, "because it would be of little help to me, and the whole comedy has become boring."[27] Instead he sent out a pile of applications for an assistantship. "Soon I will have honored all physicists from the North Sea to the southern tip of Italy with my offer!" he commented to Mileva.[28] This venture did not bear fruit. Aside from a few confirmations of receipt, he did not receive a single reply.

He declared defiantly to Mileva in December 1901, "We shall remain students (*horrible dictu*) as long as we live, and shall not give a damn about the world."[29] Since "once again, several job hunts of mine were not showing any progress," he suspected that his old professor was at the root of his lack of success.[30] "I could have found one long ago had it not been for Weber's underhandedness," he conjectured in a letter to Grossmann. "All the same, I leave no stone unturned and do not give up my sense of humor."[31]

Even Hermann Einstein, who was quite worried about his son's future, intervened in the job hunt. "Please forgive a father who is so bold as to turn to you, esteemed Herr Professor, in the interest of his son," he wrote to the famous physicist Wilhelm Ostwald in Leipzig. "My son feels profoundly unhappy with his present lack of position, and his idea that he has gone off the tracks with his career & is now out of touch gets more and more entrenched each day."[32]

His wealthy relatives had stopped funding him once he had his teaching diploma in hand, and his parents were unable to offer him money to live on; consequently, the young man gradually began to realize that he was in dire financial straits. When Mileva was pregnant, he reached an "irrevocable decision": "I will look *immediately* for a position, no matter how humble," he told her. "The moment I have obtained such a position I'll marry you and take you to me."[33]

After at least one abortive attempt of which we are aware—"we regret to inform you that you were not chosen for this position"—he took refuge for a few weeks as a substitute teacher in Winterthur. In early September 1901 he was finally offered a position as private tutor in Schaffhausen, preparing a young English pupil for his final examinations under the aegis of mathematics teacher Jakob Nüesch. He got along quite well with the pupil, but soon had a falling-out with his host family. When he demanded to be allowed to eat his meals elsewhere and have them paid for by the employer because the dinner-table talk at home bored him to tears, the situation turned ugly. He wrote to Mileva on December 12 that he had lashed out at Nüesch, declaring to him, "Very well, as you like, I have to give in for the time being. I'll know how to find living conditions that suit me better." Einstein added parenthetically to Mileva, "Imagine what nerve, in my position."[34]

Einstein's pigheadedness prevailed and he "attained [his] goal" after irritating Nüesch no end. "They are now foaming at their mouths with rage against me, but I am now as free as the next man."[35] In any case, the whole arrangement fell apart, and on February 4, 1902, he wrote to his new friend Conrad Habicht, whom he had met in Schaffhausen, "I sailed away from N[üesch] with spectacular effect."[36]

This "sailing away" turned out to be far less reckless than it may have appeared at first. Habicht received the card from Bern, where Marcel Grossmann had been working behind the scenes to pull strings for his "old luckless friend"[37] and had described Einstein's plight to his own father. His father was a friend of Friedrich Haller, the director of the Swiss patent office in Bern, and recommended the young teacher for the next available opening at the Federal Swiss Bureau of Intellectual Property. Einstein assumed that he would get the job, and moved to Bern to look for an apartment.

In December 1901 he applied for the post. He had already endured the drawn-out process of attaining the requisite Swiss citizenship on February 21 of that year. The Zurich city council had gone to great lengths to report on his comings and goings, even hiring a private detective to tail him. The detective concluded that Einstein was a "very eager, industrious, and very solid man," and added that he was a "teetotaler."[38] Eventually Albert appeared before the city council, which accepted him as a new citizen without further ado. He was careful not to reveal anything to the Swiss authorities about his affair with Mileva and the resultant pregnancy. His anxieties about being required to serve in the Swiss military turned out to be unfounded, since his varicose veins and flat and sweaty feet rendered him unfit for military service. Mileva, who failed her exams for the final time in July 1901, had been living with her parents in Neusatz, Serbia, since the end of the year. There she gave birth to their daughter "Lieserl" in January 1902, after a pregnancy that was quite difficult in its final stages.

The young papa promptly asked, "Is she healthy and does she already cry properly? What kind of little eyes does she have? Whom of us two does she resemble more?" His combination of glee and childlike curiosity prompted him to remark, "I would like once to produce a Lieserl myself; it must be so interesting! She certainly can cry already, but she'll learn to laugh much later."[39]

Right after he arrived in Bern, an "ancient, exquisitely cozy city, in which one can live exactly as in Zurich," he "saw to it that an advertisement will be published in the local gazette."[40] The notice, which would have a decided impact on the new chapter of his life, was printed the next day:

Private Lessons in
MATHEMATICS AND PHYSICS
for students and pupils
given most thoroughly by
ALBERT EINSTEIN, holder of the fed.
polyt. teacher's diploma,
GERECHTIGKEITSGASSE 32, 1st FLOOR.
Trial lessons free.[41]

It was desperation, pure and simple, that had led him to post this announcement, but the outcome was most favorable, since the response to his advertisement brought him his next set of angels. The first individual to indicate his interest in studying with Einstein was Maurice Solovine, a Jew from Romania. Solovine was a tall man, four years older than Einstein, whose small eyes peered out morosely from under his high forehead. A student at the University of Bern, majoring in physics and philosophy, Solovine was looking for someone with whom he could discuss fundamental questions that encompassed these two subjects, such as whether there is a real world, and if so, whether we are capable of ascertaining that.

Shortly thereafter, Conrad Habicht joined the duo. Habicht, an astute Swiss three years older than Einstein, looked the part of the young intellectual more than the other two, with his slight build, thin-rimmed glasses, and slicked-back hair. The three mustached men abandoned the tutoring arrangement that had prompted the formation of this group (and would have ensured Einstein a modest income), and opted instead to form a discussion group that met on a regular basis. They boisterously dubbed their new group the "Olympia Academy." It is difficult to determine Solovine's and Habicht's roles in these discussions. The two may have served merely as sounding boards for Einstein's ideas, but it is also possible that they were active in contributing ideas of their own.

The trio spent long evening meetings huddled earnestly over "bologna sausage, Swiss cheese, fruit, some honey, and one or two cups of tea," or over mocha and hard-boiled eggs, with a great deal of "merriment." Often their discussions spilled over to the next day at Café Bollwerk, a dark, smoky grotto of a place. These discussions complemented Einstein's evolving view of the world, a view that would begin to topple traditional notions of space, time, and matter in 1905. He could not have found a better extension of his boyhood readings.

Habicht had a pewter platter engraved for Einstein, with the playful inscription "Albert Knight of Steissbein, President of the Olympia Academy." The trio's impressive array of readings included books by Ernst Mach—*Analysis of the Sensations* and *The Development of Mechanics*—as well as works that zealously opposed any form of metaphysical absolutism in religion or magic according to the positivism at the close of the late nineteenth century. These works saw the "positive" in certainty and indisputable facts—in experience and empiricism.

Karl Pearson's *Grammar of Science* and Richard Avenarius's *Critique of Pure Experience* enhanced Einstein's understanding of the world that can be experienced empirically. The Olympians seem to have found the most profound exploration of scientific methods in John Stuart Mill's *System of Logic*—from measurement and experimentation to theories of the most universal natural laws.

The works of Ernst Mach, David Hume, and Henri Poincaré formed the core of the Academy's reading list. Hume questioned the principle of cause and effect, and the general validity of the law of causality: "That the sun will not rise tomorrow is no less intelligible a proposition, and implies no more contradiction, than the affirmation that it will rise."[42] According to Hume, the certainty that it will rise is based solely on experience. It will happen that way because it always has. The law of causality applied, claimed Hume, solely within one's own realm of experience—in others, things could look quite different. The only certainty lay in mathematical relationships. The fact that three times four equals twelve is ultimately based on a stipulation that can be framed differently, but the fact that within this stipulation three times four yields the same as two times six is an indisputable logical consequence. Hence, Hume regarded combining experience with empirically verified causes as the loftiest goal of human knowledge. This idea goes to the core of Einstein's theory of relativity, which makes no statements about what space and time are but does describe how they behave in relation to each other.

The trio discovered key ideas on this issue in the work of the French mathematician and philosopher Henri Poincaré, particularly in his *Science and Hypothesis*. In 1900, Poincaré summarized his findings at a major philosophical conference after announcing, "There is no absolute space, and we only conceive of relative motion."[43] He was just a step away from his breakthrough, and he went as far to call even space and time ways of looking at the world—as opposed to Kant's given framework of things and events. "There is no absolute time," he declared. "When we say that two periods are equal, the statement has no meaning, and can only acquire a meaning by a convention. . . . Not only have we no direct intuition of the equality of two periods, but we have not even direct intuition of the simultaneity of two events occurring in two different places."[44] Einstein could scarcely have put it better.

Habicht, Solovine, and Einstein grappled with the most significant epistemological problems of their era, but they also made sure to include world literature in their readings, notably Sophocles' *Antigone* and Cervantes' *Don Quixote*. The latter was one of Einstein's favorite books. Plato's *Dialogues* and Baruch Spinoza's *Ethics* were also discussed; the *Ethics* became a key text in Einstein's later reflections on "cosmic religion."

Was any other physicist of his era better equipped to reach the top? Einstein was young and not set in his ways. He had a comprehensive knowledge of physics and philosophy. He had developed a practical outlook in the family business and in the laboratory at school and college. And he possessed a combination of egocentrism, instinct, and intuition that enabled him to find his way. In addition, he had ample training in scientific and technical inventiveness. In May 1902 Einstein was interviewed at the patent office, and in late June, thanks to Grossmann, he began his job as "technical expert third class." The staff consisted of twenty-nine clerks, of whom thirteen, including Einstein, were technical experts. His annual salary was 3,500 francs. His workweek came to forty-eight hours, eight hours a day for six days a week. Work began at 8:00 a.m. He generally wore his checked suit, which he had had tailored at the request of Friedrich Haller, the head of the patent office.

The visual, critical, and analytical thinking entailed in work with patent applications served to sharpen the mind of a scientist aiming to approach theory with very realistic thought experiments. Einstein's only free time to pursue scientific work was in the evenings and on weekends. The office and the financial freedom it afforded turned out to be a blessing—in his own words "a kind of lifesaver."[45] His academic failure was ultimately transformed into the secret of his success.

Einstein continued to instruct a private student after work. Lucien Chavan came from western Switzerland and was eleven years older than Einstein. After Chavan's death in 1942, a photograph and a precise description of his teacher were found in his notebooks: "Einstein is 5 ft. 9, broad shoulders and a slight stoop. Unusually broad short skull. Complexion is matte light brown. A garish black mustache sprouts above his large sensual mouth. Nose rather aquiline, and soft deep dark brown eyes. The voice is compelling, vibrant like the tone of a cello."[46]

And Mileva? She continued living with their daughter at her parents' home. She remained apart from Albert for a total of over a year. Not only did her professional career fall apart for good, but the scientific and intellectual partnership with her boyfriend crumbled away as well. When she returned to Switzerland shortly before the end of 1902, her world had changed completely. Albert had been far away when she had had to leave her child, and as if that were not heartbreaking enough for her, she now realized that she had been ousted from her partner's intellectual life as well. She listened attentively to the Olympia Academy sessions in their home, but usually in pensive silence, and quite possibly with growing resentment.

Einstein was true to his word. He married Mileva on January 6, 1903, a few days after her return. Conrad Habicht and Maurice Solovine served as witnesses at the ceremony. Later he would explain his decision to take this step as a "sense of duty" in the face of "inner resistance." In October he had rushed to Milan to be at the bedside of his dying father, and got there just in time for Hermann Einstein to consent to his son's marriage before succumbing to his illness.

"When the end was near, Hermann asked everyone to leave so that he could die alone," Helen Dukas told Einstein's biographer Abraham Pais. "It was a moment his son never recalled without feelings of guilt."[47]

Once they had tied the knot, the couple settled into traditional roles, and dreams of working together faded away. "Well, now I am a married man and am living a very pleasant, cozy life with my wife," Einstein reported to Besso just a few days after the wedding. "She takes excellent care of everything, cooks well, and is always cheerful."[48] Would Mileva have concurred with this description of their life together?

When she traveled back to her Serbian homeland in the summer of 1903 to see to her daughter, she was carrying a new child in her belly. Albert advised her to "brood very carefully so that something good will hatch out." He also inquired, "How is Lieserl registered? We must take great care, lest difficulties arise for the child in the future."[49] Since the fate of the little girl is unknown, this may be read as an indication that she was given up for adoption. On May 14, 1904, at Kramgasse 49 in Bern, after enduring a second difficult pregnancy, Mileva gave birth to their son, Hans Albert.

In November 1903 the couple moved to a little apartment in the Old City in Bern, just a few steps away from the famous clock tower. Today this apartment, contrary to Einstein's explicit wish not to create any holy places in his memory after his death, serves as a small museum in his honor. Day in and day out, tourists from around the world stream to the house like pilgrims. Once they have ascended the narrow staircase from the shopping area and registered in the guest book, they snap photographs of one another in front of the desk from the patent office in Bern or in front of the original high desk. Even though space and time are relative, the genius committed his theory of relativity to paper within these absolute four walls in the real spring of 1905. Einstein stood at this high desk in his stylish glen plaid suit to pose for a photograph that was sent to countless recipients.

When Conrad Habicht left Bern after his graduation in the summer of 1904, the lovely tradition of the "academic sessions of our laudable Academy" quickly petered out. At virtually the same time, Michele Besso returned to the city and followed Einstein's advice by beginning work as an "expert second class" at the same patent office where his friend had been working for the past two years. Now the two of them could be seen every day walking side by side to and from work, deep in conversation, through the arcade-lined streets of the Old City in Bern. The final building blocks of the theory of relativity came together during these discussions. Besso was the only one of Einstein's many helpers whom he would single out to thank at the end of the publication. Aside from that acknowledgment, he made no reference to sources, much to the dismay of his professional colleagues.

Every now and then he toyed with the bold idea of taking advantage of a special provision that allowed outstanding researchers to get a post-doctoral degree at the University of Bern without have attained the doctorate. However, the two essays he had already published in the *Annalen der Physik* were not deemed sufficient to fulfill the requirement. Once again his attempt to make headway in his academic career met with failure. The next three essays that he published in the journal, however, show marked scientific growth.

It can no longer be determined whether Mileva participated in this phase of his development as a scientist, but there is no indication that she did. In all likelihood, her contribution was to free him up as much as possible and to take care of the household and their child. In November 1904 the news circulated throughout the world that that year's Nobel Prize for Physics would go to Pierre and Marie Curie. This news must have cut Mileva to the quick, especially because it was the first time a woman had been awarded the highest distinction in the natural sciences.

Before the year was out, Einstein had been appointed a freelancer at the *Beiblätter zu den Annalen der Physik*, which reviewed books and articles from other professional journals. Writing for this journal enabled him to develop his writing style and bring himself up to date on the latest research, since he had had little access to scientific literature. Of the twenty-three texts he published in the *Beiblätter*, twenty-one appeared in 1905, which is known as his "miraculous year."

Of course, the year does not bear this name because of the reviews, which he submitted almost as an afterthought in a state of unparalleled productivity, but because of his series of five publications in the main publication, the *Annalen der Physik*. The field of physics would never be the same. In late May 1905 he announced the completion of four of the five studies in a letter to Conrad Habicht. This letter documents a unique moment in the history of science:

Dear Habicht!

Such a solemn air of silence has descended between us that I almost feel as if I am committing a sacrilege when I break it now with some inconsequential babble. But is this not always the fate of the exalted ones of this world?

So, what are you up to, you frozen whale, you smoked, dried, canned piece of soul, or whatever else I would like to hurl at your head, filled as I am with 70% anger and 30% pity! You have only the latter 30% to thank for my not having sent you a can full of minced onions and garlic after you so cravenly did not show up on Easter. But why have you still not sent me your dissertation? Don't you know that I am one of the 1 1/2 fellows who would read it with interest and pleasure, you wretched man? I promise you four papers in return. . . .

The fourth paper is only a rough draft at this point, and is an electrodynamics of moving bodies, which employs a modification of the theory of space and time; the purely kinematic part of this paper will surely interest you.[50]

This was the special theory of relativity.

EIGHT

SQUARING THE LIGHT

WHY EINSTEIN HAD TO DISCOVER
THE THEORY OF RELATIVITY

That night in May 1905 may have been the most important night of his life. The exact date is unknown, and other details are equally sketchy. There is no documentation, and there are no witnesses, only hearsay before and after the fact. A member of the audience took notes, in late 1922 in Kyoto when Einstein, who was by now internationally famous, gave a lecture in German in which he described the hours leading up to his moment of insight. This is our only source.

Einstein recalled that it had been a wonderful day. He had spent the afternoon with his friend and patent office colleague Michele Besso, discussing, as usual, the key issue in physics, namely, how to resolve the contradictions that had shaken the foundations of the prevailing physical view of the world in recent years. All the eminent authorities had failed to find a solution—now Einstein was capitulating as well: "I give up," he told his friend.

Then came the night. How did he spend it? He surely indulged in his old habit of smoking pipes and cigars while scribbling on countless scraps of paper, the back sides of letters, bills, anything. The main thing was for ideas to find their way to paper. Was he able to sleep, or did the excitement keep him awake? Did his one-year-old son Hans Albert cry? Did he share his ideas with Mileva? Did she help him out with his calculations? Or did he work alone, and ask her to leave his meal by the door?

Something must have happened during his conversation with Besso. Perhaps his friend had given him just the pointer he was looking for, or had simply posed the right question at the right time. "Suddenly I understood where to find the key to this problem," said Einstein, a good seventeen years later, in Japan.

The next morning, when he met up with Besso again, he called out the great news to him "before I had even said hello": "Thank you; I've completely solved my problem." Several years later this solution acquired a name: the special theory of relativity. No one knows how Einstein came upon this theory, which is as closely associated with his name as evolution with Darwin and psychoanalysis with Freud. We have no indication how the inspiration came to him, which synapses fired in his brain, what logical steps were taken, what images flowed together, or how he experienced this triumph after years of rumination. In typical Einstein fashion, as soon as one puzzle was solved, a new one took its place.

We can, however, reconstruct what happened beforehand: what he knew, what he was reacting to, what drove him, why Einstein had to discover the theory of relativity. Luck was his constant companion. In retrospect, it appears that the patent clerk was the right person in the right place at the right time.

The roots of his scientific revolution ran deep. All the components of his life experience seemed to come together to form a perfect unity. His boyhood readings, his independent studies, his physicist spouse, the Olympia Academy, the conversations with Besso. And the time was ripe for his discovery, which unified space and time into a four-dimensional spacetime construct.

As a schoolboy, Albert had learned, by reading Alexander von Humboldt's *Cosmos*, "that we explore both space and time with our big telescopes."[1] Felix Eberty's *Stars and the Earth* explained, "Thus we have here the extension of Time, which corresponds with that of Space, brought so near to our sensible perception, that time and space cannot be considered as at all different from one another."[2] Both Humboldt and Eberty were addressing a topic of heated debate at the time of Einstein's youth: time as the fourth dimension, closely linked to the three dimensions of space. It was now established that they could be viewed as one collective unit. This idea was a key element in understanding the theory of relativity, although the available evidence indicates that Einstein came upon it in a very different way. He first needed to work

his way through a maze of theoretical contradictions. Today we can envision the labyrinth from above and picture Einstein taking one tenta-
tive step after another in the plan he had mapped out. If he had one
steady guide to lead him in his odyssey, it was light, the subject that fascinated him throughout his life.

What do we see when "we explore both space and time with our big
telescopes"? We see light—the light that comes to us from distant
sources. It has taken quite a while to get here. Light needs time to cover
distance, whether it goes from the lines of this book to the retinas of our
eyes or comes all the way from a faraway star: on the path from the
sender to the recipient, from the source to the goal, light covers distance through space, for which it requires time.

Light acts like any traveler who moves at a constant rate of speed:
the farther it needs to go, the longer it takes to get there, and the more
time elapses. Its velocity is almost unimaginably fast, though not infinite. Light travels at about 300,000 kilometers per second. Kilometers
can therefore also be measured in years, and years in kilometers. One
light-year is roughly equivalent to ten billion kilometers. That is a one
followed by thirteen zeros.

Space, time, and light are consequently quite closely intercon-
nected. The Dane Ole Rømer had figured this out in the eighteenth
century while observing Jupiter's moon Io, as Einstein knew from his
boyhood readings. Lunar eclipses appear later when the planet is posi-
tioned farther from earth, and sooner when it is closer to our planet.

No matter how near or far something is, however, since time elapses
while light travels to reach us, the event we see has already occurred.
"In this sense," Aaron Bernstein wrote in his *Popular Books on Natural
Science*, "we see only the past and never the present."[3] The light of the
sun, for example, takes approximately eight minutes to reach the earth,
which means that people on earth see the sun the way it appeared eight
minutes earlier. If it were suddenly to be extinguished, the earth would
not turn dark until eight minutes had elapsed. That was considered self-
evident even in Einstein's youth.

Let us imagine that the sun appears to us as a clock, with a face and
hands that we can make out from the earth on clear days. Let us further
assume that this clock runs at precisely the same rate as a clock on earth

at the location of the observer, and that one earth day takes exactly twenty-four hours on this clock as well. Looking up at the sky, we would know what time it is. What would we see, however, if the sun displayed the actual time to us? Since it takes eight minutes for the light to travel from the sun to the earth, and hence also for the image of the clock on the sun to reach us, we would not see the time that the celestial hands were now displaying, but rather their position eight minutes earlier. At the moment that we see the clock in the sky striking twelve, its hands have already moved ahead eight minutes, which means that on the sun it is already eight minutes after twelve. To show us the correct time, the clock on the sun would have to be running eight minutes fast, from our perspective on earth. This seemingly simple calculation had been standard knowledge in physics for a long time prior to Einstein.

If we are to understand the basics of the theory of relativity, we need to keep in mind that the image of the clock and the position of its hands do not change while traveling with light. A moment in time appears frozen as it travels with the light. We might say that time is at eye level with light, and light with time. The traveling image of the clock stands still, as in a photograph. Something strange happens to time and space, something entailing light, which in effect rushes along with time through space.

That the velocity of light is finite is related to its inability to outpace time, because if it were able to do so, we could peer into the future and see events that have not yet taken place. Common sense does not accept this (although in the peculiar world of quanta, such things appear to occur). Since nothing can exceed time, which travels with light, nothing can move faster than light. The speed of light is the fastest possible tempo for motions in the universe. Einstein declared it the measure of all things, since it represents a kind of absolute, constant speed.

It is essential to grasp this idea in order to make sense of the special theory of relativity. Although it was not the road to the theory, it did provide the backdrop. One of its strangest consequences is that the faster clocks move with respect to an observer, the slower they run. As they approach the speed of light, they slow down more and more—to the point of standstill, like the image of the clock on the sun.

While still a schoolboy, Einstein undertook a thought experiment that he later described as essential to the theory of relativity. He was probably inspired to do so by an idea in Bernstein's writings. His question was this: How would the world look if one could ride on a beam of light? He was really posing the basically nonsensical question of whether one can catch up with light, and if so, what one would be able to see. Would the world freeze into a static image? Or would it always look the same, regardless of how quickly we ourselves moved?

On the long path of discovery that culminated in the theory of relativity, Einstein operated on two basic and extremely important assumptions that became the twin pillars of the special theory of relativity. The first is that the speed of light in a vacuum is constant. As long as light is not impeded, its tempo remains unchanging. The other is the so-called classical principle of relativity, which is derived from Galileo, and subsequently from Newton.

When we say that one thing moves relative to another, who or what is really "moving"? The short answer, according to Galileo, is that both are always moving "relative" to each other. If a train travels past a platform, the people waiting there see it in motion, whereas for the passengers in the train, the platform and the people waiting there are moving away from them.

This statement, which may appear trivial at first, is of great moment. It means that there is no privileged observer. Each party to the event has an equal right to claim that it is at rest and the other is moving relative to it. The second point in Galileo's principle of relativity seems quite basic as well. It emphasizes that speeds can be added. The "addition principle" is a matter of common sense. If one train is traveling at 50 miles an hour, and another one goes by in the opposite direction at 100 miles an hour, their speeds can be added together. The two are traveling at 150 miles an hour with respect to each other. If a conductor walks at 5 miles an hour inside the train moving at 100 miles an hour in the direction it is traveling, he is moving at 105 miles an hour as seen from outside. This plausible method of computation is known as the Galilean Transformation.

When the classical principle of relativity applies, each body is at rest within its own world, and from its perspective the other bodies are in motion. To people on earth, the earth is a frame of reference that is perfectly at rest, although in reality it is rotating and racing through the universe at considerable speed. The traveling train may also be seen as at rest within itself. In that sense, according to Galileo, all frames of reference are equally valid.

No mechanics can determine the motion of vehicles that are in uniform motion. Inside a train moving at a constant speed, an apple falls to the ground in just as straight a line as it does on the platform. On earth, objects fall down vertically, even though the earth is turning. That seems patently obvious to us, but in Galileo's day it was not. At least in mechanics, then, the laws of nature are independent of the state of motion, as long as this motion remains uniform. The big enigma facing scientists at that time was whether that principle also applies to electrodynamics, which entails extremely rapid motions such as those of electrons or light.

That is a key question in the special theory of relativity, and Einstein had a long way to go before he could answer it. He did not set out to redefine concepts like time and space, or to create a new view of the world. The long journey that brought him to the revolution for which he is famous today took him over cliffs of well-worn thought patterns, through thickets of physical discrepancies, and across broad expanses of prescientific certainties. It was his one remaining hope of eliminating the inconsistencies that had mounted in the classical physics of his era.

The most glaring discrepancy emerges when we juxtapose the two pillars of Einstein's theory, the constancy of the speed of light and the classical principle of relativity: At what speed do two systems move relative to each other when they zoom past each other at 75 percent of the speed of light? According to Galileo's addition rule, a value of 150 percent would have to result; that is, 1.5 times the speed of light. That is impossible, however, assuming that nothing can travel faster than the speed of light. The principle of relativity and the constancy of the speed of light, as Einstein understood them at that time, are mutually exclusive. His brilliant insight was the realization that this contradiction was linked to many other contradictions; a new theory might resolve them all.

As far as we know, Einstein went about finding this theory much like a bright young trainee in a large courthouse, who knows his way around all its divisions without having a say in any of them. One division investigates drug offenses and another traffic violations. Still others handle white-collar crimes and violent crimes. He is the only one to see that the cases of all these truth-seeking colleagues are linked.

Classical physics is divided into mechanics, thermodynamics, and electromagnetism. Each attempts to describe reality best with its own methods and theories. For over two hundred years, the dominant status of Newtonian mechanics was virtually uncontested. His laws for forces and motions laid claim to exclusive validity for all eternity.

In the nineteenth century, its interpretive sovereignty began to crumble. For one thing, the mechanistic view of the world encountered competition from the theory of heat. The members of this division, called thermodynamics, believed that everything could be explained with concepts like energy. For another, the division of electrodynamics also demanded to be heard. Electrodynamics sought to interpret the physical world with the theory of electromagnetism and the fields it comprises.

There was a Babel-like confusion of data and interpretations at the turn of the century, and only the trainee Einstein had a true overview, since he was conversant in all the divisions. There was no lack of results; it was more like a surplus of results that did not seem to fit together. More than any other employee in what we might call the bureau of truth, the young messenger saw both the trees and the forest. Einstein realized that the cases could be handled more effectively if the colleagues put their heads together. That was exactly what brought him his first triumph: instead of getting caught up with any particular division, each of which defended its turf and was ruled like a principality, he turned his attention to the less established peripheral fields.

All the studies he published in his "miracle year" 1905 address interdisciplinary issues. Each of them led to the merging of physical theories, something the authors of the literature he had read as a boy had lauded as the highest goal. By 1901, Einstein was confiding to his friend Grossmann, "It is a glorious feeling to perceive the unity of a complex of phenomena that appear as completely separate entities to direct sensory observation."[4]

He was of course not the only one to be struck by the glaring contradictions in physical theory, such as the lack of consistency between Galileo's principle of relativity and the constancy of the speed of light. However, the princes of physics tended to spruce up their own divisions instead of calling for a revision of the institution as a whole. Einstein worshipped one of these illustrious men passionately throughout his life: the Dutchman Hendrik Lorentz. "I admire this man like no other," he wrote in 1909. "I might say I love him."[5] Lorentz had begun expanding on James Clerk Maxwell's theory of electromagnetism in 1895. This theory explained all magnetic, electric, and optical phenomena uniformly. Like Maxwell and nearly all of his contemporaries, Lorentz focused on the ether, that strange medium which had been flitting through physics theory since the time of René Descartes. No one had ever seen this elusive substance, let alone directly measured it or laid hold of it in any other way. Physicists nonetheless introduced it into their theories as a makeshift solution to explain the motion of light and other electromagnetic waves as a mechanical process. They imagined it to be a massless, rigid material that filled the entire universe and penetrated all bodies. The light waves were said to be transmitted in the fixed medium like the shock waves of an earthquake in the earth's crust.

The mystery substance had made quite an impression on Einstein from the time of his youth. In the summer of 1895 he composed his first scientific essay as a gift for his generous Uncle Caesar in Brussels. Its title was "On the Investigation of the State of the Ether in a Magnetic Field." In August 1899, when he was twenty years old, he wrote to Mileva: "I am more and more convinced that the electrodynamics of moving bodies, as presented today, is not correct, and that it should be possible to present it in a simpler way. The introduction of the term 'ether' into the theories of electricity led to the notion of a medium of whose motion one can speak without being able, I believe, to associate a physical meaning with this statement."[6] He was already becoming aware of the problems entailed by the notion of this kind of "light medium."

Einstein took the issue to heart, and even began to plan experiments of his own in collaboration with his former teacher Konrad Wüest. "A good idea occurred to me in Aarau," he wrote to Mileva in September 1899, "about a way of investigating how the bodies' relative motion with respect to the luminiferous ether affects the velocity of propagation of light in transparent bodies."[7] He still believed in the ether at this point, but his outlook would soon change.

The existence of an ether gives rise to a central question: Is the ether at rest in the universe as an absolutely immobile medium, and do all celestial bodies, including the earth, fly through it? Or is it mobile and swept along by celestial bodies, perhaps even standing still on the earth like the air on a non-windy day? In August Föppl's textbook, which Einstein had read as a student, the answer to this question is called "quite possibly the most important task of today's research."

The Maxwell-Lorentz theory hinges on a light medium that simply cannot be disturbed. Like many other bright intellects of his time, Einstein had a crucial question weighing on his mind: If the ether fills all of space as a stationary medium, shouldn't the motion of the earth through the ether be subject to proof? After all, our planet circles the sun at 30 kilometers per second, which means that the immobile ether would have to produce a "wind" of the same size on the earth, which moves through it. Even if it penetrates everything, including ourselves, without our registering the fact, it would have to reveal itself to light as a kind of tailwind or headwind, depending on whether the light is emitted with or against the motion of the earth. The speed of light would consequently have to increase or decrease like the tempo of a swimmer who swims with or against the current, at least according to the classical principle of relativity, which states that velocities are additive.

When the experiments with his teacher fizzled out, Einstein turned to his favorite activity: reading. He worked his way through a thick treatise about a variety of experiments to demonstrate effects of the ether. Two Americans, Albert Michelson and Edward Morley, had conducted the most important one in 1887. They constructed a new measuring device for this express purpose, and called it an interferometer. In this instrument, which is still in use today, light is divided and sent back and forth between mirrors, which enables scientists to determine even the tiniest differences in velocity—according to whether the light moves in the direction of the ether or across it.

To their astonishment, Michelson and Morley were unable to register any effect. No matter how they twisted and turned their instrument, the speed of light always remained the same. "Even this ether-wind-weather vane of utmost sensitivity," Einstein reflected in 1920, "did not feel the ether wind."[8] The earth did not seem to move in relation to the ether. In fact, observations and optical experiments revealed the exact opposite. Einstein could not accept this blatant contradiction.

Hendrik Lorentz, who firmly believed in the light medium, hit upon an elegant solution. In the absence of any experimental evidence, he assumed that the ether wind made objects contract mechanically at high speeds. He used this "contraction hypothesis" to explain the results of all experiments flawlessly, including those of Michelson and Morley. The interferometer, he contended, was shortened in the direction of the flow of the ether by precisely the amount that would be compensated for by the addition of the ether flow to the speed of light, which is why the instrument could not measure any deviation at all. Lorentz appeared to have salvaged the concept of the ether.

However, the young hothead in Switzerland did not let up: "Should [God] really have put us into an ether storm," Einstein wondered, "and should, on the other hand, [He] have arranged the laws of nature precisely so that we can never notice the storm?"[9] His answer left no room for doubt: "As a matter of fact, the truth is that we ourselves were the ones who did invent the ether together with the ether wind."

Robert Rynasiewicz, philosopher and expert on relativity at Johns Hopkins University in Baltimore, has pointed out a further contradiction Lorentz faced. His ether at rest was the preferred frame of reference relative to which everything else moves. According to historians of physics, this was comparable to Newton's idea of absolute space, which constitutes the framework of all motions. In this view, which seems borne out in our everyday lives, all events take place on a kind of set stage that is at rest. All motions, including those of light, take place relative to this absolute space. However, this frame of reference at rest contradicts the principle of relativity Galileo had described for mechanical motions, which held that everything moves relative to everything else, and that the state of rest, rightly claimed by both the traveler in the train and the man waiting on the platform, is only relative. No frame of reference is preferred or absolute; every point of view is equal.

Thus the classical principle of relativity acquired an aspect of symmetry that has been of enormous importance for Einstein's discoveries and for modern physics as a whole to the present day: no matter how you look at it, the relationships between two moving systems remain the same.

This symmetry is undermined by the notion of a stationary ether. According to Lorentz, in a moving system, for instance on the earth, the lengths and yardsticks are shortened in what was thought to be the ether wind, but they remain equal in a system at rest. However, since velocities are given as length per time, for example as kilometers per hour, each of them brings a different perspective to the time being measured. Lorentz dealt with the effects of asymmetry by using a crutch he called "local time." Einstein would not accept that. In his ongoing role as trainee in the bureau of truth, Einstein had amassed a set of inconsistencies pertaining to physics, and needed to figure out whether both aspects of the mechanical relativity principle really could be applied to the electrodynamics world. His understanding of the Maxwell-Lorentz theory of electromagnetism seemed to conflict with the principle of relativity. Not only did it privilege a fictitious frame of reference, the ether, but it contained an additional asymmetry, which struck Einstein as absurd. Hence, the son and nephew of the electrotechnical industrialists who founded Einstein & Cie. began his legendary study of the theory of relativity in 1905 with a clarification: "It is well known that Maxwell's electrodynamics—as usually understood at present—when applied to moving bodies, leads to asymmetries that do not seem to be connected to the phenomena. Let us recall, for example, the electrodynamic interaction between a magnet and a conductor."[10] Einstein was picturing a conductor as something along the lines of the windings in dynamos, as they are used to produce current. The motion of the conductor with respect to the magnet produces an electrical current.

In Maxwell's theory, however, it makes a difference which is moving: the magnet with its magnetic field relative to a conductor that is without current and hence field-free, or the conductor with current with its electrical field relative to the magnet. Einstein wondered, to put it simply, how the magnet or the electrical conductor ought to "know" which one is moving and which is not. Oughtn't it to be inconsequential which component of a dynamo was turning to the other, much like the situation between travelers in a moving train and people waiting on the platform?

Einstein's inspiration for this idea appears to have come from August Föppl, who described a comparable thought experiment in his textbook: If a magnet and a conductor move through the ether together at the same speed, no electrical current results. What matters is not absolute motion against the ether, but the relative motion of the two toward each other, in which case, as Föppl noted, the ether appears superfluous. Einstein would eventually formulate his view in exactly this way.

By the spring of 1905 he had put together virtually all the pieces of the puzzle. Just realizing that all of the pieces belonged to the same puzzle was itself an immense achievement. However, he was still missing the key insight into how these pieces could be combined in a seamless chain of evidence. He did not want to enhance a single division, but to revolutionize the field as a whole. He desperately sought a way to bring together the many differing building blocks and form a whole picture, but he could not succeed. Finding contradictions is one thing; resolving them is quite another. When he saw that he was stumped, he told his friend Besso in frustration, "I give up."

The philosopher Rynasiewicz said that at this moment Einstein was caught in a "circle of ideas" with no apparent means of escape. Feeling as though he were trapped in a dark dungeon, Einstein himself later talked about his "years of vain searching"[11] leading up to the day in May 1905 when he was ready to concede defeat. In 1932 he reflected back, "I soon felt sure that the reason was a deep incompleteness of the theoretical system. My desire to discover and overcome this problem generated a state of psychic tension in me."[12] But how could the twenty-six-year-old resolve this tension after seven "years of vain searching"?

In all these years of research, Einstein had learned that theories that rely on unproven hypotheses, such as a luminiferous ether that cannot be measured, or a mechanical contraction by the ether stream, are weak theories. He therefore took the radical step of dismissing Lorentz's unproven assumptions, even though he deeply respected Lorentz, and declaring the hypothetical ether superfluous. He placed his faith in empiricism and relied on experimental findings.

Since "neither mechanics nor thermodynamics," Einstein later recalled, "could . . . claim exact validity,"[13] he was forced to take the courageous and visionary step of declaring the speed of light independent of the motion of the source or of the recipient. It is constant, just as the Maxwell-Lorentz theory had predicted. Michelson, Morley, and all the others had measured correctly. On the other hand, the principle of relativity had to apply. He firmly believed that this was the case, otherwise asymmetries would appear that would set physics back to its pre-Galilean Stone Age.

Now he found that he was stuck in a final contradiction, and faced the dilemma of giving up either the classical principle of relativity or the constancy of the speed of light. He could not have both within the parameters of physics as it was then understood. Possibly his admission of defeat to Besso helped him let go and see the answer more clearly than anyone else did. His insight into the intimate connection between space, time, and light enabled him to perfect his work of art.

Einstein was the first to cross the threshold to a new territory of thought. He did not want to take this step, but he felt compelled to do so because he was the only one capable of resolving any remaining contradictions, fitting all the pieces of the puzzle together, and solving the case in his search for truth.

How did he do that? Calling the existence of the ether into question was a brilliant undertaking, but hardly a unique one. Einstein revealed the truly distinctive aspect of his thought in the 1922 lecture in Kyoto: "My solution was an analysis of the concept of time."[14]

Einstein analyzed the concept of time not only as a question of physics, but also as a philosophical issue. What is time as such, if no one can measure it? What are we really doing when we measure time? What does it mean for two events to occur simultaneously? It must have been a glorious moment when he suddenly knew that he was on the right track: from one second to the next, he realized that the problem had been "completely solved."

Einstein now threw everything open to question, even basic concepts such as space, which are really not subject to analysis at all, since they have traditionally been thought of as forming the framework of events in the universe. At the beginning of the Enlightenment, Immanuel Kant had stripped space of reality and declared it a form in which we perceive things. Space accordingly exists a priori, prior to all experience, and it exists even devoid of things. Einstein added space to the perceptible realm that can be experienced by man. He shrewdly observed early on "that only the discovery of a universal formal principle could lead us to assured results."[15] This principle would need to be irrefutable, such as the impossibility of a *perpetuum mobile*, for example, according to which energy cannot be created from the void. He suddenly realized that in order to be able to show science a way out of its impasse, he had to turn Newton's cosmos on its head: he made the speed of light the natural constant and light—almost biblically—the absolute. To do so, he had to take away the absoluteness of space and time and "relativize" them. Suddenly all the contradictions vanished, and all known uniform motions, from electrons to distant stars, could be explained with a single overarching theory. Einstein's insights turned a phantom into a phenomenon that could be physically objectified. Relativity was reality.

Where did he get the audacity to state principles whose universal basis actually represented pure assertions? His boldness stemmed from his intuitive understanding of the fundamental issues. The theory he derived on the basis of this intuitive understanding, which explained everything consistently, gave the principles their firm foundation. Because Einstein was determined to stick with the principle of relativity and of the constant speed of light, he pondered the question "What is to be understood here by 'time'?" and commented, "We have to bear in mind that all our propositions involving time are always propositions about simultaneous events. If, for example, I say that 'the train arrives here at 7 o'clock,' that means, more or less, 'the pointing of the small hand of my clock to 7 and the arrival of the train are simultaneous events.'"[16]

Never before and never since have such simple sentences graced a work of such profundity. Einstein made a self-evident truth a sensation. Not only did his theory become the foundation of all of modern physics, it shook up our understanding of nature and the world.

The connections are elegantly simple and can be elucidated by means of thought experiments with observers on a platform and in a train, which was the fastest means of locomotion in Einstein's day. He pictured, for instance, two flashlights set up to the right and left along the tracks at equal distances from the observer on the platform. If the observer sees both of them flash at the same moment, the flashes occur simultaneously for him. The situation appears different to the passenger in the train, who moves ahead during the space of time it takes for the flashes to reach him. Since he is traveling away from one of the flashlights and toward the other one, the flash toward which he is traveling appears earlier to him than the one from which he is traveling away. From his perspective, he is equally entitled to say that the flashes were not simultaneous.

The same picture emerges the other way around: if the flashlights are set up on the front and back ends of the train, and flash simultaneously for the observer in the middle of the train, they light up at different times when seen from the platform. The two observers would not be able to concur about the time. The one notes a time lapse between the flashes, but the other sees the flashes occuring at one and the same moment.

The sword Einstein finally used to cut the Gordian knot after years of baffling contradictions, the eureka that enabled him to connect all the pieces of the puzzle, proved to be an idea with a sensational impact: simultaneity is relative. There is no moment that can be called "now" for the whole world.

The thought experiment was far from over, however. Let us assume that a basketball player is traveling in the train, and he bounces a ball on the floor. From the perspective of the observer in the train, the ball goes up and down vertically. But anyone on the platform watching the position of the ball in the train racing by would see it zigzag. The ball does not just go up and down, but also sideways with the train. Seen from the outside—and this is the first important point—the ball covers a greater distance than when observed from within.

Imagine, instead of a basketball player and a ball, a device in the traveling train in which a light beam races up and down between two mirrors. The same picture emerges: for the traveler, the light beam jumps up and down vertically. Seen from the platform, it describes a zigzag if the train is traveling quickly enough.

What now follows is one of the crucial ideas in the special theory of relativity. Everything remains quite simple when observed from within. The light beam hops up and down at the speed of light. Seen from the outside, however, the light covers a greater distance than when seen from within. The observer on the platform might therefore deduce that the speed of light is greater in the train, but Einstein categorically excluded this possibility. There is only one way out of this situation: since the light requires more time for a longer distance, the clocks in the train would have to lag behind from the perspective of the platform. Time passes more slowly in the moving system in comparison to the resting one. This phenomenon is called "time dilation." The faster the train goes, the more extended the zigzag appears from the outside, and the slower the clocks run. When the speed of light is reached, the light beam stops bouncing altogether—the clock in the train stands still as observed from the platform. This is exactly the situation with the sun, which we see as a clock. Its image reaches us at the speed of light and stands still. All of this functions, of course, only in a thought experiment. No real clock or real train could move at the speed of light.

The most improbable element of this is actually the principle of relativity as Einstein developed it from Galileo's model. From the perspective of the observer in the train, nothing has changed. As far as he is concerned, the beam of light continues to bounce up and down vertically, and his watch keeps running normally. He is completely justified in assuming that the traveling train, which is his frame of reference, is at rest. If, however, he now looks outside to the platform and sees a basketball bouncing or a beam of light racing up and down between two mirrors, the exact opposite effect occurs: the traveler registers a zigzag, and since the speed of light must be constant, the clocks on the platform go slower from his perspective.

The crucial point is that time dilation applies in both directions, and symmetry is maintained. As the train picks up speed, the watch of the person standing still on the platform slows down to the same degree for the traveler on the train as the watch of the traveler for the observer on the platform. This relativization is Einstein's great achievement: a period of time can vary in length according to the state of motion of an observer, not subjectively, but objectively measurable with clocks.

Newton had worked on the premise that time was absolute; there was essentially one uniform clock for the whole world. Time, he assumed, "flows uniformly."[17] Doesn't man experience it that way from the cradle to the grave? Einstein basically made light the ruler over time—and over space, since distances in space also change along with slowing clocks in a moving system. The closer the train comes to the speed of light, the shorter these distances become.

This phenomenon, which is called "length contraction," can also be explained by means of motion. Motion means distance divided by time, such as meters per second. When clocks run slower in a moving train as measured from without—that is, each second lasts longer—the meter is shortened as well. The observer in the train in turn sees the situation the other way around: for him, the meter remains a meter. He, however, sees length contraction on the platform. Shortly before the train reaches the speed of light, the meter appears contracted to only one centimeter.

Length contraction applies only in the direction of travel, but not in the vertical direction. To an outside observer, a square in a moving vehicle shrinks into an increasingly narrow rectangle as its speed increases without changing in height. At the speed of light, it becomes a line running straight up and down, as though it had lost a dimension. This contraction can also be pictured as an effect of perspective: from the perspective of the observer at rest, the moving body appears as in a distorted mirror that assumes an increasingly cylindrical shape as it speeds up. In this mirror everything becomes narrower, but not shorter.

But hadn't Hendrik Lorentz assumed the same thing when he established his "contraction hypothesis"? This circumstance remains a source of confusion even today. Lorentz believed that objects shortened because they were squeezed together in the ether stream. Einstein said, "the introduction of a luminiferous ether [would] turn out to be superfluous." Contractions of time and length do not stem from a mechanical change in the clock or measuring rod, as Lorentz had surmised, but rather reflect actual characteristics of time and space.

In Einstein's theory, Lorentz's makeshift solution to save the idea of the ether was transformed into something concrete: every system has its "own time," which means that time, as understood by physics since Newton (common sense has always regarded it in roughly this way), does not exist as a unified entity. According to the old notion, to which Humboldt and Eberty also adhered, all parts of space share the same time, regardless of what moves or how it moves. According to Einstein's new interpretation, no clock is valid for the world as a whole. The image of the clock on the sun that can be read from the earth is useful only if we assume that earth and sun together represent a fixed frame of reference.

Einstein provided no explanations for these phenomena. No one knows what light and time really are. We are not told *what* something is. The special theory of relativity merely provides a new rule for measuring the world—a perfectly logical construct that surmounts earlier contradictions.

Even though we cannot know anything about the essence of things, however, we can certainly say something about their mutual relationship and interaction. The theory supplies formulas with which we can make precise calculations regarding strange behavior in "relativistic systems." If, for example, the speed of a moving object is known, its length in a resting state can be calculated. If a spaceship were to fly by us at 90 percent of the speed of light and measure in at five meters long from our perspective, the length would be about eleven meters for the astronauts inside the spaceship. The equations used for these calculations are identical to the Lorentz transformations, which is no coincidence, and has nothing to do with the fact that Einstein had adopted the theory of the venerated Dutchman. It lies in the mathematical structure of the system, which is why Henri Poincaré and several other researchers before Einstein had arrived at exactly the same formulas. But Einstein went one crucial step further than his predecessors.

His study states: "Like every other electrodynamics, the theory to be developed is based on the kinematics of the rigid body, since assertions of each and any theory concern the relationships between rigid bodies (coordinate systems), clocks, and electromagnetic processes. Insufficient regard for this circumstance is at the root of the difficulties with which the electrodynamics of moving bodies must presently grapple."[18]

Einstein masterfully combined four elements: kinematics (the theory of motions), the rigid body (the coordinate system or the frame of reference), clocks (and consequently time), and electromagnetic processes (such as the emission and reception of light), and was thereby addressing the very junction at which problems of classical physics intersect—and are resolved.

Within a "rigid body"—a train or a spaceship, for instance—"the time of the system is at rest," as Einstein explained. It makes no difference whether the vehicle is moving from an outsider's perspective; seen from the inside, it is at rest. And within the system at rest there is simultaneity, which is why earthlings can define a "world time" and synchronize all clocks around the globe. As soon as two rigid bodies move toward each other, however, each has its *eigenzeit*, its own Now. Einstein used this brilliant trick to eliminate all the contradictions.

The fascinating question remains: How did he happen upon this idea? We need more to go on than Einstein's conversation with his friend Besso to understand how this revolutionary idea that transformed our understanding of the world came to be. Einstein reached his goal on a road that was paved with deep philosophical and epistemological inspirations as well as practical insights.

Jürgen Renn of the Max Planck Institute for the History of Science in Berlin uses the term "layers of knowledge" to describe the information Einstein drew on. He explains that it is futile to try to understand the development of science by examining Einstein. The order of events needs to be the other way around: Einstein should be viewed in the context of the development of science. The goal of Renn's research group in Berlin is to place this development in the framework of the history of knowledge. In what manner do the thought processes of an individual interact with prevailing schools of thought and firmly established worldviews? How do scientific revolutions happen? What are the roles of experience, knowledge, and one's cultural milieu? A knowledge of science is not the only factor; there is also "the intuitive knowledge that shapes our everyday thinking," says Renn.

His colleague, Peter Galison of Harvard University, has pointed out a possible link to everyday reality. Around the turn of the century, simultaneity was one of the dominant technical themes. From the punctuality of trains to the coordination of shipping to well-coordinated military procedures, people relied increasingly on precisely synchronized clocks. New applications pertaining to synchronization began flooding the patent offices. As a patent expert, Einstein therefore literally had his finger on the pulse of the times. A great deal of technical effort and expense went into coordinating clocks everywhere, first in the cities, then throughout whole countries, and eventually even internationally. Numerous books, newspapers, and professional journals were devoted to the topic. Galison believes that this development had an important role in Einstein's breakthrough.

"His innovation," says the psychologist Howard Gardner, who has described Einstein as a "perpetual child," "sprang out of his capacity to integrate spatial imagery, mathematical formalisms, empirical phenomena, *and* basic philosophical issues."[19]

However, the special theory of relativity acquired its ultimate elegance not from Einstein, but from his former teacher Hermann Minkowski. He gave it the mathematical form that it essentially still has today. "Henceforth space on its own and time on its own will decline into mere shadows, and only a kind of union between the two will preserve its independence," the mathematician announced in September 1908.[20] By extending the principle of relativity to the "postulate of the absolute world," he created concepts that outdid any inventions a science-fiction author could have come up with: the events in the world lie as points on "worldlines," which together form the "world." With Minkowski's mathematical formulation, Einstein's theory could be expressed with the degree of clarity with which even its seemingly absurd consequences can be understood—such as the twin paradox.

In this thought experiment, two twins—let us call them Tim and Tom—take different routes through spacetime. Tim goes on a journey and moves away from the earth at nearly the speed of light. Tom remains back on earth. From his perspective, Tim's clocks now run far slower, which means that Tim ages far more slowly. After a year, Tim turns back, and after one additional year, he lands back on earth. To his astonishment, he finds that his brother Tom did not age by two years, as he did, but by twenty years. The paradox seems to lie in the fact that, according to the principle of relativity, Tom's clocks also ought to have run slower for Tim and hence Tom, too, ought to be much younger than Tim. The outcome that each is younger than the other, however, is something that Einstein's theory of relativity does not allow for, either. How can this strange paradox be explained? Einstein himself grappled with this paradox for a long time. In 1919 he was still having difficulty attempting it in the elegant form that Minkowski's mathematical formulation seamlessly allows. The mathematician had also put the fourth dimension into a workable arithmetical form; as the young Einstein had read in Felix Eberty's book, "space and time are not unrelated."

Minkowski's idea and the solution to the twin paradox can best be explained by means of an analogy between space and spacetime, that is, between a three- and a four-dimensional world. Let us assume that a balloon travels exactly one meter through a room. It can move in three directions—in length, width, and height. The higher it flies, the less it shifts its position forward and backward, to the left or right. If it rises only in height, it completes the meter in only one dimension. Since it "uses up" its entire allotted distance in the vertical, it does not get anywhere in the two-dimensional surface. By the same token, if the balloon moves only on the surface—that is, if it merely rolls on the floor—it does not cover any distance upward. It makes zero vertical advance.

Applying these findings to motions in the four dimensions of spacetime transcends the normal bounds of our powers of imagination. We need only know, however, that four dimensions have four axes, just as space has three. Time as a fourth dimension rests vertically on the other three—just as in space the vertical juts out of the two-dimensional plane as the third dimension. Distances through spacetime comprise four dimensions, just as space has three. The more you go in one direction, the less is left for the others.

When a "rigid body" is at rest and does not move in any of the three dimensions, all of its motion takes place on the time axis. It simply grows older. This applies to all of us, including little Albert in his easy chair. As soon as the boy stands up and walks through the room or goes upstairs, his position in space changes. The faster he moves away from his frame of reference—be it his parents' home or the earth—and covers more distance in the three dimensions of space, the less of his motion through spacetime as a whole is left over for the dimension of time. The total distance is now made up of length, width, height, and time. Whatever goes into space is deducted from time, just as what goes into the two dimensions of the plane in the three-dimensional room comes off the vertical.

This deduction in time has no practical effect on the aforementioned young Einstein or on any people going about their daily routines. In comparison with the distances light travels, all distances in the dimensions of space, even those involving airplane travel, are so very small that we essentially move only along the time axis, and we age continually. Only if we were able to move away from our frame of reference very quickly, like the traveling twin Tim, would the elapsed time shrink to near zero, as it approached the speed of light.

Light itself, which moves as it were on the crest of time, covers its entire distance through spacetime only in the three dimensions of space—like the balloon in the room that rolls on the floor, moving exclusively in two dimensions. Nothing remains for the additional dimension—the vertical in the case of the balloon, the dimension of time in the case of light. Because light particles do not move *in* time, but *with* time, it can be said that they do not age. For them, "now" means the same thing as "forever." They always "live" in their moment. The clock on the sun stands still. Since for all practical purposes we do not move in the dimensions of space, but are at rest in space, we move only along the time axis. This is precisely the reason that we feel the passage of time. Time virtually attaches to us.

These tools enable us to understand the twin paradox, which, according to the theory of relativity, is not a paradox at all. Tom on the earth really is older than the rapid traveler Tim. This effect is a consequence of spacetime. Tom, like the man in his easy chair, moved solely through time. His worldline ran on the time axis. Tim, by contrast, traveled at a very high speed, covering a substantial portion of his distance through spacetime in the dimensions of space. His worldline deviated far from the time axis. He was able to "use up" correspondingly less time, which is why he really is younger according to their common frame of reference, the earth. If he had traveled at the speed of light, he would have used up no time at all—like the image on the clock on the sun—and, like the light, he would not have aged. Einstein's later assistant Banesh Hoffmann wrote, "In spacetime past, present, and future are all spread out before us, motionless like the words of a book."[21]

The fact that bodies move through spacetime has perplexing consequences. A spaceship that races through space has more stars ahead of it than behind it. Because it can "only" travel at the speed of light, the light from the stars behind can hardly catch up to the spaceship. It is like holding an umbrella: the faster a person with an umbrella moves, the farther he has to move his umbrella handle down into a horizontal position. At a rapid pace, virtually all the drops come from the front and none from behind. Because the earth moves at only a fraction of the speed of light, we see an equal number of stars in the sky in both directions.

Einstein had made sweeping revisions to conventional ideas of space and time on the basis of very simple kinematic observations, leading some of his colleagues to claim that he had opened a Pandora's box. Since time has a role in every physical theory, all those theories needed to be brought into line with the theory of relativity. Spacetime had led Einstein into new territory. Others followed in his path and added many significant discoveries.

Einstein's own 1905 study made inroads in modifying the laws of optics and electrodynamics according to the theory of relativity. In doing so, he was able to get to the bottom of the problem of asymmetry between the motions of a magnet and a conductor, which had exasperated scientists for many years. He regarded magnetic and electrical fields in spacetime as two sides of the same coin and unified them in the electromagnetic field. Magnets and conductors combine to create this field when they move relative to each other.

In the years to follow, hordes of physicists launched an examination of all physical laws and reworked them to conform to the special theory of relativity, which is still considered valid for all of nature today—and, in the opinion of nearly all physicists, will remain valid for all time. All of the divisions of natural science work with the new basic law that the young trainee gave them in 1905. It is as if Einstein had given nature a new perspective, and now all natural laws had to be read through the lens of the theory of relativity.

It took Einstein several weeks after submitting his study to recognize a critical consequence of his special theory of relativity. In September he sent in a three-page addendum, the fifth miraculous product of his miracle year 1905. "A consequence of the study on electrodynamics did cross my mind," he reported to the renegade Olympia member Conrad Habicht. "The relativity principle, in association with Maxwell's fundamental equations, requires that the mass be a direct measure of the energy contained in a body; light carries mass in it."[22]

Thus the most famous formula in the world was conceived: $E=mc^2$. Energy is equal to its mass times the square of the speed of light. This equation was literally explosive. It describes the enormous energetic potential of nuclear fission, which Otto Hahn would accomplish in 1939 and Lise Meitner would explain to him—and which would find its first dreadful application in Hiroshima in 1945. "A noticeable reduction of mass would have to take place in the case of radium," Einstein added in his letter to Habicht.[23] Radium was one of the few radioactive materials known at the time that emitted energy-rich rays.

Einstein did not discover the equivalence of energy and matter by giving serious consideration to the connections between the two. Quite remarkably, it was a purely logical consequence of the special theory of relativity. In order to accelerate a mass, energy must be expended. The faster the mass already is, the greater the energy needs to be to accelerate it. In order to catapult it to the speed of light, an infinitely great amount of energy would have to be expended. Therefore no object that has "rest mass" (inertia) can move at the speed of light. That can only occur in a thought experiment. Photons, on the other hand, do not have rest mass. Their entire mass is in their kinetic energy. Only particles like those can travel at the highest possible speed—on a level with time.

The formula $E=mc^2$ makes the special theory of relativity the most complex simplification in the world. The French philosopher and poet Gaston Bachelard mused, "Is so little required to 'shake' the universe of spatiality? Can a single experiment of the twentieth century annihilate . . . two or three centuries of rational thought?"[24]

In unifying energy and matter, Einstein had succeeded at something on the order of squaring light. He showed what an enormous amount of energy there is in mass. There is a simple explanation as to why no one had previously established this connection: energy is so deeply hidden within matter that it cannot be measured. "It is just like an extremely wealthy man who never spends a single penny, and no one has a clue just how rich he is," Einstein commented. The mass that a lightbulb radiates at 100 watts for 100 years is less than $\frac{1}{3000}$ of a gram.

This gives us an idea of the enormous energy contained in even small quantities of matter. "The consideration is amusing and seductive," Einstein concluded his letter to Habicht, "but for all I know, God Almighty might be laughing at the whole matter and might have been leading me around by the nose."[25]

The sun provides an impressive demonstration of the conversion of mass into energy on a daily basis: millions of tons of matter are transformed into gigantic amounts of radiation energy in this glowing ball of fire every second, which makes life on the blue planet earth possible.

Albert and Mileva Einstein celebrated this breakthrough on July 20, 1905, in a manner that was quite out of character for them. They got roaring drunk—the only known case in which Albert indulged in an excess of alcohol. Conrad Habicht received a postcard of unique historic value: "Both of us, alas, dead drunk under the table. Your poor Steissbein & wife."[26]

NINE

WHY IS THE SKY BLUE?

EINSTEIN—A CAREER

The eternal mystery of the universe is its comprehensibility," Einstein exclaimed aphoristically in 1936, with a nod to Kant.[1] The world lies before us like an open book, waiting to be read. Very few people have deciphered as many lines of that book as did Einstein. By 1936, more than thirty years of professional research lay behind him. He never sought the path of least resistance. Quite the opposite. Science was not a profession, but a calling. Einstein was the artist of science par excellence.

Einstein and two colleagues had astounded the scientific establishment in the previous year with their innovative approach to quantum mechanics, the enigma of modern physics. The so-called Einstein-Podolsky-Rosen paradox, which we will be examining later, baffles theorists even today. In retrospect, it was his final scientific essay of lasting significance, but it did not spell the end of his work. He kept on plugging doggedly away up to the last day of his life, but his sustained productive phase was over by 1935.

By contrast, his accomplishments in those early three decades were unparalleled. The two great twentieth-century revolutions in physics bear his signature—the theory of relativity in its entirety and quantum theory in part. In addition to his contributions to the foundations of modern physics, he also made essential inroads in chemistry, provided the theoretical basis for the development of the laser, and, in his insatiable curiosity, answered questions such as why rivers bend and why the sky is blue. Holding fast to his "deep conviction of the rationality of the universe,"[2] he felt confident that hardly any physical problem was beyond his grasp.

His career began with an imbalance, the professional part lagging well behind the scientific one. For years he conducted his scientific research in his spare time, in addition to his forty-eight-hour week at the patent office. Early in 1908 his colleague Jakob Laub wrote him from Würzburg: "I must tell you quite frankly that I was surprised to read that you must sit in an office for 8 hours a day."[3]

Einstein's work proved that something done well is not done overnight. Even in his youth he was familiar with the feeling of moving on two tracks. Scientifically, he knew, he was far ahead of others, but career-wise he was way behind. His profession as a physicist began even before 1905, his "miraculous year," yet he was not officially appointed professor until the fall of 1909—thirteen years after writing in a high school proficiency test in French, on the topic "My Plans for the Future," that his goal was to become a professor of "theoretical natural science." By the time he attained that goal, he had already been recommended for the Nobel Prize, but he did not receive the award until 1922. By then his third career, as an international star of science and a political figure, was already in full swing.

Einstein had no choice but to adjust to the rhythm of his daily existence. He had a wife and child to support. There was no money to be made in physics without an academic appointment. At the patent office he earned more than double the salary of a university assistant professor. In September 1904 the management converted his temporary contract to a permanent position and raised his annual salary from 3,500 to 3,900 francs.

His material security was a psychological boost for the aspiring physicist after coping with a nerve-racking shortage of funds when he first completed his studies. He simply had to wait until his workday had ended to devote his time to his real interests, as he explained to Conrad Habicht, his old friend from the Olympia Academy: "Keep in mind that besides the eight hours of work, each day also has eight hours for fooling around, and then there's also Sunday."[4]

Even before finishing work on the special theory of relativity, he completed an essay titled "A New Determination of Molecular Dimensions" on April 30, 1905. This essay demonstrated that measurable characteristics of liquids and solutions, for instance the viscosity of sugar water, allow us to calculate the precise size and number of dissolved sugar molecules. Although they were invisible, molecules and atoms became part of our physical reality. His important contribution came at a time when the brightest minds in physics were still grappling with the existence of atoms, and it laid the foundation for so-called colloid chemistry, which has become indispensable in producing industrial materials.

Just nine days later, Einstein's next study was complete. Here, too, the subject was atoms and molecules. This essay examined their motions in liquids. Back in 1827, the English botanist Robert Brown had noticed that tiny particles, such as pollen grains and particles of smoke, made clearly visible jiggling motions under the microscope. Many of Einstein's predecessors attempted to interpret this "Brownian motion." Some regarded it as the effect of individual liquid molecules hitting the suspended particles. Einstein realized that individual hits of this kind would be far too weak and too brief to trigger recognizable motions of the pollen. It would be like trying to move a billiard ball with a speck of dust. However, he was able to demonstrate that, from a purely statistical point of view, the hits in their umpteen-billionth sum would have to result in a shift of the particles. This study thus established a link between macroscopic characteristics, such as temperature, and microscopic ones, such as molecular mass, which became an indispensable tool in pharmaceutical research. Einstein had no need to peer into a microscope. He set up equations and calculated. Theory was his microscope.

This was his basic creative pattern. He did not experiment, but instead made predictions for experiments and observations. Researchers continue to be amazed even today that these predictions could be substantiated. His formulas on Brownian motion inspired a key experiment. In 1909, Jean-Baptiste Perrin would succeed in dispelling any remaining doubts about the reality of atoms and molecules, and he was awarded the Nobel Prize. The "exact determination of the real size of atoms,"[5] which had been prognosticated by Einstein, now moved into the realm of possibility. A milestone in physics—in the shadow of quantum theory and the theory of relativity.

At this point, Einstein arrived at an important decision. With an eye to professional advancement, he submitted a seventeen-page text on

"The New Determination of Molecular Dimensions" to the University of Zurich as his doctoral dissertation on July 20, 1905, just a few weeks before sending off his theory of relativity. He would never have been awarded a doctorate for the theory of relativity, which was considered far too daring and speculative. The dissertation needed a positive assessment by Professor Kleiner, the main reader. Kleiner was of course the very professor who had failed to support Einstein's previous attempt to get his doctorate a few years earlier. However, everything went well this time. Kleiner attested that the applicant had "proved that he is capable of working successfully on scientific problems."[6]

The thesis was accepted, and Einstein had it printed in accordance with university regulations, and with the addition of an acknowledgment: "Dedicated to my friend Dr. Marcel Grossman."[7] Even today it is one of the most frequently cited publications in physics. After submitting the required copies, *Herr Doktor* Einstein waited to see how the world of science would react to the other products of his creative momentum in the spring of 1905 — in particular to the "Light Quanta Hypothesis," which laid the foundation for quantum theory.

He was like a revolutionary who had penetrated straight to the core of physics and detonated a series of bombs. Those explosive devices were his four essays in the *Annalen der Physik*, the preeminent physics journal of his day. However, Einstein had come to think of science as a steamer that forges full speed ahead but cannot change course rapidly. He waited impatiently for reactions from the professional sphere, expecting to be criticized rather than praised, yet instead he heard absolutely nothing, neither agreement nor rejection — nothing whatsoever.

For the time being, the only blossoming facet of the newly minted doctor's career was his work at the patent office. In March 1906 he was promoted to "expert second class" and now earned 4,500 francs a year. The Einsteins could finally afford an apartment with their own furnishings, and they moved to Aergertenstrasse 53. Unfortunately, the new apartment was nowhere near Besso's apartment, so Einstein had to walk to work alone. He felt isolated a good deal of the time. On April 27, 1906, he wrote to his friend Solovine, "I have not been meeting with anyone since you left. Even the conversations on the way home with Besso have now come to an end; I haven't heard anything whatever from Habicht."[8]

Still, the picture was not entirely bleak: "My papers are much appreciated and are giving rise to further investigations. Professor Planck (Berlin) has recently written to me about that."[9] The letter from Planck meant so much to Einstein that his sister, Maja, continued to emphasize its importance nearly twenty years later: "After the long period of waiting, this was the first time that his paper was being read at all. The happiness of the young scholar was that much greater, since acknowledgment of his accomplishment came from one of the greatest contemporary physicists. . . . At that time Planck's interest signified infinitely much for the morale of the young physicist."[10]

This famous colleague was one of the first to grasp the significance of the theory of relativity. Many years would go by before it became part of the core knowledge of physics. In July 1907, Planck wrote to Einstein, "As long as the proponents of the principle of relativity constitute such a modest little band as is now the case, it is doubly important that they agree among themselves."[11]

In September 1907, Johannes Stark, a professor from Greifswald, invited Einstein to contribute a comprehensive survey article to the *Jahrbuch der Radioaktivität und Elektronik*, a journal he had founded. After 1919, Stark became one of the bitterest opponents of Einstein and his theory of relativity.

Einstein agreed to write the article, but, he reminded Stark, as an independent scholar, "[I am] not in a position to acquaint myself with *everything* published on this topic, because the library is closed during my free time."[12] On November 1, he wrote to Stark, "I have now finished the first part of your paper for the *Jahrbuch*; I am working diligently on the second part in my free time, which is unfortunately rather scarce."[13]

Chapter five of this second part, which Einstein finished in late November, proved what he was capable of achieving in his "very limited free time" in the text he called "Principle of Relativity and Gravitation." His incomparably bold presentation ventured into uncharted territory. "So far we have applied the principle of relativity, i.e., the assumption that the physical laws are independent of the state of motion of the reference system, only to *nonaccelerated* reference systems. Is it conceivable that the principle of relativity also applies to systems that are accelerated relative to each other?"[14]

This sober presentation might sound innocuous, but it was like a thunderbolt for physics research and continues to resonate even today. Einstein had laid the foundation for his magnum opus, the general theory of relativity, which he completed eight years later. There are no clear indications as to how long he had been pondering this generalization of his special theory of relativity, but his explanations attest to how close he had come to the heart of the problem. He predicted that gravity (like motion) would make clocks run more slowly. And he anticipated the effect that would be proved during a solar eclipse in 1919 and assure him international fame: "From this it follows that those light rays . . . are bent by the gravitational field."[15]

Einstein's thoughts were focused on the general theory of relativity even before the special theory had been established. It took physicists quite a while to understand it; many clung to the well-established electromagnetic theory of Hendrik Lorentz and did not want to move away from the concept of the ether.

Henri Poincaré, who himself was just a small but significant step behind Einstein in figuring out the special theory of relativity, virtually ignored the work of the unknown scientist from Bern until his own death in 1912. After a single encounter between the two in the fall of 1911 in Brussels, Einstein wrote to Zangger, "Poincaré was simply negative in general, and, all his acumen notwithstanding, he showed little grasp of the situation."[16]

Even so, Einstein was criticized for failing to acknowledge the effect of Poincaré's work on the development of his theory. Only once, in a letter to his friend Michele Besso in 1952, did he mention the Frenchman alongside Ernst Mach and David Hume as having had "quite an influence."[17] This was a pretty paltry tribute to a scientist who may have provided a crucial idea for his theory.

Einstein reacted in a similar manner to an experiment that two American scientists, Michelson and Morley, conducted in 1887. Their interferometer substantiated the constancy of the speed of light, which was one of the two pillars of Einstein's special theory of relativity, yet Einstein's 1905 paper failed to mention it. Even later he did not make this point clear. He contradicted himself, claiming on one occasion that he knew nothing about the experiment, and on another that even if he did, "I was not conscious that it had influenced me directly."[18] It is certain that he knew at least the summary of the experimental findings, which he had read during his independent studies in 1899.

The theory of relativity took quite a while to gain acceptance, and Einstein was getting fed up with his strange, splintered routine, going back and forth between his duties at the patent office and his glorified hobby of physics research. In January 1908 he asked his friend Marcel Grossmann for advice, "at the risk of having you ever so gently poke fun at me . . . I would like to make an attempt at a teaching position at the Technikum Winterthur," he wrote. "Don't think that I am driven to the pursuit of such an overambitious path by megalomania or some other questionable passion; this craving comes only from my ardent wish to be able to continue my private scientific work under less unfavorable conditions."[19]

It seems strange that Einstein, whose reputation as a physicist was steadily growing, was seeking employment as a teacher. In January 1908, shortly after completing his essay for Johannes Stark's almanac, he applied for a teaching position in mathematics at a high school in Zurich "while also noting that I would be ready to teach physics as well."[20] Of the twenty-one applications submitted, his did not even make the short list.

His "craving" was preceded by yet another misguided attempt to launch his academic career. In 1903, Einstein had tried to obtain his final credentials for a professorship at the University of Bern. In June 1907, he tried once again, but now he had "17 papers from the field of theoretical physics" under his belt. In late October the senior faculty decided at their faculty meeting to "deny his application until Mr. Einstein has submitted a *Habilitation* thesis."[21]

Einstein had no choice but to abide by the regulations and write the required thesis. On February 24, 1908, the faculty voted to "accept the *Habilitation* thesis and to invite Mr. Einstein to a sample lecture."[22] Four days later, Einstein presented himself for the lecture and colloquium, and one day after that he was granted the *venia docendi* — permission to teach. Einstein now qualified as a lecturer.

His run on the academic treadmill was far from over, however. In the summer semester of 1908 he offered a course at the University of Bern, which met on Tuesdays and Saturdays at 7:00 a.m. so that he would get to work on time by eight o'clock. Only three participants showed up: his two friends Michele Besso and Heinrich Schenk from the patent office and his former student, Lucien Chavan, who was now in the working world. In the winter semester, when he had moved his course to the evening hours, Max Stern was the only enrolled student to join the small group. In the summer semester, Einstein's three old friends bowed out, and only Stern remained. Einstein canceled the lecture course.

In September 1908 the young lecturer missed the first public acknowledgment of his theory of relativity at the meeting of German Natural Scientists in Cologne. Although he really wanted to participate, he eventually decided against it, because "it was imperative for me to use my short vacation for rest and recuperation."[23] He should have gone. His former mathematics professor at the "Poly," Hermann Minkowski, who was now a professor in Göttingen, introduced the mathematical formulation of the theory with a now-famous remark: "Henceforth space on its own and time on its own will decline into mere shadows."[24]

The man who had inspired these powerful words was still eking out an existence as a "patent boy." For the time being, he had given up his hopes for a university appointment. "The business with the professorship fell through, but that's all right with me," he wrote to Jakob Laub. "There are enough teachers even without me."[25]

Then everything changed. In Zurich, a nontenured teaching position in physics had just been established because the only physics professor to date, Einstein's old friend, Alfred Kleiner, had been appointed rector of the university. In the early summer of 1908, Kleiner, who was considering Einstein as a candidate, traveled to Bern. He sat in on Einstein's course, "to size up the beast," as Einstein later reported to Laub.[26]

The visit could not have gone worse. "On that occasion I really did not lecture divinely—partly because I was not prepared very well, partly because the state of having-to-be-investigated got on my nerves a bit."[27] The story made its way around physics circles, and Einstein, as fearless as ever, wrote a letter of complaint to Kleiner. "I seriously reproached him for spreading unfavorable rumors about me and thus turning my position, which is already so difficult, into a final and definitive one. For such a rumor destroys any hope of getting into the teaching profession."[28]

Kleiner decided to give Einstein a second chance. The candidate would be asked to give a lecture in Zurich in mid-February. This time "I was lucky. Contrary to my habit, I lectured well on that occasion."[29] Kleiner's recommendation praised Einstein's "extraordinary acuteness in the conception and pursuit of ideas, and a profundity aiming at the elemental."[30] One of his most cordial supporters in the process was the dean of the medical faculty, his friend Heinrich Zangger. Einstein, however, ultimately got the job only because the committee's top choice came down with an incurable form of tuberculosis. In mid May he reported to Laub, "So, now I too am an official member of the guild of whores."[31]

He stuck to his guns during the salary negotiations. They initially offered him far less than the patent office was paying him, "but I refused to accept the job under such conditions."[32] Eventually he got the 4,500 francs per year he had insisted on, plus auditor and examination fees, and his appointment was set to begin on October 15. Before Professor Einstein even appeared in front of his students for the first time, the University of Geneva granted him an honorary doctorate on July 8. Two days earlier he had given notice as of mid-October at the patent office. He was now thirty years old.

On September 21 the new professor was a guest of honor at the Congress of Natural Scientists in Salzburg, where he had the opportunity to get acquainted with the preeminent German scientists. A few decades later, scientists began referring to his well-attended lecture on the nature of light as a turning point in theoretical physics.

His new post, which entailed eight hours of teaching per week, took up more of his time than his office work had. Einstein had to develop his lectures and prepare for each unit for hours on end. To the elation of his students, he introduced a new teaching style. He cultivated the type of camaraderie with them that did not become widespread until the 1970s. "During the breaks," recalled one of his students, "he was often surrounded by students attempting to ask questions. He was patient and forthcoming, and he tried to answer them."[33] After the evening seminar he often went to the Café Terrasse with the young people and continued his discussions with them until closing time.

In the summer of 1910, Einstein turned his attention to the blueness of the sky. His interest was of course purely theoretical, and his point of reference was the recently discovered "opalescence," an intense scattering of light in liquids and gases caused by density fluctuations. As a kind of sequel to his two 1905 essays on atoms and molecules, he aimed at redetermining the dimensions of molecules in an entirely new way. He noted in passing why the sky opalesces blue during the day and shines red at dusk. Sunlight comprises all the colors of the spectrum and is therefore white. When it enters the atmosphere of the earth, it is scattered by small particles; blue light, with its shorter wavelength, is far more affected than red light. When we look up at the sky, we see the scattered blue light. The sun, however, appears to be yellow, not white, because a part of the blue is missing in its spectrum. The lower in the sky the sun is, the longer the path of its light through the atmosphere, and the more blue gets lost, and the more it thus appears to be red.

In March 1910, Einstein climbed the next rung of his career ladder. The University of Prague lured him with a full professorship and a "significantly better salary."[34] His professional career was now catching up with his scientific career at a rapid pace. The university recognized the significance of having Einstein join the faculty. Max Planck had just published a book that paid tribute to the historic theory of relativity: "This principle has brought about a revolution in our physical picture of the world, which, in extent and depth, can only be compared to that produced by the introduction of the Copernican world system."[35]

In response to a petition from his students in Zurich to keep the beloved professor there, no matter what the cost, the university offered Einstein a salary raise to 5,500 francs a year. In September, however, he traveled to Vienna, the capital of Austria-Hungary, which included Prague. The imperial officials offered him the equivalent of a little over 9,000 francs, and he accepted. On January 6, 1911, the emperor put his signature on the appointment. Three months later, the Einstein family, which had now grown to four people with the birth of their son Eduard, arrived in the city on the Vltava River. They moved into a spacious modern apartment near Palacky Bridge. A housemaid, Fanni, took care of their now solidly middle-class household.

Einstein was quite content with his working conditions. "I have a magnificent institute here," he told Grossmann.[36] He wrote to Hans Tanner (the only doctoral student he ever had) in Switzerland that his post offered him the enticing combination of "a fairly good library and few official duties . . . although Prague is not Zurich."[37] Still, he found the bureaucracy annoying ("the paper-pushing in the office is interminable"),[38] and the sanitary conditions made Prague seem like the vestibule of the Orient. The brown tap water was a health hazard.

He liked the city itself more than he liked its residents. "Prague is marvelous, so beautiful that it alone would make a big trip worthwhile," he wrote to Besso, in an attempt to entice him to move to Prague. "Only the people are so alien to me. These are not people with natural sentiments; unfeeling and a peculiar mixture of class-based condescension and servility, without any goodwill toward their fellow men. Ostentatious luxury side by side with creeping misery on the streets. Barrenness of thought without faith."[39]

He felt cut off from his friends and colleagues. Right from the start, it must have been somewhat clear that Prague would be nothing more than a way station in his career path, albeit an important one, since he was finally able to find suitable working conditions and a decent salary.

The high point of the period Einstein lived in Prague did not take place in Prague, where he was so isolated, but in Brussels. The Belgian industrialist Ernest Solvay invited eighteen of the most prominent physicists to a summit meeting to discuss the most pressing problems in science. This first gathering, in late October 1911, began a tradition of Solvay conferences. Einstein met the greatest minds of his guild. Under the chairmanship of the Dutchman Lorentz, whom Einstein called "a living work of art,"[40] the speakers and discussants in the chic Brussels Grand Hotel Metropole included Max Planck, Marie Curie, Ernest Rutherford, and Henri Poincaré.

The conference did not yield any scientific advancements of note. "Nothing positive has come out of it," Einstein concluded.[41] It did, however, mark a decisive breakthrough in his career. At the age of thirty-two, he had arrived at the Olympus of physics, becoming a full-fledged member of the pantheon of science. There were no national borders here. On this highest level, fewer than a dozen researchers from around the world could converse with one another.

Einstein began to pave the way for his departure from Prague even before the meeting. Just one day after he had taken his oath of office on August 23, 1911, he entered into negotiations with the University of

Utrecht to secure an academic appointment. In September, Heinrich Zangger visited him from Zurich and discussed the possibility of bringing him to the ETH. This institution, formerly called the "Poly" (Polytechnic), had just been upgraded and granted all academic privileges.

To his Swiss colleagues, Zangger was trying to allay any doubts about Einstein's teaching abilities.

> He is not a teacher for mentally lazy gentlemen who merely want to fill a copybook with their notes and then learn it by heart for an exam; he is not a smooth talker, but anyone wishing to learn honestly how, deep down, to construct his physical ideas, carefully examine all premises, see the pitfalls and problems, review the reliability of his reflections, will find Einstein a first-class teacher, because all of that emerges impressively in his lecture, which compels intellectual participation and unrolls the whole extent of the problem.[42]

Shortly before Christmas, Einstein traveled from Prague to Switzerland, and departed soon after with an oral agreement for an appointment. On January 22, 1912, the Education Council recommended the appointment to the Council of Ministers, and in early February, Einstein had cause for celebration: "Two days ago I was appointed to the Polytechnic in Zurich (hallelujah!)."[43] He would be starting his duties in the summer. The teaching obligations were minimal, and the annual salary was set at 10,000 francs, plus a 1,000-franc bonus from the federal government.

Just after Easter 1912, Einstein traveled to Berlin to "talk shop," since he lacked suitable conversational partners in Prague. There he met up with two participants in the Solvay conference, Max Planck and Walther Nernst, and also with the world-famous chemist Fritz Haber and the young astronomer Erwin Freundlich from the observatory in Babelsberg, near Potsdam. Haber and Planck became the driving force behind the push to bring Einstein to Berlin. Freundlich had resolved to check the deflection of light by gravitation in the proximity of the sun that Einstein had predicted (although it was of course incorrect at the time). After 1920 he was in charge of the construction of the "Einstein Tower" near Potsdam, which was designed by the architect Erich

Mendelsohn. He would later become the director of the solar observatory, the purpose of which was to confirm the general theory of relativity. In all likelihood, Einstein had a second, life-altering tête-à-tête during his visit to Berlin: He met up with the daughter of his uncle Rudolph, his divorced cousin Elsa Löwenthal—née Einstein.

Prague fit logically into the rhythm of his development, just as the third stage of his career in Zurich would later. He returned to the institute of his student years as a renowned physicist with a full professorship and a magnificent salary. Before his career path brought him to Berlin, where he spent the height of his career, he had to tie up some loose ends in Zurich. The work on his major study, the general theory of relativity, was entering its crucial phase. His accustomed milieu and his friends, particularly Marcel Grossmann, proved invaluable.

Although Einstein was quite content in Zurich, the same old pattern was now repeated for the third time: he again entered into negotiations for an even better job. On July 12, 1913, Planck and Nernst visited him and made him an offer he could not refuse. Einstein had known that sooner or later these emissaries from Berlin would approach him with an offer, but their actual proposal astonished him nonetheless. The venerable Prussian Academy of Sciences had concluded its deliberations on July 3 with the decision to make Einstein an offer. The membership proposal stated, "In sum, it can be said that among the important problems, which are so abundant in modern physics, there is hardly one in which Einstein did not take a position in a remarkable manner."[44]

In accordance with custom, the members of the "Physical-Mathematical Class" voted on him with white and black balls. The results were overwhelmingly positive. Twenty-one white balls were cast in favor of the candidate, with only one black ball indicating rejection. (We do not know who cast the dissenting vote.) His salary was set at 12,000 marks a year, half of which was financed by the academy's own funds, and the other half by the industrialist Leopold Koppel. As if that were not enough, if he accepted the position, he would also be made a full professor at the Friedrich-Wilhelm University without any teaching obligations. He would have complete discretion over whether and when he would give lectures or offer seminars. What is more, he would be made director of an institute of theoretical physics to be created specifically for him within the newly established Kaiser Wilhelm Society, the forerunner of the Max Planck Society.

Even though this position would be a sinecure, essentially devoid of duties, and even though Nernst described Berlin as "the place in which eight of the twelve people work who have understood your theory of relativity," Einstein asked for time to mull over the offer. He would communicate his decision to Nernst and Planck the following day. When they returned from the excursion to Mount Rigi, he would be waiting for them at the train station, and if the answer was yes, he would be carrying a bunch of red flowers. He left no documentation of the thoughts that went through his head while he slept on that decision, but the prospect of moving closer to his cousin Elsa was a crucial factor in his deliberations. "She was the main reason for my going to Berlin, you know," he confessed to Heinrich Zangger in June, 1914.[45] The next day, he awaited his new academy colleagues at the train station carrying red flowers.

At first he told only his closest friends about his new position. The first time it comes up in the *Collected Papers* is in a letter to Elsa: "Next spring at the latest, I'll come to Berlin for good."[46] A few days later he said, "Seeing you regularly will be the nicest thing that awaits me there!"[47] To his colleague Jakob Laub, who was now a physics professor in Argentina, he confided, "At Easter I leave for Berlin as an Academy man without any obligations, rather like a living mummy. I am really looking forward to this difficult profession!"[48]

On November 22, 1913, the Prussian Academy made an official announcement that Einstein had been chosen as a member. Einstein sent a modest thank-you letter: "When I reflect upon the fact that each working day demonstrates to me the weakness of my thinking, then I can only accept the high distinction intended for me with a certain trepidation."[49] When he was leaving Zurich, he confided to his old colleague Louis Kollros, "The gentlemen in Berlin are gambling on me as if I were a prize hen. As for myself, I don't even know whether I'm going to lay another egg."[50]

The prospect of being able to continue working on his general theory of relativity without official duties must have made his acceptance much easier, but that was not why the Berliners had recruited him. His new colleagues expected him to make important contributions to atomic theory. They had the foresight to attribute a great potential to it at this very early stage, not only for science, but also for technology, industry, and new products. They hoped that Einstein would initiate an interdisciplinary research project and enhance the theoretical foundation of chemistry. However, nothing ever came of this interdisciplinary cooperation. After hearing Einstein's inaugural address in the newly constructed Unter den Linden academy on July 2, 1914—an address that, contrary to expectations, focused on the theory of relativity—Max Planck could "not resist the temptation to record my objection."[51]

When Einstein arrived in Berlin in April 1914, he could not help noticing that he had attained a level of prestige far beyond his years. Here he was, a man of thirty-five, surrounded in the academy by older men whose etiquette and clothing style seemed to outweigh any other priorities. On May 4 he wrote to his old mathematics professor Adolf Hurwitz, "In its habitude the Academy entirely resembles any faculty. It seems that most of the members restrict themselves to displaying a certain peacocklike grandeur *in writing*." He noted, however, that "against expectation I am managing to settle in well; only a certain drill with regard to attire, etc., to which I must submit myself on the order of some uncles so as not to be counted among the rejects of the local human race, disturbs my peace of mind a bit."[52]

Now he held the position in which he became world-famous; he would not give it up until forced to do so after nineteen years. There was virtually no way of advancing any further. On October 1, he became the director of the Kaiser Wilhelm Institute of Theoretical Physics, in accordance with the agreement. It did not have a building, however, and therefore took the form of a home office in his apartment. Still, he received an annual compensation of 5,000 marks and secretarial support. Until he left Berlin, the situation remained the same: "his" institute existed without any additional staff, primarily to award research funds. Even so, he was grateful, and wrote to Zangger in 1917, "Without these local colleagues I would surely have remained an 'unappreciated genius.' "[53]

He turned down other offers in 1918, such as an extremely generous joint appointment from the ETH and the University of Zurich. Instead, he accepted a teaching post that would bring him to Zurich for four to six weeks each year. Apart from that, he agreed to a guest professorship in Leiden in 1920 and a comparable position at Christ Church College in Oxford in 1930. His third, unintended career as a politically active star of science had begun. In 1929 he summarized the consequences upon turning forty, in his standard doggerel form:

> Wherever I go and wherever I stay,
> There's always a picture of me on display.
> On top of the desk, or out in the hall,
> Tied round a neck, or hung on the wall.
>
> Women and men, they play a strange game,
> Asking, beseeching: "Please sign your name."
> From the erudite fellow they brook not a quibble
> But firmly insist on a piece of his scribble.
>
> Sometimes, surrounded by all this good cheer,
> I'm puzzled by some of the things that I hear,
> And wonder, my mind for a moment not hazy,
> If I and not they could really be crazy.[54]

A flurry of prizes and distinctions followed in rapid succession. On December 31, 1920, he became the youngest member ever to be elected to the "Peace Class" of the order "Pour le Mérite." On November 9, 1922, he was awarded the Nobel Prize in physics for the year 1921. At his belated Nobel speech in July 1923 before a gathering of Nordic scientists in Göteborg, the pattern of his inaugural lecture in Berlin was repeated. Einstein did not speak about quantum theory, for which contribution he won the prize, but about "Fundamental Ideas and Problems of the Theory of Relativity."

Einstein faced one final professional decision in August 1932, when he accepted an appointment at the new Institute for Advanced Study in Princeton, where he planned to conduct research for half of each year and return to Berlin for the other half. Once he could no longer go back to Germany, he decided to remain in Princeton and turn down other offers from England and France. He now embarked on the final stage of his career.

The Americans really did get the famous professor in a state that was "rather like a living mummy." He no longer produced anything of lasting value other than his contribution to the Einstein-Podolsky-Rosen paradox. Over the course of more than three decades he had found that "the world of our sense experiences is comprehensible. The fact that it is comprehensible is a miracle."[55]

Others paid the price for his career. He left the saddest victims behind in Europe: his two sons, Hans Albert and Eduard, and their mother, Mileva. Their world remained closed off to his senses forever.

TEN

"DEAR BOYS . . . YOUR PAPA"

THE DRAMA OF THE BRILLIANT FATHER

Berlin, Anhalt Train Station, July 29, 1914, Wednesday evening at about nine o'clock. An unpleasant summer day was coming to an end. In the afternoon the thermometers had reached just sixteen degrees Celsius under a dismal, rainy sky. A bracing wind was blowing from the northwest. At the kiosk, the fluttering front pages of the newspapers were wedged fast to their stands.

"The Danger of a World War," screamed the headlines and the train station vendors. "The Battle for Serbia!" After more than forty largely peaceful years, Europe was now poised at the precipice of the worst armed encounter of its history. And on the platform, at the night train heading south, the thirty-five-year-old Albert Einstein was standing, enduring one of the bitterest defeats of his life.

In front of him was his family, whom he had forced to leave, and his friend Michele Besso, who stood by their side. Besso had come from Switzerland for the express purpose of picking up Mileva and the children. It was still broad daylight. A glance over at his wife—"we parted rancorously"—then the father bent down to his boys, to Hans Albert, called Adu, ten years old, and to little Eduard, called Tete, Tetel, or Teddy, who had turned four the previous day. He gave each of them "a last kiss."[1]

The black locomotive started hissing along, and the train with his friend and family eventually disappeared in the distance. The young superstar of German physics stayed behind on the platform, accompanied by his new colleague, the chemist Fritz Haber. Einstein had been in the German capital for just four months. Haber had spent the last few days trying to mediate between the couple—in vain.

When the two scientists left the imposing main hall of the train station and walked out into the wind on Askanischer Platz, Einstein broke down and wept. "I cried yesterday," he wrote the following day to his cousin and lover Elsa, for whose sake he was making his family move. Elsa was on a vacation in the Alps with her two daughters. "I bawled like a little boy yesterday afternoon and yesterday evening after they had gone."[2]

Four days earlier he had told her about the "meeting with Haber": "It lasted three hours. The way to a divorce has also been smoothed. Now you have proof that I can make a sacrifice for you."[3] And as if he needed additional evidence that he had made his decision between the women, he went to Elsa's apartment after the "final meeting": "Tonight I'm sleeping in your bed! It is peculiar how confusedly sentimental one is. It is just a bed like any other, as though you had not yet ever slept in it. And yet I find it comforting that I may lay myself in it, somewhat like a tender confidence."[4]

Einstein shed tears of loss and guilt, for his "boys" and for himself. "Such an affair is a bit similar to a murder!"[5] he had written ruefully to Elsa three days before taking leave of his family at the Anhalt train station. Mileva "perceives my conduct as a crime against her and the children," but he consoled himself with the thought, "You, d[ear] little Elsa, will now become my wife and become convinced that it is not at all so hard to live by my side."[6]

A man leaving his wife for another woman is a fairly commonplace occurrence, but Einstein wanted to make sure everyone understood that the woman he had abandoned was the guilty party. He recounted the same story to all his friends, describing her "barbaric nature," which, he claimed, made it impossible for him to stay.[7] "Life without my wife is a veritable rebirth for me personally," he wrote to his friend Heinrich Zangger in Zurich in the spring of 1915. "It feels as if I had ten years of prison behind me."[8]

Two weeks before the farewell, after he had moved out of the apartment they shared, he told Mileva, "After all that has happened, a comradely relationship with you is out of the question." He assured her of "proper comportment on my part, such as I would exercise toward any woman as a stranger."[9] We do not know exactly what had taken place. He did not bother his friends with the details. Mileva also kept mum. Two letters that she wrote him in early 1914 and late 1918 are the only available sources.

There is not a single letter from Elsa to Albert in the *Collected Papers*, although his replies indicate that there ought to be a big stack of them. To what extent was she an active partner in destroying his marriage? How much was Einstein influenced and pressured by her and her family? It is as though we were listening in on a dialogue in which we could only hear one of the two voices and have to imagine the rest. The story of Albert, Mileva, and their sons after the separation also needs to be reconstructed solely on the basis of his letters to them and a few scattered passages in his other correspondence that discuss their relationship.

He wrote to Elsa that his colleague Haber "also understands entirely that I could not live with Miza. It was not her ugliness, but her obstinacy, lack of accommodation and flexibility and tenderness that had ruled out a fusion. He does not consider me a hard inhuman person but loves me just as much as before."[10] He even appealed to Mileva's girlfriend Helene Savić to justify his position. "Separation from Miza was a matter of life and death for me. Our life together had become impossible," he wrote in September 1916. "She *is* and always *will remain* an amputated limb to me."[11]

What does that mean? Was he feeling phantom pain? He did not even mention Mileva's feelings, nor did he express any self-criticism. Instead, the relationship as a whole fell into disrepute from its beginnings. The former object of his affections was pronounced a "living plague that had made my life so hard since my youth is gone."[12] However, the love letters that Robert Schulmann of the Einstein Papers Project tracked down in the 1980s tell a different story about the early years of their relationship.

Whatever had "happened" and was so "hard" about living with him was not the crucial point. Elsa had to find that out for herself, but at first she would "have to perform wonders of tact and restraint so that you are not looked upon as a kind of murderess; appearances are very much against us."[13] On the same day he wrote, "Now after all my ruminations and work I shall find a dear little wife at home who receives me cheerfully and contentedly."[14] Mileva, the mother of his sons, who had been a failure at her chosen profession in large part because of motherhood, had not been able to offer him this carefree and uncritical cheerfulness, which he required to free himself up for his science.

Just four days later, Einstein was already backing down with respect to Elsa and taking back his promise to marry her. "It is not a lack of true affection which scares me away again and again from marriage! Is it a fear of the comfortable life, of nice furniture, of the odium that I burden myself with, or even of becoming some sort of contented bourgeois?"[15] One last time, the youthful rebel in him was resisting the inevitable. His postscript read, "Best regards to the ex-stepchildren!"

Einstein was now living in a metropolis. He tried to steer clear of its hectic atmosphere as much as he could, even though this was a time of war. "Despite being in Berlin, I am living in tolerable solitude," he wrote to Zangger.[16] In May 1915 he confessed to his friend, "My life here is ideally pleasant if you disregard things that actually have nothing in the least to do with me."[17] A year later he stated, "I am living peacefully and contentedly in my quiet cell, into which no newspapers penetrate."[18] Fortunately for us, his most important friends were far away, resulting in copious correspondence that conveys a sense of Einstein's private life during these years. His letters read like a series of written monologues. "I am musing serenely along in my peaceful meditations," he told his friend Paul Ehrenfest.[19]

He had no institute, barely any duties, not even marital ones, and little contact with the outside world. He hid himself away, lived in meager conditions, ate canned food, immersed himself in science, and often worked throughout the night. Any disturbance at all was one too many. After all, he was just about to turn the world of physics upside down once again with his general theory of relativity. In the final weeks before he finished it up, between late September and early November 1915, he went through the most intensive work phase of his life. Anyone who crossed his path stood in danger of getting a taste of his ruthlessness. Quite a lot of porcelain was smashed during this period. His relationship with his sons, particularly Hans Albert, who was old enough to recognize what was amiss, suffered a permanent blow. "The emotional price that Einstein must pay for creative isolation is indeed high," the editors of his Collected Papers note.[20]

In late 1914 he moved out of the apartment in the suburb of Dahlem, which Mileva had rented for the family, and into a place of his own on Wittelsbacherstrasse in the center of the western section of Berlin. It took only a couple of minutes to walk from this new apartment to Elsa, her daughters, and her parents (his aunt and uncle). Haberlandstrasse 5 became his own address about three years later. The temporary bachelor was taking a breather before mastering the art of succumbing to the conveniences of a comfortable bourgeois life without becoming bourgeois himself.

He had told Mileva on August 18, 1914, that "I do not intend to demand the divorce from you, but only that you stay in Switzerland with the children."[21] Elsa was not satisfied with this state of affairs; she wanted her conquest all for herself. For five years she fought to make Albert her husband, five years of back-and-forth that entailed his threatening Mileva with divorce, promising to marry Elsa, and backing down.

At least Einstein could rest assured that his wife and sons were safe. On August 2, 1914, when Germany declared war on Russia, and World War I was beginning its disastrous course, Einstein was entering a private war of the kind waged by countless men when they leave their families: a battle over money and the children. He wrote to Mileva, "I have kept only very little for myself, namely the blue sofa, the rustic table, two beds (originating from my mother's household), the desk, the small chest of drawers from my grandparents' household, unfortunately also the electrical lamp you want, not knowing that you are attached to it."[22] Even for prominent people, the little things loom large when a marriage is at stake.

Even though Einstein had a steady income, his financial obligations to Mileva brought him to the brink of bankruptcy. "I would have transferred even more money to you, but *I have no more myself* to be able to manage at all without help."[23] Like so many others of his gender, he learned that living separately can be more costly than living together, and that buying one's freedom by going through a divorce, which in Einstein's case was finalized in 1919, can be financially ruinous.

Their haggling over money, however, was nothing compared to the battle over the children. From that teary July 29 on, Einstein and his sons would be on an emotional roller coaster, an up and down between loving affection and harsh rejection, with the father seeking to lay the blame for these problems on his ex-wife. "You have taken my children away from me and are ensuring that their attitude toward their father is vitiated."[24] Had he deserved this treatment?

"I have carried these children around innumerable times day and night, taken them out in their pram, played with them, romped around and joked with them," he rhapsodized to Elsa. "They used to shout with joy when I came."[25] And now? "Greetings from me to the children do not seem to be relayed," he reproached Mileva on December 12, 1914, "otherwise they would have sent their regards to me at least once in this long period."[26] Many fathers who leave their families experience this tangle of censure and fear, helplessness and obstinacy.

"The upkeep of all of you has been generously provided for," Mileva read, "and I find your constant attempts to lay hold of everything that is in my possession absolutely disgraceful. Had I known you twelve years ago as I know you now, I would have considered my responsibilities toward you at that time quite differently."[27]

Children and money, money and children—a nonstop back-and-forth. In the same letter he demanded "that no pressures be exerted on the child aimed at giving him a distorted image of me."[28] He added, "If I see, however, that Albert's letters are prompted, then I shall refrain from sustaining a regular correspondence out of consideration for the children." He did not realize that he was only making things worse with threats of that kind. The latent fear of losing his children was transformed, as it is with so many fathers, into passionate hatred of the mother forced to bring up the children herself without his participation. The other aspect of this disastrous entanglement in the father-son relationship was the boys' constant quest for love from their larger-than-life father. "Dear Papa," Hans Albert wrote in April 1915, when he was ten years old, "Today we told each other our dreams. Tete suddenly said, 'I dreamed that Papa was here!' "[29]

In the summer of 1915, Einstein spent an eight-week vacation with Elsa and her daughters in the little town of Sellin, on the island of Rügen. The vagabond got to see the sea for the first time in his life. "It is wonderful here," he wrote to his friend Zangger. "Never have I been able to rest so well since adulthood."[30] Thus fortified, he went to Zurich in early September for his first reunion with the family he had left. This visit was preceded by a crisis that would recur quite frequently. "My son has written me a very curt postcard in which he decidedly rejects going on a tour with me," he had angrily written to Zangger back in July.[31] This card is one of the supplementary documents in Volume Ten of the *Collected Papers* that provide new insights into Einstein's difficult and guilt-ridden relationship with his children. "Dear Papa," Hans Albert wrote, "You should contact Mama about such things, because I'm not the only one to decide here. But if you're so unfriendly to her, I don't want to go with you either."[32]

Evidently Einstein had attempted to make arrangements directly with Hans Albert for a "trip with just you alone for a fortnight. Then I'll also tell you many fine and interesting things about science and much else."[33] Apparently the trip was not quite so harmonious. "Twice I was together with the children," he wrote to Elsa. "After that, standstill. Cause: Mother's fear of the little ones becoming too dependent on me."[34] Two days later he added, "I saw the children only twice; *she* appears to have become distrustful."[35]

Still, Einstein now knew that his friend Zangger would "attend to my children."[36] In October he was pleased to learn "that my boy has come to see you. Do ask him whether he has received my long letter." Neither a letter nor a possible reply has been located. "But if I see that the woman is thwarting all of my efforts to remain in contact with the boy, I shall use legal means after all to enforce that the boy spend a holiday month with me every year."[37] However, he also saw the basis for part of the problem: "I noticed distinctly that in front of the others Albert and his mother were uncomfortable that I was in Zurich and did not stay with them. I understand that."[38]

Early November brought the relief he was hoping for: his son got in touch with him. "I was already afraid you didn't want to write to me at all anymore," he confessed to Hans Albert.[39] He owed the resumption of their correspondence to his friends Anna and Michele Besso, who lovingly lent their support to Mileva and the children. "We acknowledge the legitimacy of your wish," they wrote to him around October 30, "to communicate with your children without disturbance, but also the legitimacy of your wife's reservations that this not take place close to your Berlin relatives."[40] In the separation agreement, Mileva had been assured "that Mrs. Einstein would *never* have to yield the children to Einstein's relatives."[41]

The Bessos felt that Mileva was right, and concluded, "The discord resulting *from this* can only eventually place the child's emotional harmony at risk. If you enforced it, then in both of our views the time would inevitably come when the suffering caused by it, also for you, would far outweigh the present conveniences. For *the children,* no contact would without a doubt be better than contact *in Berlin.*"[42]

Moreover, his friends insisted on "a written confirmation so as to exclude the possibility that a means of financial pressure remain that could be used toward achieving some unforeseen purpose."[43] That did the trick. Einstein heeded the adults like a child. He gave in and softened his tone. "Your letter honestly pleased me," he told Mileva on November 15, 1915, "because I can draw from it that you don't want to undermine my relations with the boys." His greatest concern was her influence on "his" boys, whom he wanted to see grow up to be like their father.[44]

The letter provides some insight into how cleverly and patiently she attempted to clarify her relationship to her "dear Albert." Had the circumstances been different, Einstein would probably have reacted to it far more indignantly. She asked him not to make any appointments with the children, but to ensure that "matters relating to the children first be arranged with *me.*" She reasoned, "believe me that if Albert had the feeling that what is being demanded of him is done with the consent of both parents, he would much sooner succeed in calmly appreciating you than if he has the feeling that you were working as an enemy of this little world I have built up here for the children in which they are living and which they love."[45]

Einstein promptly wrote to Hans Albert, "I would like to come to Switzerland around New Year's."[46] A scant two weeks later their relationship took another turn for the worse. The father was treating his son like an adult. He threatened, carped, chastised, and overreacted crossly, feeling that he had been slighted—like a child, but not the way a child could grasp. "I see from your long delay and from the unfriendliness of your letter that my visit would bring you little joy. Therefore I consider it not right that I sit in the train 2x20 hours without the result of making someone happy. I'll come to visit you again only when you ask me to do so yourself."[47]

Again Besso was obliged to intervene and stand up for Mileva: "It is more difficult for her to understand you than you her—and it always has been more difficult for her, if only because the role as the wife of a genius *never* is easy."[48] One day later he sent off a letter to Mileva: "A string of misunderstandings has prevailed."[49] Einstein laid on the charm and did all he could to get her to "entrust [Hans Albert] to me from time to time. For my influence is limited to the intellectual and the aesthetic. I want to teach him mainly to think, judge, and appreciate objectively"—as though the mother could not or would not impart all of that to the boy.[50] "Your letter, which just arrived," he wrote on December 10, 1915, "prompts me to travel to Switzerland now after all. For there is a faint chance that I'll please Albert by coming. Tell him this and see to it that he receives me fairly cheerfully."[51] If the boy really was happy, he must have been crushed to learn shortly before Christmas, "It's so difficult to come across the border. . . . That's why I cannot come to see you now. At Easter, however, I definitely will visit you."[52]

Before this Easter visit rolled around, Einstein was again stirring up trouble for the "little world" of his distant family in February 1916: "I propose herewith to turn our now tested separation into a divorce."[53] One month later he declared, "For you it involves a mere formality, for me, however, an imperative duty. Try to imagine yourself in my position for once. Elsa has two daughters, the elder of whom is eighteen years old, that is, of marriageable age. This child . . . must suffer from the rumors that are circulating with regard to my relationship with her mother. This weighs on me and ought to be redressed by a formal marriage."[54]

The burden, as he saw it, was not Mileva's but his own, because the daughter of her rival was "suffering." What she was enduring as an abandoned woman and a single parent was a matter of indifference to him. In an apparent attempt to console the woman he had betrayed, he added cynically, "I shall never give up the state of living alone, which has manifested itself as an indescribable blessing."[55]

We do not know what Mileva said in response to this comment, but from Einstein's next letter we can deduce that her resistance was gradually crumbling, and she was heeding the dictates of reason. "I have prodded myself into action now," he wrote, "and, with your consent in principle, have discussed the matter of the divorce with a lawyer."[56] However, he added a threatening remark about the upcoming visit: "I hope confidently that this time you will not withhold the boys from me almost entirely again. If you repeat what you did the last time, I shall not return to Zurich so soon."[57]

When he got together with his sons there one week later, he sent her "my compliments on the good condition of our boys," and thanked her for their "proper upbringing" and "that you have not alienated me from the children."[58] As we know from his postcards to Elsa, the hike with Hans Albert really did take place, and Einstein was pleased with their rapport. "The boy delights me, especially with his clever questions and his undemanding way. No discord exists between us."[59]

Einstein pushed his children to perform; he placed great importance on their becoming "well-rounded individuals," and stressed intellectual rather than emotional maturity.[60] "You still make so many writing errors," he nagged Hans Albert.[61] Einstein later confided to Hans Albert about his brother Tete, "Digging deep does not appear to be his passion, but there have to be people who simply take pleasure in God's creation—perhaps that is the true purpose of the latter."[62] He had already given up hope that they would follow in his footsteps. "I cannot see them as my temporal successors," he wrote to Elsa in October 1920. "They have large chubby hands and for all their intelligence something indefinably four-footed about them."[63]

Now that Einstein was in Zurich, Mileva evidently wanted to discuss the divorce arrangements with him in person. He dodged the issue, however, and delegated it to the lawyers. Hans Albert tried to stand by his mother. A couple of days after Einstein's harmonious excursion with his son, he wrote to his cousin, full of indignation, "He urged me to call on his mother. When I resolutely declined, he became stubborn and refused to return in the afternoon. That is how it remained; and I saw neither of the children since and also did not arrange any more meetings. I should see the children only when they are not also under their mother's influence; only in this way are serious conflicts avoided."[64]

Mileva suffered a severe nervous breakdown as a result of the unrelenting pressure. Einstein categorically ruled out the possibility of visiting her "after the bad experiences at Easter"—"partly on an inalterable resolve, partly also to spare her the agitation." Instead of showing concern for her condition, he insinuated that she was just trying to pull the wool over their friends' eyes. "From your letter it appears that my wife really does seem to be seriously ill," he wrote to Besso. "I personally have the suspicion that the woman is leading both of you kindhearted men"—the other one being Zangger—"down the garden path." In case his friend had failed to notice, "You have no idea of the natural craftiness of such a woman."[65] He wrote to Zangger, "Whenever my wife confided in any one of my friends, I almost always had to give him up for lost."[66]

Besso assured his friend that in his opinion there was no way that Mileva was feigning illness. "The woman's sufferings had left their mark clearly on her appearance for a long time already; she did not let herself go either, but overburdened herself with too much work."[67] Einstein promptly replied, "As concerns my wife, please consider the following. She has a worry-free life, has her two fine boys with her, lives in a lovely area, can use her time freely, and basks in the aura of abandoned innocence."[68]

Evidently he was counting on the solidarity of his male friend in his attitude toward Mileva. That Besso and Zangger were siding with Mileva and asking him to listen to reason alarmed him. "Dear Michele, We have understood each other well for 20 years. And now I see you developing a bitterness toward me for the sake of a woman who has nothing to do with you. Resist it! She would not be worth it, even if she were a hundred thousand times more in the right!"[69]

On July 25, 1916, he wrote to Hans Albert, "I cannot get away right now, because I have a lot of work to do."[70] The very same day, however, he told Zangger the real reason: "I am very afraid that my wife will express the wish that I visit her. . . . On this occasion I could be forced to make promises regarding the children that result in the boys being taken away from me, also in case of the woman's death."[71] Still, he demonstrated concern since his friend had told him off: "I am very sorry for the woman, and I also believe that her difficult experiences with me and through me are at least partly to blame for her serious illness."[72]

In the final analysis, his assessment of the situation in a letter to Besso in late August reveals what a somber turn his thoughts had taken: "If it is cerebral tuberculosis, as seems probable, a quick end would be better than long suffering." He would not be shedding any tears for her.

Something else was bothering him: "My Albert is not writing to me. I believe that his negative attitude toward me has fallen below the freezing point." In a sincere combination of frustration and paternal pride he added, "In his place, under the given circumstances, I also would probably have reacted in the same way."[73]

On September 6, he finally gave in. "From now on, I'll not trouble her any more with the divorce," he announced to Besso. "The corresponding battle with my relatives has been fought. I have learned to withstand tears."[74] Mileva enjoyed her final victory against Elsa. Einstein's main problem was not solved, however. "I am writing you now for the third time without receiving a reply from you," twelve-year-old Hans Albert read in a letter from his father. "Don't you remember your father anymore? Are we not even going to see each other again?"[75]

Two weeks later the ice was broken once again. His entreaties had been heard. Hans Albert got in touch with his father. "Although I am over here," Einstein implored his sons to realize in his reply the very same day, "you do have a father who loves you more than anything else and who is constantly thinking of you and caring about you."[76] In November he was back to his old strident self. "It is not through joys and pleasantness that a decent fellow develops," he informed his eldest, "but through suffering and injustice. Your father's path was also not always strewn with roses like now, but rather more with thorns!"[77]

An alarming message from Besso in early December 1916 concerning Mileva's health also indicates that we are missing many of Einstein's letters: "She has to lie still again and is understandably discouraged by the return of the attacks after about five weeks' respite. It seems that the recent deterioration coincides temporally with a letter (from you?) that little Albert had supposedly received and that he did not want to show her."[78]

Einstein's letter to Hans Albert in January 1917 shows that Besso was right: "I am glad that now Mama is feeling better again; you may always show her what I write you."[79] Now there was new cause for concern, however: since the beginning of the year, six-year-old Eduard had been bedridden with a severe case of pneumonia and a high fever. His father's reaction: "Bearing it, not bewailing it, is the solution. We care for the sick and console ourselves with the healthy."[80]

Einstein sided with his older son. "I draw strength from the idea of taking Albert out of school and teaching him myself and, where I fall short, supplementing it with private tutoring," he confided to Besso on March 9, 1917. "I think that I could offer the boy very much, not just intellectually."[81]

However, he considered his younger son's prognosis grim: "My little boy's condition depresses me very much. It's impossible that he will become a fully developed person." As he had earlier in the case of Eduard's mother, he now came to terms with the possibility that Tete would die. "Who knows whether it wouldn't be better for him if he could depart from us before he really came to know life! I am to blame for his being and reproach myself, for the first time in my life."[82]

In a rather bizarre amateurish diagnosis, he connected Tete's illness to a "glandular swelling that my wife had at the time" when their youngest child was conceived. He claimed not to have known about "scrofula," a form of tuberculosis of the lymph nodes, with "hereditary risks for the children," back then.[83] "The genetic makeup of our children is not flawless anyway," he explained to Mileva.[84] He put it more bluntly in a letter to Zangger: "I begot children with a physically and morally inferior person. . . . I would be very unhappy if I believed that I could have begotten valuable progeny with another woman."[85] For decades to follow, he would blame himself for Tetel's "lamentable condition." Finally, on August 4, 1948, when he learned of Mileva's death in Switzerland, he confided to Hans Albert, "If I had been fully informed, he would not be in the world."[86] "To keep something alive that is not viable beyond the years of fertility is undermining civilized humanity," Einstein lamented in 1917. "So it would be urgently necessary that physicians conducted a kind of inquisition for us with the right and duty to castrate without leniency in order to sanitize the future."[87] This statement is chilling indeed.

Like many of his contemporaries, Einstein was positively obsessed with genetics. Along with psychoanalysis, genetics was becoming one of the trendiest topics of his era. In 1900, three researchers had rediscovered the laws of heredity proposed by the Augustinian monk Gregor Mendel. The next year a *mutation* would be described for the first time. In 1906 the term *genetics* was coined, and in 1909 the term *gene* entered the language.

Einstein kept returning to this touchy subject. He was constantly on the lookout for similarities between himself and his sons, particularly his firstborn son, Hans Albert. He also made a genetic connection between Mileva's precarious state of mind and the mental illness of her sister Zorka and his son Eduard. "Unfortunately all indications are that the severe hereditary flaw will have a pronounced effect on him," he wrote to Besso in 1932. "I have seen it coming slowly but surely since Tetel's childhood. The outward causes and influences are of minor significance in cases like these, compared to secretory causes, which are hard to identify."[88]

Although the role of hereditary defects in shaping our personalities is generally accepted, we can easily imagine that the boy suffered significant emotional damage when he had to move as an infant to Prague with his unhappy mother. And how had all the marital strife affected his psychological well-being? The teary farewell at the Anhalt train station must have been a traumatic experience. Tete's father turned his back on him just as his lifelong odyssey through hospitals, sanatoriums, and psychiatric clinics was beginning—an odyssey full of upswings and relapses, releases and readmissions. It is the sad story of a talented son of a genius and his father's unremitting guilt about the hereditary and parenting conditions that gave rise to his son's condition.

On the advice of friends, the Einsteins sent their frail son to the mountains to convalesce. "Of course, I approve of his being brought to high altitude for a year. It is only human!" Einstein wrote to Zangger. "I am inwardly convinced that it would be in the public interest to imitate the Spartans' method."[89] Nine months later he had had enough. "I definitely want Tete to be brought back down again from up there in the New Year. I am opposed to having such a child spend his youth in such a disinfector."[90]

In early 1917, Einstein's own "cranky body"[91] began to exact its revenge for years of living an unhealthy lifestyle. In August 1913 he had boasted to Elsa, "I have firmly decided to bite the dust with a minimum of medical assistance when my time has come, and up to then to sin to my wicked heart's desire. Diet: smoke like a chimney, work like a horse, eat without thinking and choosing, go for a walk *only* in really pleasant company, and thus only rarely unfortunately, sleep irregularly, etc."[92]

Now he had to pay the price. His doctor diagnosed gallstones and recommended that he recuperate in Tarasp, a spa in Switzerland. For the time being, he was helped by a "spa treatment, strict diet. You surely know that our Zangger has arranged for the appropriate fodder for me. I'm feeling significantly better; no more pain, better appearance."[93] In the course of the year 1917 his health problems even seemed to subside. "The constant outdoor air and the good care, in conjunction with the comfortable, peaceful existence, are having their effect," he wrote from Switzerland.[94]

While he was there, he had a visit from Hans Albert, who was now thirteen years old. The visit did not go quite the way he had hoped. "He has developed well, but often behaves a bit roughly toward me out of ingrained habit."[95] His friend Zangger, who, like Besso, was looking out for the boy's best interests, had warned him, "You should not subject Albert to the disappointment of expecting you and then not coming for the third time."[96]

In the second half of the year, Einstein's condition took a turn for the worse. "My stomach has become so sensitive and weak," he reported to Hans Albert in early December, "that I have to follow a diet like that of a small child. If this persists, that's the end of our traveling; because someone must always cook something separately for me."[97] On Christmas Eve he reported, "I am staying in bed for 4–6 weeks because of my stomach ulcer."[98] Evidently he was now suffering from jaundice in addition to the disorder of his digestive tract. It would take not four weeks for him to recover, but four years—which did not stop him, however, from working at a frantic pace.

In January 1918 he was given a conclusive diagnosis. "My ailment involves a stubborn ulcer at my stomach exit, which is healing only very reluctantly," he wrote to Hans Albert. Now that Mileva had suffered several severe relapses, Hans Albert was the only remaining healthy member of the family. "There is no doubt that I have had this illness longer than you have been in this world. . . . I am probably never going to become truly healthy again, but am going to have to eat a kind of infant food for the rest of my life." Then he told his son that Elsa would be preparing this food for him. "My cousin does a splendid job in seeing that I have enough food to 'peck' at."[99]

In September 1918, Einstein, who was severely stricken by his illness, conceded defeat; he gave up his bachelor apartment and moved to Haberlandstrasse to be cared for by Elsa. His lover became a caretaker who tenderly nursed him back to health. His weight had dropped below 143 pounds, which was 55 pounds lighter than his normal weight. One of the features of the special diet Elsa prepared for his "queasy insides" was sweetened rice pudding.[100] It was no secret to Mileva and his mother, Pauline, that the way to their Albert's heart was through his stomach. Now Elsa was exacting her price.

Einstein gave in to her pressure and agreed to marry her. He abandoned his resolution not to remarry. His decision was driven not by feelings of love, but by the realization that he needed to be nursed back to good health. Her family members were no innocent bystanders in this matter. "The attempts to force me into marriage," he wrote to Zangger, "come from my cousin's parents and are mainly attributable to vanity."[101]

Einstein let two years go by before making an offer to Mileva on January 31, 1918—from one sickbed to another: "The endeavor finally to put my private affairs in some state of order prompts me to suggest the divorce to you for the second time."[102] He wanted to buy his freedom, and talked up the "huge sacrifices" he was prepared to proffer. He offered her the entire money for the Nobel Prize he was firmly counting on, which converted to 180,000 Swiss francs—but made sure to reiterate his threat to her: "If you do not consent to the divorce, from now on, not a cent above 6,000 M per year will be sent to Switzerland."[103]

Mileva's reply shows how embittered she had become: "Exactly two years ago, you pushed me over the brink into this misery with such letters. . . . D[ear] Albert, Why do you torment me so endlessly? I would never have thought it possible that anyone to whom a woman had devoted her love and her youth, and to whom she had given the gift of children, could do such painful things as you have done to me. . . . You ought not to have subjected me to the relentless, insensitive, and callous protests about Tete either."[104] Another letter at this time reveals, however, how calmly and prudently she reacted: "You will understand that with my current illness it is difficult for me to come to a decision. I do not have an overview of the situation—I must first accustom myself to the idea, also for the children's sake. I understand that you want an unhampered future; I don't know whether it is necessary for you and your work, but I don't want to stand in your way and obstruct your happiness."[105]

But Mileva stuck to her guns concerning the financial arrangements. She regarded Albert's generous offer as the minimum to which she was entitled in a divorce—enough to live on comfortably if she handled her finances competently. Mileva and Albert were in agreement that their children needed to be well provided for. Money was again an issue when Einstein instigated another conflict. Beleaguered by the situation, he picked a fight with his friends in Switzerland.

"I refuse to tolerate constantly being used like a gullible schoolboy," he barked at Besso in early January, after his friends had apparently criticized him about sending insufficient funds to the family. "We are faced here with a vicious circle unless I put an abrupt end to it."[106] And shortly after that, he confided to Hans Albert, who was almost fourteen, "My friends in Zurich, who are irresponsible in this regard, are to blame for this calamity."[107]

He was quite annoyed by the escalating costs of Tete's stay in Arosa, and told Tete's brother that he was "firmly convinced that it's wrong to pamper him up there for so long." He also groused about "the misfortune that an enormous amount of money is being spent so that all my savings are being used up."[108]

In any case he had a "gripe" against Zangger, who criticized his threats of divorce, calling them a "knife at the throat without advance warning."[109] Einstein complained to Besso's wife, Anna, in early March "that all are united in making life unnecessarily hard for me."[110] He tried to win them over in the same immature manner he had used with Mileva: "Think of the two young girls, whose prospects of getting married are being hampered considerably under the present circumstances, through my fault."[111]

Einstein actually suggested removing his older son from Mileva's custody because of her illness and sending him to his sister Maja's house. When he also mentioned his new living arrangements with his nursemaid-lover—"Think of the difficulties I have at every turn because, owing to my illness, I am compelled to live in the same flat as Elsa"[112]—Anna Besso-Winteler blew her top.

"If Elsa had not intended to make herself so vulnerable," she replied, "she ought not to have run after you so conspicuously. A mother with children ought to know what she is doing. . . . The fact that you are ill is dire fate, but I do not understand how this should be a reason for marriage. . . . It was very misguided of Elsa—also of you— to want to attack these persons"—she meant her husband and Zanger—"both of whose intellects I esteem very highly (each in his own way). I will not let myself be blindfolded with open eyes."[113]

Oddly, when Einstein had finally had enough of his brother-in-law's sister's attacks a few months later, he vented his frustrations to Mileva: "She has written me such impertinent letters that I have put an end to further correspondence and am never going to be able to have dealings with her again."[114] He was even insensitive enough to lash out unapologetically at Besso: "Never before has anybody been so insolent to me, and I hope that no one ever will again in the future!"[115]

Mileva apparently agreed to his offer if he raised her annual payment without delay from 6,000 to 8,000 francs. "I am very willing to comply with your wishes," came his swift reply.[116] They bickered about the details until late 1918. It would be quite interesting to read Mileva's letter that prompted Albert's rejoinder in May 1918: "Only about death can we be *secure*, not about possessions of any kind. There's no changing this. . . . At all events, the best that I can bequeath to my boys, or that they can inherit from me, is not money, but a good mind, a positive outlook, and an unblemished . . . name that is known everywhere on Earth where science-loving people live."[117]

Shortly thereafter, Einstein attempted to seal the deal with a perfect proposition: "Securities for 40,000 M are being transferred to the Swiss Bank Corporation, Zurich, for you one of these days. I request, now, that you send the contract and file the divorce."[118] On the face of it, he seemed to take the matter lightly. "Dear Michele," he wrote in July 1918 to his friend, with whom he had since reconciled. "I have received your letter with the original divorce advice—Till Eulenspiegel."[119] The financial burden was eating him up inside. In one of the few dreams he recorded on paper, he imagined slitting his throat with his razor.

In June, Einstein finally had a small personal reason to celebrate. Although, as he wrote to Besso, he was "very sorry that I can't see my boys,"[120] he was overjoyed that Eduard had written to him for the first time. "My dear Tete!" he replied promptly. "I am very proud that now my second boy also can write already!"[121]

During his summer vacation at the artist's retreat in Ahrenshoop on the Baltic Sea—"I am lying on the shore like a crocodile, allowing myself to be roasted by the sun, never see a newspaper, and do not give a hoot about the so-called world"[122]—he waxed absurdly dramatic in his letter to the fourteen-year-old Hans Albert: "How much simpler it would have been for you to come to me if you had wanted. . . . So you see that you are reproaching me unjustifiably; maybe later one day you'll think that it would have been better if you had worried more about me at this time."[123]

Nonetheless, he continued to be "very happy about the nice letters my boys are writing me." He went on to tell Besso, "Your advice about my remarriage is well intentioned." Apparently his friend had warned him not to make the same mistake twice. "I'm not going to follow it, though. Because if I ever were to decide to leave a second wife as well, I wouldn't allow myself to be held down by anything."[124]

Writing to Zangger from the same vacation spot, he told his friend about his older son's career interests, adopting an arrogant tone that was not exactly suited to the role of a father: "Albert is already starting to think quite amusingly, oddly enough, about technical questions. In the end, though, I am glad for any mental quickness, even if it clings to narrow-minded views. Perhaps he will realize the superfluousness of the many conveniences sometime, after all."[125]

Biographer Peter Michelmore, who spoke to Hans Albert in person, describes the dramatic scene in Zurich: "Hans Albert was hostile to his father. Now a husky boy of fifteen and fiercely independent, he told his father he had definitely made up his mind to become an engineer. . . . 'I think it's a disgusting idea,' Einstein said. 'I'm still going to be an engineer,' the boy insisted. Einstein strode away saying that he never wanted to see his older son again."[126]

Hans Albert wisely chose not to accede to his father's wish to follow in his footsteps as a natural scientist, and pursued a technical route instead. It took years for his father to realize, "I am actually glad that neither of you dedicated yourselves to science, because it is a hard thing, full of difficult and futile work."[127] Eduard, by contrast, regarded his father as an idol whose ideals he could never attain. "The young one is a fine little fellow, with shy, bashful movements, not at all Einsteinian,"[128] Zangger reported to his friend in Berlin. This boy played music, read, and wrote poetry.

Eduard hoped to show at least flashes of brilliance when he later wrote highly artistic poetry and dabbled in literature to come to grips with his medical and psychological issues—as though he had to play the genius that he could not really be. Einstein felt uncertain and was inclined to rebuff his son because of the boy's sickliness, but he began acting proud and hopeful. Einstein's vacillating behavior toward his children inflicted permanent damage on them. As we have learned from the field of psychology, if a parent constantly alternates between rejection and love, a child may never be able to develop trust in relationships. Quite possibly this emotional roller coaster also caused Tete's emotional pendulum to swing to an extreme against his father. Gushy admiration alternating with brusque rejection, coupled with bouts of complete indifference, was testimony to his despairing inner rift.

On November 9, 1918, there was an epistolary handshake about the divorce: "I agree fully to the suggested procedure," Albert explained to Mileva. "See that you accelerate our divorce so that the 40,000 M are transferred to your name."[129]

One day before Christmas Eve, the secretary of the Royal District Court Berlin-Schöneberg entered the professor's deposition into the official record in the matter of *Einstein v. Einstein*. The opening words of document number 1286/1918 read as follows: "It is correct that I committed adultery. I have been living together with my cousin, the widow Elsa Einstein, divorced Löwenthal, for about 4½ years and have been continuing these intimate relations since then. My wife, the Plaintiff, has known since the . . . summer of 1914 that intimate relations exist between me and my cousin. She has made her displeasure known to me."[130] The divorce was granted on February 14, 1919, in the absence of both parties.

At this time illness and suffering took center stage in the drama of the brilliant father, who now faced a difficult test as a son as well: "My mother unfortunately is lying mortally ill in Lucerne," he wrote to Hans Albert in June 1919. "She is sure to die within a year and is suffering dreadfully."[131] In the summer of 1919 he traveled to Switzerland, where he visited his mother in the sanatorium. After his departure, he assured her that the "three days' sojourn in Zurich, which I spent very satisfactorily with Albert," were quite pleasant.[132] From Zurich, he fled to Schaffhausen to visit his friend Conrad Habicht. "It was high time, too, for Thursday evening the lioness came back to her den."[133]

He finally made a crass suggestion to the ailing "lioness": "It is necessary for you to move to German territory as soon as possible."[134] Owing to inflation and the dismal exchange rate — "our money is growing ranker . . . by the minute"[135] — Einstein was barely able to support his family in Switzerland on his German salary despite the lavish provisions of his job. He even urged her to "try to get a roof over your head by apartment *swapping* with a Baden family wanting to go to Switzerland."[136]

Over the next few years, this topic became an ongoing theme. When Hans Albert was old enough to enroll at a university, Einstein advised his wife and children to move to Darmstadt. "There is a good technical college there, and not only could you all live much better there than in Zurich, but also save substantial amounts of money, whereas now nearly my entire income goes to giving you a dismal standard of living in Zurich."[137] In the summer of 1938 he was still asking Mileva "to keep in mind that the monthly allowance I give Tetel would easily suffice for both of you in Yugoslavia."[138]

Rampant inflation was draining Einstein's wallet to the point that he became dependent on the generosity of his uncle and father-in-law Rudolph. However, Mileva rejected the idea of moving, and insisted on staying in her city of choice, Zurich. Eventually he relented: "It seems that a kind of vagrant existence is our destiny. Under prevailing conditions I can understand you well. So we'll postpone the moving problem by half a year."[139]

On November 16, 1919, when all the fuss about Einstein had just gotten under way, he wrote to Mileva, "I'm very much longing for a few lines from Albert and Tete. Tell them so!"[140] By late February 1920, contact between them had been restored. Hans Albert addressed his father tenderly in a letter as "Big Beast."[141] "I miss you thoroughly, too," Einstein replied.[142] And he made a request to both boys: "Both of you write me soon about school and whatever else you are up to."[143]

His financial distress found repeated expression in his letters to his sons as well. "Who knows whether one fine day I won't be forced to look for a job abroad after all," he pondered in a letter to Hans Albert on April 5, 1920. He was hoping that they would get together during their fall vacation: "You really should come to visit me then, with Tete if possible. A trip to Switzerland is too expensive for me."[144] In July he sent his sons practical tips for their travel: "Make inquiries at the German consulate, d. Albert, about what is required for you to obtain your travel permit. But also tell them you are the son of the Berlin resident Prof. Einstein."[145]

However, when Mileva steadfastly refused to hand over her children to his relatives (which is how she saw the situation), he was furious: "In the future, you should really stop forbidding at least [Hans] Albert to come to Berlin. It's simply ridiculous to treat a virtual grown-up like such a child. My wife will keep her distance from Albert; I could even take meals with him alone if he likes. But these are silly trivialities. One shouldn't have to make such a fuss for you old women."[146]

On August 1, he wrote to Tete, who was ten years old, "It hurts me often, too, that I have so little of both of you, but I am a very busy man and can't get away from here much. . . . The two of us were so rarely together that I hardly know you at all, even though I am your father. I'm sure you have only a quite vague idea of me too."[147]

In the summer of 1921, after returning from his first major trip to the United States, he finally had the chance to spend a vacation with the boys in northern Germany. "Albert is a wonderful fellow," he wrote to Mileva in Zurich. "Tete is clever, but of course still a bit of an embryo."[148] On a postcard, the little poet apologized to his mother, "Dear Mama! I am giving up sending you a poem every time; it is unbelievably hot, and the heat has quite a detrimental effect on my mind."[149]

After the vacation, Einstein wrote to her in regard to "our dear boys": "I am grateful to you for having raised them with a friendly attitude toward me." He could not help remarking "that the younger daughter"—Ilse, whom he would have liked to marry instead of Elsa—"is a very good, modest girl who cannot be held responsible for the deeds of her elders."[150]

This pleasant concord did not, however, last very long. Evidently Hans Albert refused to play the intermediary between his divorced parents. He remained loyal to his mother, and "an unpleasant and awkward scene" ensued. Of all the people Einstein could have confided in, he chose twelve-year-old Tete: "You can imagine how terrible this issue with Albert is for me; but no father can be treated the way Albert's last letter treated me. This letter is full of distrust, lack of respect, and a nasty attitude toward me. I truly did not deserve that and cannot put up with it, even if it would cause me great pain if there were to be an estrangement between me and Albert."[151]

The letters indicate that apart from an occasional thaw, the climate between Albert senior and Albert junior remained frosty over the next few years. However, while he "generally" considered the older boy a "wonderful fellow," he arrived at a grim diagnosis for the younger one with his layman's knowledge of psychology. "Tete's situation is far more complicated," he wrote to Mileva in 1925. "His intellectual abilities may be even stronger, but he seems to lack equilibrium and a sense of responsibility (egoism too strong). Too little personal interaction with others and too much ambition, which gives rise to a feeling of isolation and a kind of anxiety, along with inhibitions of other sorts. He is an interesting little fellow, but he will not have an easy time in life."[152]

In a letter to his sister, Maja, he characterized his sons as a classic study in contrasts: "Tete has shot up in height, almost taller than me and Albert, a bookworm, intellectual, messy, unreliable. . . . Albert is somewhat rough around the edges, a bit tyrannical, very competent and clever, orderly, responsible, reliable, not exactly good-natured and considerate. . . . But when it comes to our family, the boys have a deeply embedded thorn in their sides."[153]

Communications between Berlin and Zurich kept getting testier, at least as far as we can determine from the tone of the extant half of the correspondence, namely Einstein's letters: "I am very sorry that Tetel's illness has gotten so much worse, but I will not come to Zurich anyway, because I am quite certain that you will not be nice to me."[154] Again, the talk was of finances. Mileva had bought several apartment houses with the Nobel Prize money, which ought to have enabled her to live well. Einstein insinuated that she was not good at handling money, and he was probably right.

"Instead," Einstein went on to say, "you have availed yourself of every opportunity to keep getting money out of me. It is up to you to restore a sense of trust between us. All three of you can do that by sending me a legally binding declaration that either states that the Nobel Prize is to be applied to the children's portion of the inheritance or that you will not contest my will. . . . I know that you have not even told the children that you have received from me the total amount of the Nobel Prize. Quite the opposite: in Zurich you always tried to give the impression that you and the children have been deprived by me."[155] He was probably already contemplating leaving part of his estate to his stepdaughters. After his death, Margot would inherit more than the two sons put together. She was left $20,000 plus the house and its contents. Eduard received $15,000, and Hans Albert $10,000.

Einstein's dialogue (which we are in the unfortunate position of reading as a lopsided monologue) is quite revealing, but falls short of providing all the answers we hope to find. However, in the endless series of hanging files in the basement of the Einstein Papers Project in Pasadena, California, a new document in the rounded penmanship of a woman suddenly turned up amid the somewhat wobbly writing that typified Einstein's letters. Mileva was writing to Albert and Elsa—and in doing so adding a new dimension to our understanding of Einstein. Her letter is undated, and its recipients are addressed with abbreviations:

> D[ear]. A[lbert]. Thank you for your letter, and thank Elsa for her long letter. Your refusal to spend some time here made me sad. This issue has been bothering me constantly over the past few days, and you surely will not be upset if I write a few words about it. I hope that you are not forgetting completely that you have a very dear boy who is quite ill. If you only knew how often he has longed for you over this long year and a half of terrible suffering, you would drop everything and come to him. He kept asked me anxiously, "Do you think he might come?" and after a while, when you didn't come, he lapsed into silence.[156]

As if this were not heartrending enough, she added her ailing child's lament about his father's absence, which he purportedly expressed in hushed tones "about six weeks ago": "Well, what can he do with a son like me; I'm sick, after all." Then Mileva dealt him the final blow. "Since then he has not mentioned you a single time and does not join in any discussion about you. I think that his poor needy heart is deeply wounded because he is convinced that you do not want to see him, which only adds to his suffering."[157]

She offered him a helping hand: "Perhaps no one needs to find out that you are here; you could come in the evening, for instance, and take a curtained car to Teddi." The closing line of the letter delivered a punch: "After all, you give your help to anyone who asks for it—why not to your own child?"[158]

Mileva even directed a desperate appeal to Elsa, who, she declared, had always been "almost sympathetic and compassionate" to her: "D[ear]. E[lsa]. Because you are a woman and a mother, perhaps you are in a better position to understand how much I am suffering from the fact that Teddi has wanted to see his father for such a long time and this wish cannot come true for him."[159] There is no record of Einstein's reaction to this letter, but subsequent letters indicate that, at least in writing, he was growing closer to Tete while moving further apart from Hans Albert. Although the older son was now past the age of twenty, the conflict with him reads like a quarrel with a teenager.

The crisis escalated in 1925 when Hans Albert turned twenty-one and his father tried to interfere in his love life. In Einstein's estimation, as he confided to Mileva, "he is obviously quite inhibited when it comes to the other gender. . . . I want to try giving him some unobtrusive instruction as best I can."[160] The young man would not put up with that, however. And Einstein was anything but "unobtrusive." In his private war with his son, Einstein let loose a virtual barrage against Hans Albert's girlfriend and future wife Frieda Knecht—as though he needed to reenact precisely the manner in which his own mother had once tried to torpedo his relationship with Mileva. Since Mileva had apparently noticed the connection, he was forced to mollify her: "I am not trying to compare you to Albert's wife."[161]

He tried to make clear to Hans Albert, however, that "It all comes from the fact that *she* was the one to make the first move, and now you regard her as the embodiment of femininity. . . . Never send Miss Knecht to me, or bring her along with you, because as things stand, I would simply not be able to bear it. But if you should feel the need to break up with her, *don't stand on ceremony with me*; trust me to help you. The day *will* come."[162]

For years on end, his letters were full of tirades against Frieda. Einstein considered her too old for his son (she was nine years older), and also genetically inferior because she was short (barely five feet tall). Einstein worked himself into a frenzy over her alleged genetic shortcomings. He consulted the directors of the hospital in Berlin-Neukölln and the Burghölzli psychiatric facility near Zurich, trying to do his part, as he wrote to Hans Albert, "to avert the disaster" of Frieda having children. "I have nothing personal against Miss Knecht; a woman cannot be expected to have as much of a sense of responsibility, but you need to."[163]

He told Tete, who was only sixteen, "The deterioration of the human race is surely a bad thing, one of the worst possible things." Then the emotionally ill younger son also got his comeuppance: "Do you believe that your father sinned? Perhaps. In that case, forgive me for your existence. . . . In order for God not to regret his creation, it is in need of some luxury."[164]

Forgive me for your existence. Now that he had forfeited Hans Albert as his conversational partner, Einstein unfortunately increasingly turned to Hans Albert's fragile brother when he felt miserable and was in need of a confidant. "I see a real kindred spirit in him," he wrote to Mileva.[165] When Tete was just fifteen, he was told, "We became an animal on two feet by way of the monkey, a fleeting consciousness heavily enforced with atavistic instincts."[166] This was not just general philosophizing; the family's dirty laundry was also being aired. Einstein found himself in an emotional vacuum, and feebly played each son against the other.

"Albert is a good-natured, psychologically ingenuous individual," he wrote to his teenaged son Tete. "That woman is a sly, egoistic, disagreeable person who dominates him. She is more sophisticated than he, whose intelligence is thoroughly exhausted by every aspect of his job. He may well have an inkling that his emotional wings have been clipped."[167]

At Christmas in 1927, Eduard received a long letter from his father in which Einstein, evidently referring to an equally lengthy communication, launched into a psychological diatribe: "When I say that I wish you would show more of your animalistic side, I mean that natural feelings and actions should not be shortchanged by cerebral ones. The animal I have in mind is not the tiger, who is not an animal with social traits, and does not have our best instincts and feelings."[168] It is not hard to imagine how the young man feverishly devoured lines like these from his famous father. Didn't Einstein know that he was dancing on an emotional volcano? Did he lack the ability to realize how he was behaving? Or did he simply want to communicate to his son that he took him seriously the way he did an adult?

"I certainly accept your axiom that 'life has no ultimate purpose apart from life.' . . . Life in the service of an idea can be good if this idea is life-giving and emancipates the individual from the fetters of the self without propelling him into a different bondage. Science and art *can* work this way, but they can also lead to enslavement or unhealthy pampering and overrefinement. But I would dispute the notion that these efforts lead to an inability to cope with life. After all, even water is a poison if you drown in it."[169]

Water as the poison of the drowning. Einstein advised his son not to become a professional writer: "Creative work with literary matters is as a main occupation an absurdity, like an animal that eats nothing but lilies."[170] He declared his solidarity with his son's criticism of the university (Eduard had begun the study of medicine): "That is what happened to me as a student. You feel as passive as a stuffed goose."[171] He warned him against switching to psychiatry: "Only those who are extraordinarily strong can afford to leave the prescribed professional paths without coming to grief."[172] He criticized his son's "ditties" ("You have a very strict father") because they were "for my taste somewhat too clever and insufficiently natural and unaffected." He attempted to share his son's interests: "When you come to visit me, you will have to inform me about psychoanalysis; I promise to . . . concentrate on what you are saying and keep a serious look on my face."[173] He was pleased "to find that we are kindred spirits,"[174] and endeavored to interfere in Tete's unhappy love life just as much as in that of his brother: "You seem to be locked in a battle with the eternal feminine. Just don't get stuck with your older girlfriend, who is too sophisticated for you. Find an uncomplicated girlfriend instead, someone who is more of a friendly game to you."[175]

He once even tried his hand at solidarity with the suffering of his son: "I know the feeling I get when I appear to be nervous. It always happens when I see my actions or goals 'unravel': Wanting many things that are mutually exclusive, but nothing strongly enough to lead to un-interrupted action. This state can originate from without, for instance when I receive too many letters that I ought to reply to. . . . What is left is a kind of depression that is somewhat disengaged from what triggered it and impedes self-assured, firm action."[176]

He seems to have meant well, but Eduard knew psychiatry as both a student and a patient. Kindred spirit of his father or not—he was also mentally ill. Einstein's letter to Tete after the latter's psychiatric diagnosis of "manifest schizophrenia" proves that a show of empathy does not automatically result in sensitivity: "Being overwhelmed does not have to result in fatalism; there is no need to keep banging your head against the wall. . . . I notice in your letter that you are not nearly as calm as you used to be. . . . Don't make the same mistake that many anguished people make when they take themselves too seriously and think everything revolves around them."[177]

Did he really believe that advice of that sort would help someone who is psychologically ill? The next week he wrote to his son, who was passionately interested in psychology, "In a sense, you really ought to be glad that you have symptoms: The best way to gain a profound understanding of something is to experience it yourself." In another letter, he remarked, "I'm nutty too, but only in *my* way."[178]

While he was making these intense and dangerously clumsy overtures to his younger son, he was having it out with his older one. In his desperation he even tried to forge an alliance with Mileva. "You will start to notice that there could hardly be a more pleasant divorced man than I. I am devoted, not in the sense that a young girl pins her hopes on, but still devoted."[179] His attitude toward Mileva appeared quite different in a letter to his sister, Maja; he called Mileva "a big pig."[180]

"This business with Albert is the worst part. Sometimes I wonder whether he is normal," he wrote to Mileva.[181] He also told her: "[Hans Albert] recently wrote me an outrageous letter manipulated by female power."[182] He did not even discount the possibility that his son was acting as a spirit of discord: "He gives you the feeling that I am conspiring with him, but that is not at all the case. . . . I never told him that he should not bother about you and did nothing whatever to encourage him to diminish his natural gratitude to his truly good mother who is willing to make great sacrifices for him. . . . This girl has a bad influence on him."[183]

As time went on, he became more and more adamant on this issue: "Her miserable influence is apparent in how Albert has fallen apart in this brief period of time. . . . This mistake will be his undoing."[184] He could not stop thinking about the question of their having children: "I will be satisfied if he doesn't start the risky process of having children."[185] Even once he had met his future daughter-in-law, he did not let up altogether. "For someone other than for him, she is not as bad as I had expected. *She* is somewhat egoistic and egocentric, and not *very* tactful. But there are far worse people. . . . In short, if they remain childless, I have come to terms with the situation."[186]

He eventually offered his son what amounted to a truce: "Once you tell me that you have definitely decided not to have a child with Miss Knecht, I will seriously accept your decision to marry, even though I feel sorry for both of you."[187] However, Einstein was not able to stop the "wretchedness." In 1927, Hans Albert married Frieda. In 1930, grandson Bernhard Caesar was born. Einstein found that he had been "quite disrespectfully promoted to a grandfather."[188] However, the grandpa would grow quite fond of little "Hadu" and think of him as his own child; eventually he even bequeathed this boy, who would be carrying on the family name, his beloved violin Lina. Today this instrument is in the possession of the violinist Paul Einstein, a son of Bernhard Caesar and great-grandson of Albert Einstein.

Adversity came in the form of material as well as biological inheritance—and more was involved than simply a violin. When he failed to make his case with genes, he tried to gain control of the situation with money. His own history of illness may have been a factor governing his behavior, "since they have diagnosed a severe heart condition (enlargement of the heart with elevated blood pressure and low pulse rate). It remains to be seen whether there will be any substantial improvement," he told Hans Albert. "In any case I will be bedridden for months."[189]

When he carried his valise a few hundred yards uphill through deep snow during a visit to Switzerland in March 1928, he suffered a circulatory collapse. He returned to Berlin in a fragile state of health and had to spend several months in bed adhering to a strict diet. His new doctor,

János Plesch, whose patients included many celebrities, diagnosed peri-carditis. At the age of forty-nine, Einstein was in such bad shape that he feared for his life. "I came close to kicking the bucket," he wrote to Besso in early 1929, "which one ought not to let drag on too long."[190] From this point on, he began to focus on the arrangements he needed to make in case he did not survive. As far as his material legacy was con-cerned, he evidently had some ideas that his first family did not share.

Einstein had a furious falling-out with all three of them. He could not stomach the idea that while the rest of the world revered him, his own children turned against him. In the fall of 1932 he wrote to Hans Albert, who was acting as the family spokesman: "You need to under-stand that my trust has gradually eroded, and that I have no desire to go to Zurich as she requested. I can no longer turn to Tetel, considering his condition. Talk it over with your mother, and if possible with Tetel, and then do what you think is right. . . . I explained everything to you again and again—to no avail—when there was still time to avert an impending disaster. Now that it cannot be changed, it has to be endured."[191] This was quite uncharitable for a man who preached humanism and the idea that "only a life lived for others is a life worth-while."[192]

In late August he had given Eduard a piece of his mind as well: "Dear Tete! Your two letters were sketchy and did not provide any ex-plicit answer to my question. I assume that you have thought over the matter with Mama and Albert. The result was that minor misgivings on my part turned into justified suspicion. I am sorry, but I have learned to accept the facts with my eyes open, no matter what—or who—is con-cerned. Under the circumstances, getting together would not be uplift-ing for the two of us, but harmful. Everyone does what he can, or, rather, must. That is fate. Keeping that in mind, regards from your Papa."[193]

These lines also show the degree of flustered impatience that guided his hand while writing these letters. The following description by a family friend provides some insight into the cause: "In the boy's re-lationship with his father there was passionate admiration mingled with unexpected rebellion, which was like a secret resentment. Perhaps it was an impotent gesture of revolt in reaction to an exaggerated idealiza-tion," Elsa's friend Antonina Vallentin recalled. "Suddenly letters began to arrive from Switzerland, incoherent letters in which the desire to affirm a weak personality through grand language alternated with outbursts of despair; they were pathetic, unhinged letters. Einstein was deeply upset by them. He could not understand these figments of an overexcited imagination. Suddenly young Einstein's passion for his father turned to hatred, expressed itself in bitter recriminations and vehement accusations—the feverish confessions of a sick mind."[194]

Again Einstein implored Eduard, his only source of support in the family, although he refused to let him visit: "It is especially important to me not to have such a crushing disappointment from you. It is easier for me to handle disappointments from others." He went on to say, "You write that I am aloof from you, but aloofness is my fate, and that is as it should be."[195]

A typed copy of an undated letter from Hans Albert to his father, apparently written in 1932, provides an indication of what was making him lash out so impatiently:

> You yourself will have to admit that of all the people who are or have been closest to you, I am the one who has given you the least trouble, at least from a financial point of view. I am coming to realize that I was really an idiot to make the effort to earn my living as quickly as possible. While all the weak and ill are pampered and cared for, it is evidently not enough for you that I take care of myself; you apparently do not even consider me worthy of getting at least a modest token from you after your death, after you were stolen from me in life. If your fatherly feelings toward me have really hit rock bottom, you need to tell me so.[196]

You were stolen from me in life. In November 1932, Hans Albert received a fitting reply: "You accuse me of not being a good father. The truth is that emotional ties have always meant more to me than family ties. I felt the full force of this as I was reading your loveless and brutal letter, which is as alien to my disposition as the war dances of some black tribe."[197] And he wrote to Mileva in late April 1933, "I will never forgive Albert for the way he treated me, even though I am realize that that woman is largely responsible for his behavior."[198] All of this was taking place during the turbulent days when Einstein gave up his German citizenship after Hitler's takeover and lost his German home for good.

We have no record of what happened next. Did Mileva shake him up when he finally did visit her and Eduard again in May 1933 in Zurich? Did their friends intervene once again? Was there an additional letter from Hans Albert that opened his eyes? Or did Einstein realize on his own that now that he could (or would) no longer return to his homeland, Germany, he might lose his family as well? In any case, he made an about-face in light of the jarring shock. On May 30, 1933, three weeks before books—including his own—were burned on Bebelplatz in downtown Berlin, he wrote from Oxford: "Dear Albert! I realize that I have been unfair to you de facto by misjudging the motivation behind your bad behavior to me. The main reason was a feeling of undeserved neglect and my need to take revenge at some point."[199] Two weeks later he confessed to Mileva, "My defiant attitude prevented me from noticing how much I have suffered from the personal quarrel with all of you."[200]

Once he had emigrated, Einstein's life was accompanied by illness and the deaths of those close to him. In May 1934, Elsa traveled to Paris to be with her dying daughter Ilse, who was being cared for by her sister, Margot. Although Elsa beseeched her husband to come with her, he refused to go. Ilse was suffering from tuberculosis. Like many at that time, her faith in the power of psychoanalysis ran so deep that she refused any medical treatment and eventually suffered a horrible death. Accompanied by Margot and Margot's husband, Elsa returned to Princeton a broken woman. Shortly after moving to 112 Mercer Street, Einstein's final residence, she came down with a severe illness.

On January 4, 1937, Einstein told Hans Albert, "About 14 days ago, my wife died after a prolonged period of suffering. So much is happening all at once."[201] In the same year, Hans Albert and his family immigrated to the United States. Einstein was now quite preoccupied with the subject of his own demise. "At our age, the devil doesn't give you much time off!" he wrote to Zangger in 1938.[202] No sooner had he "finally seen my grandchildren *in natura* and heard their little voices" in 1938 than the family had to weather yet another blow: the second grandson, Klaus Martin, who was born in 1933, died of diphtheria. "Dear Children!" Einstein wrote to Hans Albert and Frieda in January

1939. "Now, at such a young age, you are having to go through the hardest thing that can happen to loving parents."[203]

The next year there was finally a pleasant piece of news: "I am glad that you have the courage to take on another child in these uncertain times. I think you are right to do so."[204] His son and daughter-in-law adopted a baby, Evelyn Einstein, who now lives outside of Berkeley in poor health. She is the possible biological daughter of Albert Einstein, the product of an affair with a dancer in New York. "Little Eveli is growing," Hans Albert reported to his mother in July 1942.[205] A few days earlier he had told his mother, "She is now 16 months old and her little legs and healthy appetite are as strong as the boys' were."[206]

At the same time, Einstein's state of health was declining. "They found that I have an intestinal ulcer," he informed Hans Albert in June 1947.[207] A subsequent letter went into detail: "Whenever I eat somewhere, everything has to be prepared specially, and I get sick anyway. I also had the flu; I feel terrible and am in worse spirits than usual."[208] He wrote to Mileva in September: "I would not like to get to the point that I am no longer good for anything, but everyone has to take it as it comes."[209] As it turns out, he was not the one facing death; Mileva was.

When his son wanted to travel to Switzerland in June 1948 to take care of his mother, his father reminded him, "Just keep in mind that savings are better used to improve the lot of those who are left behind and still strong enough to live than to squander them on such a hopeless case."[210] Hans Albert did not go. Einstein even objected to his daughter-in-law going. He told Hans Albert one week later, "I don't like the idea of Frieda going to Zurich, mainly for psychological reasons. She is one of the women who robbed Mileva (in her mind, anyway) of the men in her life."[211]

Mileva Einstein died alone and bewildered in a hospital in Zurich. Her ex-husband was informed about her death in a telegram from Tetel's guardian. Evidently Hans Albert raked him over the coals. Einstein replied, "I fully understand your emotional reaction in this very sad matter, but think it is better that you did not go there. The death of our closest relatives reopens old childhood wounds. There is not much I can do to help you; it is meant for each of us to deal with our share alone."[212]

No sooner was Mileva buried than a new disaster loomed. Shortly after Christmas 1948, Hans Albert received an alarming letter from his father. "I am writing to you today because I am scheduled for a major operation. . . . If I don't come out of it, it is not before my time. I am writing to you because it would not be nice for you to read it in the newspaper."[213] On March 2, 1949, Einstein was able to report with satisfaction, "Malformations were removed, which greatly improved my function."[214] When he turned seventy a few days later, he had "a hailstorm of a birthday, which was like being buried alive."[215]

Einstein was now sharing his house with three women: Margot, Maja, and his secretary, Helen Dukas. His stepdaughter had separated from her husband shortly after her arrival in 1934. His sister had come to the United States in 1939 and left her husband, Paul, behind in Europe. In late 1944 she still believed that she would be returning. "It is so nice," she wrote to Tetel, "that I can hope to see everything I cherish once again."[216]

This plan did not work out. Maja, whose white hair now made her seem like the spitting image of her famous brother, suffered a stroke, and was confined to her bed in 1946. Margot took loving care of her, and Albert sat down at her bedside every evening and read her literary and philosophical works. "Every day I look forward to this hour," she wrote to a friend, "and have the satisfaction of seeing that he too looks forward to it."[217]

Her death in the early summer of 1951 was difficult for Einstein to bear. Now he was nearly alone in the world. He was also making preparations for his own death. "Monday afternoon at 4 o'clock my sister passed away. . . . Tuesday morning at 11 a.m. we brought her to the crematorium in Trenton without the presence of friends. . . . I have decided that this is how I want it too."[218] In December 1954 he announced that his end was near. "I have severe persistent anemia that keeps me housebound," he told Hans Albert.[219] He still had about a quarter of a year to live. After long years of quarreling, he was finally reconciled with his son and heir, who had constant reminders of his father while living as "Professor Albert Einstein" in Berkeley.

For Hans Albert's fiftieth birthday in May 1954, Einstein had written to him, "It is a joy for me to have a son who has inherited the chief trait of my personality: the ability to rise above mere existence by sacrificing oneself through the years for an impersonal goal. This is the best, indeed the only way in which we can make ourselves independent from personal fate and from other human beings."[220]

His son Eduard—"My beloved rascal!"—was experiencing a far darker side of life. In 1932, even before Einstein had to emigrate, he had written off his younger son for good. "He is something of a hypochondriac anyway"[221] and "will always have to remain a problem child."[222] In September 1932, Besso was still imploring him to pay more attention to Tete: "Take the boy along with you on one of your big trips. Once you have given him freely of your time for six months of your life . . . both of you will know once and for all what you share."[223]

In October 1932, Einstein extended this offer to his son, who alternated between living with his mother and residing at a psychiatric clinic: "Don't be upset about having to undergo this cloistered medical treatment. . . . And stop worrying about the issues concerning my will. . . . I hope to be able to take you with me to America next year."[224] He never did. During his visit in Zurich in May 1933, about which no details are available, Einstein saw Mileva and Eduard for the last time. He later described his son's condition to his sister, Maja: "He is depressed and keeps losing the thread of the conversation."[225]

When Tete visited Einstein's sister and brother-in-law in Italy in 1934, accompanied by a male nurse, Maja was horrified by the changes in him. She wrote to a friend that he was bloated and was obsessed with theories of his own devising. "There also hangs over him a leaden melancholy, which makes it even sadder when his old sunny smile very seldom flashes like lightning across his face and disappears. . . . He suffers terribly, poor, poor boy."[226]

Einstein stayed in touch with his younger son from time to time at first, but there is no record of any replies to his letters. In 1937 he wrote to Tete, "I am now living the life of a lonely old man, a gradual preparation for my departure."[227] He still had eighteen years ahead of him at that time. In June of the same year, Besso described his encounter with Einstein's son:

Last night . . . choral preludes . . . played by Eduard Einstein with wonderful tenderness. He has physical problems: he is overweight and fearful of going out, and has not left the apartment for a year—friends who come to visit do not always get to see him. But then again, three times recently he gave complete, well-structured, and highly original lectures in the area of psychology. The presentation was a bit rough, reminiscent of those old organs you had to bang with your fists—one syllable after another pushed out of him as though he were caught behind a curtain of overwhelming bashfulness, yet with the kind of force you associate with playing an old organ.[228]

Eduard Einstein, though still a medical student, was now spending quite a bit of time in the Burghölzli mental institution to treat his schizophrenia, first as an outpatient, then as a resident. Between his institutional stays, he lived with his mother, and, after her death in 1948, in family settings for a few years as well—first with a pastor's family, then with an attorney's widow. Pastor Freimüller and his wife even temporarily succeeded in integrating him into their local community and finding him a job addressing envelopes. The journalist and Einstein biographer Carl Seelig took loving care of the invalid; Seelig was also looking after the writer Robert Walser, who was a patient in Burghölzli. "The face of your son has something tormented and brooding, but also a serene smile and a trustfulness that are speedily charming," Seelig wrote to Einstein.[229]

Even though Tete had broken off his own psychoanalysis after a matter of days, he was so well versed in the form of therapy that was common at the time that he made his treatment more difficult by his constant, often condescending interference. The registration form of Burghölzli patient file no. 27445 of January 10, 1933, stated, "Even while he was still an outpatient here (Fall 1932), . . . he held forth at length about psychoanalytic theory. He explained that his romantic relationship with an older woman who had invited him for intercourse too late and under false psychological premises had repressed him, made him fearful, etc., and he would now have to work through these problems. His language was pregnant with metaphors and muddled language, but he stuck to his theories adamantly and obstinately."[230]

In the patient file there is also a notation by Hans W. Maier, the director of the Burghölzli, that would surely have met with Einstein's approval if he had had a chance to read it, since he believed so strongly in the influence of heredity. "It is certain that a schizophrenic heredity comes from his mother, whose sister is institutionalized with catatonia."[231] As Einstein had been saying all along, "His mother has a schizoid personality."[232]

The Swiss author Thomas Huonker reports that Eduard was subjected to electroshock therapy in 1944, which was broken off after six cycles. The following year, Eduard attempted suicide for the first time. Before that he had undergone several unsuccessful rounds of insulin shock therapy. Even more drastic measures had been considered: "Are you actually considering an operation on his brain or the induced-sleep cure?" Einstein asked Mileva. "I say: hands off."[233]

"It's a crying shame that the boy has to spend his life without hope of a normal existence," he lamented in 1940 in a letter to Besso. "Since the insulin treatment definitely did not work, I no longer see the point in medical help. I have a very low opinion of this profession, and when all is said and done, I find it better to leave nature alone."[234]

Sometime around 1946, he finally realized once and for all that his son would never be able to lead an independent life again. For this reason, he wrote to Mileva, he considered "it absolutely necessary for Tetel to be declared legally incompetent."[235] Dr. Heinrich Meili, a friend of the family, became Eduard's guardian for the rest of his life. A 1953 report stated that the patient was currently "conducting studies on how to build a fool's paradise based on scientific research. He thinks there must be a way to create hybrid plants to produce trees that bear loaves of bread."[236]

Einstein eventually broke off all contact with his son. One year before his death, he wrote to Carl Seelig to explain his reason for doing so: "It is based on an inhibition that I am not fully capable of analyzing. But it has to do with my belief that I would awaken painful feelings of various kinds if I entered into his vision in any way."[237]

After his father's death, the bloated, chain-smoking Teddy deteriorated rapidly. The patient file in March 1957 states that he "skulks around the building and scares away visitors with his hobo look."[238]

Two years before his death, a newspaper article described the life of "Inmate Third Class" in Burghölzli, who now bore a startling resemblance to his father since he had grown a mustache. The article was titled "Forgotten in Zurich": "He wore a blue overgarment and wooden shoes because he had been working in the field. . . . He explained that he would have liked to play the piano, but he understood that it would disturb the other residents. He did not like to work in the field, but he understood that it would do him good. He would like to sleep alone, but understood that this was not possible. . . . He said that he was used to life here; this is just the way things are."[239]

Eduard's October 1965 obituary reads, "Son of the deceased Prof. Albert Einstein." There is no mention of Mileva. His medical file contains an extensive collection of his poems. One bears the title "Lonely End":

> Imagine how I will die alone
> Fading away in silence
> Not carving my existence in the bark of a tree.
>
> Things I have sown
> The winds have scattered away.
> Things I gathered together
> The brook quickly washed away.
>
> Imagine how I will die alone
> And how the shame
> Took away my hold on life,
> Took everything from me.

ELEVEN

ANATOMY OF A DISCOVERY

HOW EINSTEIN FOUND THE GENERAL THEORY OF RELATIVITY

Papa, why are you so famous?" His father thought it over for a minute, then answered his nine-year-old son: "When a blind beetle crawls over the surface of a curved branch, it doesn't notice that the track it has covered is indeed curved. I was lucky enough to notice what the beetle didn't notice."[1] Einstein was explaining the essence of the general theory of relativity to his younger son. What the branch is to the beetle, spacetime is to the physicist: imperceptibly curved.

And where was Papa's laboratory? Einstein tapped his temple with his finger and said, "Here." He constructed his experiments inside his skull. He rode atop imaginary beams of light, traveled on superfast trains, whizzed in elevators through the cosmos, made blind beetles crawl, and established a new order with signs and numbers. He accomplished these feats with images that blended into filmic sequences in his head. "No scientist thinks in formulas," Einstein stated categorically.[2]

What would it be like to watch this brain at work? The organ still exists, thanks to pathologist Thomas Harvey. Of course, seeking out his memory in fixed tissue sections would be like asking Einstein's violin to have his hands play another Mozart sonata, yet there are researchers who attempt to watch Einstein think after the fact, in a manner of speaking. They find traces of him in a notebook full of cryptic symbols and equations, which Einstein started in 1912 when he lived in Zurich. The book is impenetrable for normal mortals. For those experts who have scrutinized the eighty-four-page manuscript down to every last period and comma, however, it reveals an important element of the twisted path Einstein traveled to find his general theory of relativity.

Two men in Minneapolis, Jürgen Renn and Michel Janssen, have devoted their energies to studying the magic formulas. Each of them is examining a folder with photocopies of the notebook pages. The trustees of Einstein's estate keep the original at the Einstein Archives in Jerusalem. Renn and Janssen barely glance at the picturesque, snow-covered university campus outside the picture window, preferring to talk a blue streak within the cozy confines of their scientific incubator. These two kindred spirits re-create the Einstein laboratory in action, discussing mysterious matters like "metric tensors," "general covariance," "second derivatives," and "gamma kappa 1, dx kappa dx1," all of which resulted from a series of trials and errors, setbacks and triumphs.

When Jürgen Renn and Michel Janssen get together for their research sessions (they switch off between Minneapolis and Berlin), Einstein awakens to new life. It is as though the über-physicist is sitting in the room right with them, conversing with his students who are struggling to understand him—as a scientist and as a human being. They think his thoughts, speak his words, follow his arguments, and try out his tricks, and when the bare outline of his view of life in all its simple elegance is restored in their minds' eyes, they succumb to the magic of his meticulous writing on the graph paper.

They would, without hesitation, be able to pick out his thin, regular penmanship, which slants slightly to the right, from among all the handwritings in the world. They no longer need to rely on the manuscript. After fifteen years of research, they can recite every one of the equations in their sleep. Contemplation of his thought processes has swallowed up more of their time than those processes took in the first place. The Zurich notebook has been studied by as many as eight experts at a time, but Renn, a director at the Max Planck Institute for the History of Science in Berlin, and Janssen, of the University of Minnesota, are essentially the core team. Renn and Janssen's findings cast doubt on the legends that the cult of genius has created and reinforced.

They owe their research to the happy coincidence that the first director of the Einstein Papers Project, John Stachel, whose job it was to help organize the physicist's papers, was himself a theorist of relativity. When his expert eye lit on the notebook, which he spotted among thousands of pages and folders full of handwritten scientific notations, he instantly recognized the enormous value of this work, which on the surface had seemed so sketchy. He realized he was holding a laboratory journal in his hands—notes of the experiments in Einstein's head.

Einstein once said that, compared with the general theory of relativity, the special theory had been mere "child's play." Without the special theory, of course, he could never have developed the general theory. The work he completed in 1905 did away with the absolute character of space and time. He relativized them by making the speed of light the measure of all (moving) things. He enabled the mathematician Minkowski to unify them into four-dimensional spacetime. He gave new meaning to the simultaneity of events by running a train image through his head. He recognized electrical and magnetic fields as two sides of the same coin. He declared that it was superfluous to introduce the idea of the ether as a carrier for light and all other electromagnetic waves. And he discovered the interrelationship between mass and energy and summarized it in the formula $E=mc^2$.

Although these achievements would have been enough to earn Einstein his place of honor on the Mount Olympus of science, he quickly realized that he had accomplished only half of what he had set out to do.

Although the special theory of relativity was bound to be figured out sometime during the early twentieth century, by another physicist if not by Einstein, Einstein went it alone with his general theory for quite some time. Even Max Planck thought nothing of this new intellectual venture: "As an old friend, I have to talk you out of doing this, because you will not succeed, and if you do, no one will believe you."[3] Einstein's standard reaction to challenges like these was an added surge of determination. Nothing would stop him now.

Until this point he had demonstrated that his principle of relativity applied only to uniform motions. The same physical laws applied to travelers in a train going at a constant tempo as to observers on the platform. Clocks slow down and distances are shortened relative to one another. Now he sought to extend the principle of relativity to nonuniform motions as well—when the conductor accelerates or applies the brakes—and thereby generalize the theory of relativity, hence the name. This was exactly where he would fail, however, as Janssen, Renn, and their colleagues were surprised to discover.

Besides increases in velocity by propulsion, such as in vehicles and airplanes, there is acceleration by centrifugal force. As any child knows from the playground, if a disk is rotated, everyone who is not standing in its center is pressed to the edge and accelerated like a projectile. If you jump into a swimming pool from a diving board, you experience the most common form of nonuniform motion, falling due to gravity. The new theory had to account for these forms of accelerated motion by impact of force and gravitation.

Einstein had to clear up several issues, the first involving symmetry between moving systems. As long as two vehicles move at a constant speed relative to each other, their relationship to each other can be described mathematically. Lorentz transformations allow each of them to calculate how the other's clocks run and how its yardsticks are shortened. Can that also be achieved if the two systems accelerate toward or away from each other? In other words: Are systems moving toward one another nonuniformly related to one another in a way that allows the same formulas to be applied from one to the other? Are they—in the language of physics—"generally covariant"? If so, can the principle of relativity, which Einstein had expanded using Galileo's model, be applied to centrifugal forces resulting from rotation as well? What about gravity, which has no role in the special theory of relativity? What is the essence of this strange force of acceleration called gravity, which not only makes apples fall to the ground, but also keeps celestial bodies on their tracks? What connection is there between gravity and spacetime that causes the attraction of masses? How is matter (which, according to the formula $E=mc^2$, is also energy) connected to spacetime?

Every answer gives rise to new questions; every breakthrough is followed by a renewed beginning. This is how science functions. Young researchers of a given generation typically strive to solve the problems raised in the work of their predecessors. Kepler followed on the heels of Copernicus, as did Newton with Galileo, and Maxwell with Faraday. In the case of the theory of relativity, Einstein functioned as his own forebear and successor. He was like a marathon runner who decides to have another go at it after reaching his goal.

Newton's law of inertia states that an object remains in uniform motion until an external force is applied to it and forces it into a different motion. For cars, it is the force of the motor. The mass of the object determines how strong it needs to be. Larger cars require more powerful engines or more energy to achieve the same acceleration. That is the essence of Newton's law: the more extensive the "inertial mass" of an object, the more force it takes to get it moving and to accelerate it by a specific amount.

In the case of gravity, however, something odd occurs, as Galileo's experiments with falling bodies demonstrated impressively: all objects fall equally quickly, regardless of their weight. All it takes to re-create Galileo's experiments is two stones of different sizes, dropped from a bridge into a river. They hit the water simultaneously. If air resistance is eliminated, for instance in a vacuum, the feather falls with the same acceleration as the stone—independent of its weight. How is that possible? How can one and the same force accelerate two objects with different masses to the same degree? How does this differ from cars, which need more force to accelerate as they get heavier? Until Einstein came along, this was one of the deepest mysteries in nature.

Although Newton did not find any explanation for this phenomenon, his law of gravitation provided some insight into the matter. No one had ever been able to say what this mysterious force might be, but it clearly accorded with Newton's law that the heavier an object, the more powerful the force of gravity. The resistance of an inertial mass to an external force increases as that force increases. Moreover, the resistance is exactly equal to the exerting force. Newton contended that gravity provides a perfect counterbalance to inertia, and hence all bodies in a vacuum fall to the ground with the same acceleration regardless of their weight.

Curiously, in the nearly two centuries following his death, no one was able to interpret this exact correspondence between inertial and gravitational mass. Because Newton's law could be applied so wonderfully to both celestial mechanics (which he considered divine) and to man-made machines, the concurrence was taken to be an odd coincidence of nature.

One man, however, called this coincidence into question: Albert Einstein. Not only did he fundamentally distrust all untested assumptions, as a good scientist should, but he also sought theories that *explained* natural phenomena rather than merely *describing* them. What are the actual effects of earth in space, pulling all things toward it with exactly the strength of its gravity and keeping the moon on its track at the same time? And how could this—mechanical—force be transmitted if the objects did not touch? What does force really mean in this context?

An additional oddity compounded Einstein's mistrust. Newton's law of gravitation includes an effect that cannot be reconciled with the special theory of relativity: two bodies attract each other directly, without any time lag. According to Einstein, however, that kind of long-range effect ought never to occur. Nothing can travel faster than light, not even gravity, and certainly not infinitely fast.

Einstein retreated to the laboratory of his imagination to ponder these questions. In the Einstein train of 1905, no physical experiment could ascertain whether it was traveling. In both the train moving at a constant speed and the stationary train, an apple falls to the ground vertically. The passengers are also unable to feel the speed of the train. If the conductor accelerates, however, and the train speeds up, the change in motion is noticeable, and the apple falls down diagonally against the direction of travel. The passengers also feel the acceleration as a force that acts on them and presses them into their seats. This everyday situation—what do we feel when in motion?—is one of the points of departure for the general theory of relativity.

In 1907, Einstein had contemplated a thought experiment in which acceleration was not discernible. He would later call this thought, which established a link between acceleration and gravity, the "happiest thought of my life." Einstein used a stark image to explain the effect of gravity: "an observer in free fall from the roof of a house."[4] Can a person feel gravity while falling? In a later thought experiment, Einstein enclosed the observer in a windowless elevator racing down to the ground. Within the elevator, the person who is falling can perceive neither the headwind nor his fall. But the crucial point for Einstein is that the man in the elevator is not able to register the field of gravity in which he experiences the fall, even without his prison. He does not feel the acceleration, either, because he is floating within his frame of reference, namely the elevator.

"The extremely strange and confirmed experience that all bodies in the same gravitational field fall with the same acceleration immediately attains, through this idea, a deep physical meaning," Einstein commented later. "If the observer lets go of any bodies, they remain relative to him, in a state of rest or uniform motion."[5]

His image captures man in a state of weightlessness, a phenomenon that has become familiar to us since the beginnings of manned space travel. If a space traveler lets go of a tube of toothpaste, it remains in the spot at which it left the astronaut's hand. No external force pulls the tube in any direction. If the astronaut touches it, however, it floats through the spaceship until it comes to a stop somewhere. Weightless flotation was unknown in Einstein's day, but he pictured it correctly. Even in a free-falling elevator, if you let go of objects, they float—in principle exactly the way that feathers and stones fall at the same level in a vacuum. The gravitational field is eliminated from the standpoint of both falling and floating objects.

Einstein's observer and any objects that float with him are essentially on a level with gravity. To clarify the special theory of relativity, we can compare this situation with the image of the clock on the sun, the moment at which time travels. Photons do not age. As gravitation occurs, the falling person does not feel the "force" that makes him race to the ground. "He has every right," Einstein said, "to consider himself in a state of rest and his vicinity as free of fields as far as gravitation is concerned."[6] Something must be acting equally on all objects, independent of their mass, and not, as Newton believed, growing larger as the mass increases. But what could it be?

Einstein sought principles to form the foundation of his theory. For the special theory of relativity he used the principle of relativity he had adopted from Galileo and expanded, and the principle of the constancy of the speed of light. His "happiest thought" had touched on the first principle of the general theory of relativity, the so-called equivalence principle, which states that gravity and acceleration have the identical effect. Newton had assumed that gravity on the one side generated acceleration on the other. What God has joined together, Einstein now declared to Newton, let no man put asunder. Gravitational and inertial mass are identical and not two separate entities with coincidentally identical values. This conclusion elevates an apparent coincidence to a general principle. Once again, a brilliant idea had come to Einstein's rescue. In this case, however, the equivalence principle did not ultimately hold up in its original form.

Einstein-trackers Renn and Janssen wend their way into Einstein's thought process, and notice that the new principle's minor error had major consequences. Janssen explains, "Einstein believed he could develop a new theory of gravitation and in doing so generalize the principle of relativity." He failed in this effort, however, because the matter is far more complex.

"By clinging instinctively to this intuitive but flawed principle, Einstein could begin moving ahead," says Renn. "Even though he had to revise it later, it was an illuminating first step." Einstein's initial assessment may have been wrong, but it set him on the right track and enabled him to come up with a better replacement. He realized that the close affinity between inertia and gravitation had to be quite significant. The proverbial genius of Einstein was composed in good part of instinct and intuition.

The equivalence principle was not the only thing to point him in the right direction. John Stachel has drawn attention to an often-overlooked thought experiment in Einstein's laboratory. Einstein, who was working on the phenomenon known as Ehrenfest's paradox, placed his observer onto the center of a rotating disk. Since its rim moves relative to the center, clocks move slower there than those in the middle, according to the special theory of relativity, and it has indeed been established that clocks at the equator run a bit slower than those at the North Pole. This is seemingly paradoxical, since both are on the earth. The clocks at the rim of the disk are subject to acceleration, an effect we have all experienced when riding on carousels. Einstein may have figured that if, as the equivalence principle shows, acceleration is equal to gravity, gravity must slow down clocks. This was his first discovery on the way to the general theory of relativity, namely that gravitation influences time.

He drew an important conclusion from this discovery, namely that the farther clocks are from the center of a mass and the less gravity acts on them, the faster they run. This insight enabled him to make an experimentally verifiable prediction: From the earth, people ought to be able to measure the effect on the sun, which is far heavier, but how can the ticking of clocks on earth be compared to the corresponding rhythms on the sun, where there are no clocks? Researchers rely instead on oscillations within atoms, which give atomic clocks their high degree of precision. The frequency of the light that atoms emit depends on the rate of oscillation. Frequency, in turn, determines the color of the light in the spectrum. The slower the oscillation (and hence the "clock"), and the lower the frequency, the more the light shifts toward red. These differences in frequency can be measured quite precisely with spectrographs. The sun's far greater mass and gravitational pull would cause clocks to run more slowly, with the light shifted more toward red than is the case on earth.

Long before he had completed the general theory of relativity, Einstein predicted a second important effect: a redshift by gravitation. Of course, the projected value between the earth and the sun is so small—only a $\frac{1}{500,000}$ deviation—that the effect could not be proved until many years after Einstein's prediction, and even then not on the sun, but on a far heavier celestial body in the proximity of the star Sirius.

In the laboratory of his mind, Einstein realized that gravitation must have an additional effect. In a thought experiment, he sent a beam of light through a small hole perpendicular to the motion of a free-falling elevator. Not only does this beam hit the opposite wall higher up, since the elevator keeps on moving, but since the motion is accelerated, the elevator moves away from the beam faster and faster, making it appear curved to the observer in the falling elevator. If, however, acceleration and gravity are equivalent, gravity has to have the same effect on light. Hence the general theory of relativity allowed for an additional prognosis long before its definitive formulation: Large masses deflect light beams. As long as he was still lacking the correct equations, Einstein was unable to indicate the exact value of the curvature of the light, but the groundwork had been laid for the eventual verification in 1919.

The Ehrenfest paradox of the disk, centrifugal forces, and slower clocks at the outer edge yielded yet another, extremely significant insight for Einstein. According to the special theory of relativity, not only do clocks run slower in uniformly moving systems, but the length of a yardstick is also shortened. This conclusion has sweeping implications. If the length of the yardstick is shortened by motion, more yardsticks would have to fit along the outer edge of the disk.

As every schoolchild knows, the laws of Euclidean geometry state that the ratio between the radius and the circumference of a circle is twice the number pi. Since length contraction affects only the direction

Albert Einstein at the age of three, 1882. Hebrew University of Jerusalem Albert Einstein Archives, courtesy AIP Emilio Segrè Visual Archives.

Einstein's parents: Pauline and Hermann Einstein. Hebrew University of Jerusalem Albert Einstein Archives, courtesy AIP Emilio Segrè Visual Archives.

Albert with his sister, Maja, circa 1893. Hebrew University of Jerusalem Albert Einstein Archives, courtesy AIP Emilio Segrè Visual Archives.

Einstein's first wife, Mileva, with their sons, Eduard (left) and Hans Albert;
Berlin, 1914. Hebrew University of Jerusalem Albert Einstein Archives,
courtesy AIP Emilio Segrè Visual Archives.

This photo from late October 1920 was taken during "Magnet-Woche" (literally "Magnet Week"), a scientific meeting on various topics related to magnetism and the behavior of solids at low temperatures. Left to right: Albert Einstein, Paul Ehrenfest, Paul Langevin, Heike Kammerlingh-Onnes, and Pierre Weiss at Ehrenfest's home in Leiden. Present but not pictured were Hendrik A. Lorentz and Johannes P. Kuenen. From *Einstein: His Life and Times,* by Philipp Frank (New York: A. A. Knopf, 1947). Alfred A. Knopf, courtesy AIP Emilio Segrè Visual Archives.

With Niels Bohr, 1925. Photograph by Paul Ehrenfest, courtesy AIP Emilio Segrè Visual Archives.

Einstein collaborating with Wolfgang Pauli in Leiden, 1926. Photograph by Paul Ehrenfest, courtesy AIP Emilio Segrè Visual Archives.

With the Indian writer Rabindranath Tagore, Berlin, 1930. Courtesy AIP Emilio Segrè Visual Archives.

Einstein with his second wife, Elsa, at their summer house in Caputh, circa 1930. Courtesy AIP Emilio Segrè Visual Archives.

The "Doktor-Quartett" at the house of Felix Ehrenhaft in Vienna. The event was a "musical evening for Prof. Einstein" on October 13, 1931. Felix Ehrenhaft is standing behind Einstein, who is third from right. Alfred A. Knopf, courtesy AIP Emilio Segrè Visual Archives.

Einstein with three other Nobel Prize winners at Max von Laue's home in Berlin, November 1931. Left to right: Einstein, Planck, Millikan, von Laue; from *Schopfer des neuen Weltbides: grosse Physiker unserer Zeit* (Bonn: Athenaum-Verlag, 1952). Deutscher Verlag, courtesy AIP Emilio Segrè Visual Archives, Brittle Books Collection.

Bicycling in California, February 1933. Courtesy of the Leo Baeck Institute, New York.

Einstein attending a dinner in honor of Franz Oppenheimer, German
sociologist and economist, hosted by the New York Zionist Club at the
Hotel Park Central, November 24, 1935. Left to right: Einstein, M. Fertig,
F. Oppenheimer. Courtesy AIP Emilio Segrè Visual Archives.

Einstein with Helen Dukas (left) and Margot Einstein, taking the U.S. oath of allegiance, October 1, 1940. Courtesy AIP Emilio Segrè Visual Archives.

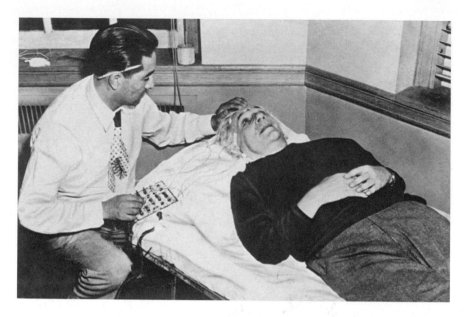

Einstein as research subject: measuring his brain activity, 1950. Can genius be "explained?" © Rowohlt Verlag GmbH, Reinbek / Germany.

of the motion, but not the one perpendicular to it, the distance between the middle and the outer rim—the radius of the disk—remains
unchanged. If, however, the circumference of the disk changes, but not
its radius, Euclidean geometry loses its validity as well—owing to acceleration or, as the equivalence principle tells us, owing to gravitation.
That was the next important discovery on the way to the general theory
of relativity: gravitation influences not only time, but also geometry.
Janssen explains that while rotation gave Einstein a great deal of trouble, it also led him to his key idea.

He now proceeded on the assumption that gravitation would
change the structure of spacetime and that the changes could be described mathematically, but that classical post-Euclidean geometry was
of no help to him in formulating this description. He also knew, however, that his new theory had to encompass and not contradict the principles of Newtonian physics, which had been verified a thousand times
over in experiments. But how can you ensure that the old principles are
preserved while implementing new ones?

"Einstein took an astonishing liberty," says Jürgen Renn. "He reverted to the position of a traditionalist." Instead of thinking up an entirely new system, as Newton had with his differential equations, he put
his faith in the treasure trove of long-established physical knowledge.
He tried to build up his theory with essentially the same type of field
equations that James Clerk Maxwell had developed fifty years earlier
for electromagnetism. Instead of having to invent new procedures,
then, he was starting from a well-established foundation. Once again,
his intuition served him well—and so did his luck.

Einstein would support and maintain this reliable framework of
long-established theories through many ups and downs. He put his
faith instinctively in the idea that gravitation could also be described by
field equations—with the help of fields that were familiar from electromagnetism. Fields of this kind do not exert direct forces but rather alter
the characteristics of space in such a way that objects follow them—the
way iron filings follow the "field lines" of a magnet. Ultimately, Einstein's assumption proved correct. The problem, however, could not be
grasped as easily as he first anticipated.

For a long time, he tried to formulate Maxwellian-style field equations that would satisfy each of his conditions, but his attempts came to

naught. Evidently something was fundamentally wrong, and his fever-ish attempts to figure out what was amiss failed. Finally, in 1912, he got a bright idea—again in view of the special theory of relativity. He pon-dered what the actual source of gravitation might be. Newton's answer was: mass exerts gravity. According to Einstein's formula $E=mc^2$, mass is bundled energy, which means that a gravitational field that deflects masses from their tracks must itself contain energy—and hence mass. Einstein realized he was dealing with a process that had yet to be de-scribed in physics. The field contributes to its own source. Gravity pro-duces gravity, and gravitation brings about an increase in gravitation.

"In a certain sense, the field pulls itself up on its own," Janssen ex-plains. Gravitational field and mass go hand in hand—the yin and yang of gravity. Masses stipulate their forms to gravitational fields. At the same time, the form of gravitational fields tells the masses how to move.

It dawned on Einstein that simple field equations would no longer do for this complex dynamic. If everything that moves perpetually influ-ences all motions, the matter gets trickier. Isaac Newton's mathematics, and, for that matter, the mathematics of the special theory of relativity, were simple by comparison. The erstwhile child prodigy now made the frightening discovery that he lacked the requisite mathematical tools. He had not taken higher mathematics at the university seriously enough, and his deficiencies in that area were coming back to haunt him. He was like a car mechanic who had saved on equipment only to find out that he was now unable to complete a crucial repair.

Once again, Einstein found himself in the right place at the right time. The ETH in Zurich, his old "Poly," appointed him full professor. His third period of residence in Zurich began in 1912, and with it the era of the Zurich notebook. As luck would have it, one of his new col-leagues on the mathematics faculty was Professor Grossmann, the same Marcel Grossmann whose impeccable notes Einstein had once used to catch up on the mathematics lectures he had cut, and whose paternal assistance had landed him his lifesaving job at the patent office.

Einstein recalled the subject matter of the lectures he had neglected as a student. Hadn't there been discussions about the great mathemati-cian Carl Friedrich Gauss, who had jettisoned Euclid's axiom of the parallel lines and invented a geometry of curved surfaces? One glance at the globe makes the distinction clear: on even surfaces, squares have four right angles on their four sides. The shortest distance between two points is a straight line, and the edges run parallel. By contrast, the lines of longitude and latitude are not parallel and the corners not right-angled. Their curvature makes them form smaller and larger angles. The effect is minimal at the equator and most dramatic at the poles.

The shortest distance between two points on a curved surface is called a "geodesic line." The curvature makes it longer than the most direct line between the two points. A straight path would go diagonally across the inside of the globe—which, however, is not part of its world and is not described in Gaussian geometry. Geodesic lines as the bent shortest distances between two points play an important role in the general theory of relativity.

Einstein remembered that Gauss had created his new geometry for curvatures of this kind. However, the Gaussian system, in accordance with common sense, pictured only three dimensions: height, length, and width. Einstein was now dealing with four dimensions in space-time. He was stumped, and wrote to his friend, "Grossmann, you have to help me, or I'll go crazy." Sure enough, his friend came to his aid. For quite some time, mathematicians had been exploring arenas that initially existed only in the cosmos of their mathematics. After Gauss, they played with dimensions that no one can picture, which proved to be a major stroke of luck for Einstein. Thanks to Grossmann, he had the requisite mathematical tools at his disposal.

The German mathematician Bernhard Riemann had expanded the Gaussian system in the mid-nineteenth century using purely mathematical considerations. In doing so, he created a complex set of mathematical tools for non-Euclidean geometries in more than three dimensions. Riemann's geometry was to Euclid's like a mountainous expanse on the globe to a flat surface. His "metric tensor" was exactly what Einstein needed.

A glance at the atlas shows what he was dealing with. The map of the world is a flat rendition of a curved surface, the globe. Everyone knows how distorted the proportions are on maps of this kind: Africa seems comparatively small, and Greenland disproportionately large. The correct distances between two points can be determined only by conversion factors, two for the north-south direction and two additional ones for the east-west direction, collectively designated as "components of the metric tensor." The transformation of distances on the map into real distances describes the "metric field."

Einstein was tackling a far more complex issue. Though the atlas needs only to deal with the two-dimensional surface of the earth with its rivers and coastal lines in adapting a spherical form for flat paper, Einstein was faced with the task of representing a four-dimensional curved object—spacetime. His friend Marcel put the tools right into his hands. On page 14L of the Zurich notebook, Einstein acknowledged the source of his breakthrough: "Grossmann." He then proceeded to test several ideas and set out on a search for field equations. However, calculations with the "Riemann tensor" proved to be devilishly difficult.

Renn, Janssen, and their colleagues made the crucial discovery that Einstein had gone about this process quite differently from the way he later depicted it. Even today, the scientific myth stubbornly persists that he thought up his theories in a dreamlike state and then considered them correct solely on the basis of their formal mathematical beauty. "In truth he acted quite differently, namely like a good physicist," says Janssen. He knew the data and laws, and was guided by the physical reality. "The crazy part is only that Einstein later disregarded that," Renn adds.

Renn and Janssen have adopted two distinct perspectives, as befits their individual temperaments, to carry out their scientific and historical detective work. The researcher from Berlin takes the bird's-eye view, and his Dutch partner in Minneapolis the worm's-eye view. Renn peers down from far above, while Janssen moves down to be cheek by jowl with the equations. This is the only way to ensure that the mass of symbols will not make them lose sight of the total picture, and they will be able to understand Einstein's method of action in the rhythm of its individual stages.

"From a distance you can see," says Renn, "that Einstein attempted both a mathematical and a physical strategy." He did not simply start with purely mathematical formulas and try them out by tacking on physical reality. He also proceeded from basic physical assumptions and then sought the suitable mathematics. "If we look at his procedure up close," says Janssen, "we see how he kept getting tangled up in one or the other strategy." Ultimately, however, physics came to his rescue.

Einstein once said, "Mathematics is the only perfect method of leading yourself around by the nose." Elsewhere he remarked, "One thing is certain: never in my life have I tortured myself anything like this, and I have learned to have the utmost respect for the most subtle aspects of mathematics, which, in my ignorance, I used to regard as a pure luxury!" His commentary in the notebook indicates how hard he was struggling. "Should disappear," he noted at the bottom of page 14L; "too complicated" at the close of page 17L.[7]

Janssen points out an important advancement in the course of the calculations: "Einstein's numbers now served double duty," he says. "They comprised the characteristics not only of gravitation, but also of spacetime." This removed an additional hurdle, but not the final one. If you want to describe a physical characteristic like gravity in a geometrical manner, you need equations with physics on the one side and mathematics on the other side of the equal sign, with measurable reality equaling the geometrical term.

Janssen goes beyond reading Einstein's hieroglyphs the way a musician reads notes. Like a mind reader, he recognizes the author's motives every step of the way. He also dares to sound off about his great role model, throwing in words like "nonsense," "prejudices," and "self-deception" to describe Einstein's thinking, because he is seeking the real Einstein behind the miracle worker who recorded the greatest work of any scientist with no apparent effort, and all on his own. Janssen does not mince words when exposing his idol's weaknesses. "Einstein muddled up mathematics and physics," he says.

Einstein set his sights higher than developing a new theory of gravitation. He sought to generalize his theory of relativity at the same time. Observers in each of the systems that are accelerating toward each other have to be able to deduce the situation of the other using the same formulas. "But that is like comparing apples and pears," says Janssen.

Jürgen Renn opens to page 19L of the notebook, which says, in Einstein's handwriting, "Recalculation of the Surface Tensor"—one of the few lines in comprehensible language. "This gamma kappa i, dx kappa dx i, is a second derivative with respect to x," Janssen says, "and then there are also second derivatives with respect to y, z, and t. He has to get rid of those if the tensor is to reduce in the limit to Newton's gravitation theory." Renn runs his finger over the equations. "Here he seems to be quite happy. At the end of the calculations he writes, 'Result certain. Valid for coordinates that satisfy the equation delta psi equals zero.' " Two thinkers on the trail of Einstein, in their element.

"He probably got that trick from the mathematical literature," says Janssen. "On page 20L we made one of the most exciting discoveries. This is where we located the crucial element, the correct tensor for him to complete the general theory of relativity in 1915." However, back in 1913 Einstein had failed to notice what a treasure he was holding in his hands. "From the perspective of modern research this was a breathtaking result," says Renn. Einstein had two long years ahead of him before getting back to this place. "Why did he dismiss the correct formulas?" Janssen wonders. "This question has driven our research for over fifteen years."

On page 22L, the name Grossmann appears once again, which means that Einstein was still sticking with a mathematical strategy. "This strategy kept getting him in trouble," says Renn. But Einstein did not give up. On the bottom of page 24R came the point where he opted for a physical strategy. Now he had the exact same problem in reverse: the physics was right, but he could no longer tell whether the equations satisfied the mathematical condition of covariance. Nonetheless, he published an "Outline of a General Theory of Relativity" with Grossmann. Was he really unaware that he was wrong, or did he simply not wish to admit it?

Janssen has spent many years working on the so-called Einstein-Besso manuscript, which contains calculations that Einstein made with his friend in 1913. The manuscript clearly indicates that Einstein was collaborating with Besso, since both of their handwritings appear in the paper. The fifty-two loose pages supply the missing link that enables us to understand Einstein's work method. The two of them evaluated whether the equations he generated in 1912 could be used to determine accurately the shift of the perihelion of Mercury—that is, whether that stubbornly persistent ugly little glitch in Newtonian celestial mechanics could be eliminated. Their complex calculations filled up page after page, but Einstein's theory failed. The friends' calculations yielded only half the sum that the astronomers have reported time and again.

However, Renn and Janssen believe that this difficulty did not dampen Einstein's enthusiasm. He was so convinced of the correctness of his work that he could not be dissuaded. Moreover, Einstein and Besso dealt with the question of whether the theory (which was still incorrect at that time) also applied to rotations like those on the circling disk. The manuscript contains a sloppy calculation in which Einstein quite simply lost a couple of minus signs. But he was so sure of being right that errors in calculation did not bother him. After half a page he noted with satisfaction, "is correct."

Einstein researchers long assumed that he had failed to notice the error or did not regard it as such. In 1998, however, Robert Schulmann came upon an additional fourteen pages of the manuscript in Switzerland that prove the opposite. The notes clearly demonstrate that in late 1913 Einstein knew his calculations were false. Besso had told him so. Did he simply suppress this knowledge? Or was he too impatient to finish his work at long last?

His devotion to the scientific confirmation of his theory demonstrates the degree of his confidence in his work. Just a few days before the outbreak of World War I in the summer of 1914, a research expedition led by the Potsdam astronomer Erwin Freundlich went to Russia at Einstein's instigation. Their aim was to measure the deflection of the light of distant stars during a solar eclipse on the Crimean Peninsula. (Five years later, Sir Arthur Eddington would achieve this aim elsewhere.) The Prussian Academy made 2,000 marks available to its new member Einstein for this purpose—the only time the academy financed his research. The expedition was brought to a halt, however, by the turmoil of the war. Freundlich and his colleagues were interned as enemy aliens, which proved fortunate for Einstein, since his theories were still yielding incorrect values. The correct readings would only have set him back—and made him look foolish.

In October 1914 he published a comprehensive survey article, the title of which demonstrates how he continued to believe in his breakthrough. It was no longer called "outline," but, pompously, "Formal Foundation of the General Theory of Relativity." In the introduction, he claimed that this theory relativized every motion, constant and accelerated, and had the same implication for rotation. "A simple recheck could have shown him that that was not true," said Janssen.

A rude shock awaited him in 1915. At first everything seemed to be going along smoothly. In the summer, Einstein had introduced his new work in Göttingen. Sitting in the audience was David Hilbert, the most important living mathematician of the era. Einstein wrote to Heinrich Zangger, "I held six lectures in Göttingen in which I was able to convince Hilbert of the general theory of relativity. I am quite enchanted with the latter, a man of astonishing energy and independence in all things."[8] He did not know that he would soon be facing his greatest rival. Hilbert, who had listened to his lectures with rapt attention, would engage him in a head-to-head race for the correct general theory of relativity. Hilbert had correctly noted that Einstein's theory was wrong—even though Einstein considered it right. "Einstein was still quite sure of himself," said Janssen.

Einstein appears to have gotten the devastating news in the late summer of 1915. For the first time since he had separated from his wife, he was visiting her and his sons in Zurich, and he met up with Besso there as well. His steadfast companion from his days in Bern reminded him again about the problem with rotation that he had broached two years earlier—and gave him valuable advice. "We assume," says Renn, "that Einstein finally repeated the incorrect calculation. To his horror, he realized that his equations did not work." On September 30, he reported to Freundlich, the astronomer who had happily returned from Russia, about a "blatant contradiction" in his theory.[9]

"Either the equations are numerically incorrect," he confessed, "or I am applying the equations in a fundamentally incorrect way. I do not believe that I myself am in the position to find the error. . . . I must depend on a fellow human being with unspoiled brain matter to find the error."[10] Had everything been in vain? Eight years of grueling work a waste of time? To make matters worse, Einstein now knew that this "fellow human being" in Göttingen was trying to get his theory from him at the eleventh hour. The mathematician had once quipped that physics was too difficult to leave to the physicists. This idle banter had turned serious.

Weeks of almost maniacal work ensued—days and nights of obsession, hours in the fever of fanatic brooding, minutes of utter despair, seconds of a quiet hurrah. A battle of the Titans—smack in the middle of World War I. Einstein, the defender, was pitted against Hilbert, the attacker. Göttingen, the stronghold of mathematics, versus Berlin, the international capital of physics. Through it all, the two of them exchanged friendly letters, both feeling out the situation, lying in wait, and preparing to pounce. Einstein quipped that he knew Hilbert had found "a hair in my soup."[11] Hilbert wrote to Einstein that he would soon provide "my axiomatic solution to your great problem."[12] Einstein maintained a civil tone in his reply: "The hints you give . . . awaken the greatest of expectations."[13]

What had happened in the brief period between the end of September and the beginning of November 1915, two weeks of which he even spent as a guest speaker in Leiden, Holland? How did Einstein get to his solution? Which path did he take through the thicket of his own theory to find the clearing where mathematics and physics merged? The standard version has Einstein realizing that he had reached an impasse with his physics and reverted to mathematics. This explanation seems quite plausible, especially because he kept claiming so himself. Jürgen Renn and Michel Janssen believe, however, that he was fooling his readers—and himself. Perhaps all the acclaim had blinded him to the truth. Doesn't pure mathematics produce far more elegant results than messy physics, which deals with physical reality? Surely a scientist who has restricted his investigative tools to the royal discipline of mathematics to hear the knock from the end of the universe is considered the greater genius.

But why would Einstein suddenly want to position himself as a man of mathematics? Was this an unconscious compensation for his earlier carelessness? His new credo of the union of mathematical elegance and cosmic harmony would give him quite a lot of trouble. "Moreover," says Janssen, "it has had a very damaging effect on modern physics. People still think today that this cannot be a bad strategy, since it is how Einstein found his theories."

Janssen and Renn conclude that in reality, Einstein remained true to his original area of expertise, physics, and did not deviate from this path, where he knew that he was on firm ground. By doing so, he had found his way back to the old familiar formulas he had derived mathematically and then discarded in 1913. Now he was able to appreciate their value. Had he gone about it the other way around, according to Renn, he would not have had a chance.

In a letter to the physicist Arnold Sommerfeld, Einstein reported that a "disastrous prejudice" had caused him to make an error in the definition of the gravitational field. Janssen has shown that correcting this error almost automatically leads to the right equations. "But he could not simply cut the knot; he had to untie it."

On Thursday, November 4, 1915, Einstein appeared before the Prussian Academy and explained to his colleagues in a rare moment of candor that he had lost his "trust in the field equations I had derived." He added, "In this pursuit I arrived at the demand of general covariance, a demand from which I parted, though with a heavy heart, three years ago when I worked together with my friend Grossmann."[14]

He presented his new field equations to the gray-haired gentlemen, who did not know what to make of them. Hadn't this fellow Einstein said the same thing before? They left his lecture muttering their disapproval. In all likelihood, not a single one of them could imagine what the speaker had been going through for the past few weeks. It was nearly impossible for them to grasp the fact that although he had made enormous progress, he was not quite at his goal, and that he appeared before them so precipitately because he felt Hilbert breathing down his neck.

But Einstein knew it, and he sat right back down to his work. One week later he repeated the same presentation to the academy. Again he had new field equations, but still not the definitive ones. The following Thursday, November 18, he yet again stood before the group of professors. This time he had a surprise in store for them. He had subjected his theory to an important test and calculated the perihelion of Mercury with his new equations.

When he saw the result, his heart began to race. It was as if the heavens were acting in accordance with his formulas. The value he had worked out corresponded precisely to the deviation that proceeded from the increasingly exact measurements astronomers had been making for over a hundred years. He wrote breathlessly to Besso, "Perihelion motions explained quantitatively. The role of gravitation in the structure of matter. You will be astonished."[15]

"I was beside myself with joy and excitement for days," Einstein confessed to Paul Ehrenfest.[16] He announced gleefully to Heinrich Zangger that his success was "glorious."[17] He had just a few final corrections to insert and the field equations would reach their definitive form. On the very day that he introduced his perihelion calculations, however, he received a dismaying letter from Hilbert. The mathematician had formulated equations that were bafflingly similar to his own. On Saturday, November 20, Hilbert submitted his study to the Göttingen Society for Sciences with the presumptuous title *The Foundations of Physics*. Einstein again set out to put the finishing touches on his equations. On Thursday, November 25, 1915, he introduced his reworked theory to his academy colleagues for the fourth time in a month, with one difference: the gentlemen were now hearing the field equations of the general theory of relativity, which are still valid today.

Einstein had thus attained the pinnacle of his inspiration and won out over Newton for the second time. For high speeds and strong gravitational fields, his mathematical framework offered a congruence between the results and the physical facts that was superior to Newton's. To the dismay of many of his colleagues, he had derived his results virtually devoid of measurements and experiments. In contrast to the spirit of classical physics, his theory was not based on experience. Consequently, the majority of physicists wrote off this theory as a speculative mathematical game.

However, this was exactly the way Einstein succeeded in conquering the macroworld. Physics and mathematics celebrated a royal union. Gravitation is geometry. Like space and time, it is an integral part of the world stage on which events unfold—but not a stage of the sort Newton pictured, which is independent of the events played out on it. In response to a reporter's request to provide a brief description of his theory, Einstein later said, "It was formerly believed that if all material things disappeared out of the universe, time and space would be left. According to the relativity theory, however, time and space disappear together with the things."[18]

Space is not an empty vessel in which God has placed things any which way. Events set the stage, and the stage sets events. Matter and hence energy establish spacetime with their actions and motions. Without them there would be no space and no time—perhaps not even a world.

The composition of the universe in all its motions is unique. There is no alternative, which means that Einstein was simply doing away with one of the most mysterious of all phenomena in the universe, namely gravity. There is no "force," as people have imagined since Newton, that two bodies exert upon each other with their mass. The curvature of spacetime created by the masses contained within it produces gravitation. These masses determine the form of gravitational fields. At the same time, the form of the gravitational fields tells the masses how they are to move.

The accumulation of mass (and hence of energy) alters the geometry of spacetime. All bodies move through it on geodesic lines. They keep pursuing the shortest route along all the bends and curves as long as no external force acts on them. Examples would be rocket propulsion accelerating them, or the earth's surface stopping their free fall. The structure of spacetime is the sole factor determining the direction in which they move.

Geodesic lines in four-dimensional spacetime are the magical paths of the general theory of relativity. They can best be understood when trimmed down by one dimension. Let us imagine that we are looking out the window of a fourth-floor apartment and watching children playing with marbles on the lawn. The ground is uneven, but from where we are sitting, we cannot make out the little hills and valleys. From our vantage point, it looks as though the marbles are by turns attracted and repelled by inexplicable forces.

Newton saw the world in much this way. He was as blind to the additional dimension as the beetle in Einstein's explanation for his son Eduard. Einstein enabled the blind beetle to see. He went down into the yard and saw how things really were. The marbles are diverted by the unevenness in the ground, and the geometry of curved surfaces is responsible for the paths they take.

Although we continue to use Newton's concept of the force of attraction today, it is not really a force in the usual sense that is keeping us on the ground. Rather, the structure of spacetime constantly pulls us toward the center of the earth, and we are stopped from falling only by the firm ground below us. This also explains why the mass of a body is of no significance, and why, as Galileo had discovered, all bodies fall to earth at the same speed. The curvature is identical for all bodies.

The phenomenon of weightlessness makes the connection clear. If an astronaut were to leave his or her space station as it is orbiting around the earth, the astronaut, though weighing far less than the spaceship, would continue to fly next to the spaceship on the geodesic line that applies to both of them. The situation is altogether different when a force, such as propulsion, acts on bodies. This energy produces an acceleration that makes bodies leave their trajectories—and hence their geodesic lines. As long as the force continues to act on them, they move on nongeodesic lines across spacetime. Only when the propulsion stops do the bodies return to their new trajectory, which follows only the dictates of spacetime.

The forced deviation is always evident, since the acceleration that results from propulsion can be felt. What we do not feel, however—and here the weakness of the original equivalence principle is revealed—is acceleration during a free fall. Someone falling from a roof follows a geodesic line, in contrast to Einstein at his desk chair, who would feel his own deviation from that line on his backside.

Once again we see dramatic consequences for the nature of time. The more a body deviates from its geodesic line, the smaller the share of space it covers in its spacetime—as in the twin paradox of the traveling Tim in relation to his stay-at-home brother Tom. It therefore has more time remaining, which means that in the gravitational field, clocks run more slowly—exactly as Einstein had supposed.

This colossal idea, the implications of which were incomparably more momentous than those in the first study of relativity, showed that matter can influence time, even bringing it to a standstill at its most extreme. All at once, an evolution of the world is conceivable, thus laying the foundation for the big bang theory—a beginning of the world—that was formulated concretely in 1947. With the big bang, space, time, and the universe as a whole arose.

And Hilbert? Issues concerning the laws of nature lay outside his usual realm of interest. Nonetheless, his *Foundations of Physics* was published in March 1916 with the same equations that Einstein had introduced on November 25. Since the mathematician had submitted the publication five days before Einstein's report to the academy, Hilbert actually deserved the victor's laurels. Many researchers continued to believe that Einstein had cribbed from Hilbert. The object of this accusation naturally saw the matter the other way around. He described the situation with the theory, as he had introduced it in Göttingen in the summer, to his friend Zangger: "However, only one colleague has really understood it, and he is seeking to 'partake' [*nostrifizieren*] in it . . . in a clever way." In other words, he meant to modify it to make it look like his own work. Einstein was right.[19]

Hilbert, in short, was not one of his angels. In 1997, Jürgen Renn, John Stachel, and their colleague Leo Corry were finally able to shed light on the case. They found the galleys of Hilbert's article, which leave no doubt that Hilbert had inserted the correct equations later, in December 1915. Einstein did not plagiarize; his competitor had amended his text. In any case, Hilbert apologized to Einstein with the rather flimsy excuse that he had forgotten Einstein's lecture in Göttingen.

The injured party was gracious. "Objectively it is a shame," he wrote back, "when two real fellows who have extricated themselves somewhat from this shabby world do not afford each other mutual pleasure."[20]

That was not the end of the story. In March 1916, Einstein published a major survey article that replaced the publication of 1914 which contained the false equations. "He had not learned his lesson," said Janssen. "He still believed that general covariance was the same thing as general relativity."

Einstein had created a magnificent theory of gravitation; there is no doubt about that. He would go on to lay the foundation for modern cosmology with this as its basis. By 1917 he had reached a pinnacle that surpassed his wildest dreams. He had actually found generally covariant equations. But he had not reached the goal he had set out to achieve ten years earlier. He did not succeed in relativizing all motions, since he could not solve the rotation problem.

Newton had used a bucket of water to demonstrate absolute space (an approach that worked against a general relativity) in an experiment that anyone can perform at home in the kitchen. As long as the bucket and water stand still, they are at rest. If the bucket is suspended on a rope and rotated, the water stands still at first, and its surface stays flat. If the bucket keeps turning, the water begins to move until it rotates along with the bucket. Its surface becomes concave and the water ascends up the sides of the bucket to the rim. If the bucket is stopped, the water not only continues to rotate, but temporarily retains its concave form, with the water up at the rim of the bucket.

The point of Newton's argumentation is that in the second and the fourth stages—that is, shortly after the beginning of the rotation of the bucket and just after its end—the bucket and water are in relative motion to each other. Seen from the outside, the bucket sometimes moves against the water, and at other times the water against the bucket. The surface of the water, however, is flat in the one case and concave in the other. This difference proves, said Newton, that the form of the water surface cannot be caused by relative rotation, in which case they would be equal, but by absolute rotation, that is, rotation with respect to absolute space—which in this case corresponds to the space around the earth.

Generations of physicists have been grappling to interpret this seemingly banal experiment. In the late nineteenth century, the Viennese physicist Ernst Mach proposed an odd solution. Mach claimed that the crux of the matter was not the relative rotation of the water against the bucket, but of the rotation of both relative to the rest of the universe. Einstein's field equations seem to reflect that, too. However, he continued to confuse the mathematical characteristic of covariance, which is satisfied, with the physical one of relativity, which does not apply.

Even after 1915, Einstein tried everything possible to master the problem of rotation. "Finally he threw in the towel," says Janssen. The walls of the bucket and the water in it could not move as freely as gravitation dictated; they are constrained by the forces that hold the bucket together and those that hold the water in the bucket. They cannot follow their trajectory, and consequently they deviate from their geodesic line.

The man falling from the roof past Einstein's office window at the patent office in Bern could also contend that he is at rest and that Einstein and the entire earth are rushing toward him in a gravitational field. However, unlike the situation in the train with a constant velocity, the two observers in their systems accelerating toward each other are not identical, their situations not symmetrical. To put it simply: one is sitting, and the other is falling.

"In this sense the motion of the water in Newton's bucket is absolute," says Janssen. A belated victory for Newton—keeping in mind, of course, that absolute space does not stand still, the way he had imagined it, but forms a constantly evolving, dynamic entity, which made Janssen feel justified in contending that "the general theory of relativity is a misnomer." This is a remarkable statement, especially in light of the misapprehensions and trouble this name would subsequently cause. "Some of the adversaries who later attacked Einstein's theory of relativity were using sound physical arguments," says Janssen. "Einstein's friends were well aware of this, but they kept quiet about it, because the arguments came from German National anti-Semites." Janssen's compatriot Willem de Sitter explained to Einstein why the theory unsettled him. "It is not merely and not primarily a fear of ghosts that makes your theory somewhat unpalatable to me; most [predominant] for me is that you make time absolute again. Your hypothesis violates the principle of special relativity."[21]

In 1920, Einstein wrote an essay for the British journal *Nature*. In this essay, which was never published, he explained that the situation resembled the special theory of relativity, where he had united the magnetic and electric fields. Maxwell and Faraday had established that a magnet lends a particular quality to the space that surrounds it. In the same way, celestial bodies determine the geometric makeup of space. Moreover, inertia and gravitation are not identical, but two sides of the same coin, the "gravitoinertial field." Which of the two components initiates the motions of bodies depends upon the motion of the observer.

In the same essay, Einstein made a remarkable admission: "For this reason, my opinion in 1905 was that one should no longer talk about the ether in physics. But this judgment was too radical."[22] He concluded, "Therefore, one can say the ether has been resurrected in the theory of general relativity" because ultimately "the concepts of 'space' and 'ether' flow into each other."[23]

Years earlier, when he was an unknown scientist, he had simply done away with the ether in the face of bitter opposition by veteran physicists. Now, surprisingly, he was restoring it to favor in a new guise. If spacetime can be curved, thereby exhibiting characteristics of a substance, what is the big difference? His adversaries, who did not bother with concepts like spacetime geometry and banded together to assail Einstein after 1919, saw this as a victory. Not every blind beetle wants to learn how to see.

TWELVE

LAMBDA LIVES

EINSTEIN, "CHIEF ENGINEER OF THE UNIVERSE"

Einstein did not need to look up a single time to explore every last detail of the universe, but the rest of us have to supplement the power of the mind with telescopes, antennas, and satellites in our quest to see beyond the range of human travel.

The road to the stars is steep. The outlines of cubes and cupolas can be made out behind a curve. "Turn out the light," says the man. "Any artificial lighting will interfere with our work." The world is instantly enveloped in blackness, with the stars close enough to pluck, yet farther than the eye can see. There are hundreds of billions in every one of the hundreds of millions of galaxies—the cosmos, the greatest possible negation of nothingness. Somewhere in one of these dots, called the Milky Way, on a planet that circles around one of the countless suns, John Beckman, a professor of astronomy, is following his idol, Albert Einstein, to the end of the world and to the beginning of time. Alpha and omega, united in the field equations of spacetime and gravitation.

"The farther we look into the cosmos, the deeper we see into the past," says the professor when he reaches his workplace: the Observatorio del Teide, the highest astronomical observatory in Europe, on the Canary Island of Tenerife. Domes, houses, and cabins, snow-white by day, silky gray by night. In the pale light, the ensemble, surrounded by solidified lava, looks like a human frontier settlement on the moon or Mars. On the horizon, the peaks of the Teide volcano rise 3,770 meters, making it the highest mountain in Spain.

Beckman talks in terms of light years, billions of light years. This concept is mind-boggling, yet he throws it into the conversation the way others might say "two dozen eggs" or "seven yards of fabric." No sooner has he sat down in one of the heated cabins than this British man in his late fifties, who is employed by the Spanish government, begins to talk, aflutter with excitement, about the beginnings of modern cosmology and about Einstein's legacy, about predictions and confirmations, instinct and error, and the kind of astronomy he and his colleagues research today. They consider the "chief engineer of the universe," as a letter writer once called Einstein, the prophet of their guild.

One of the first words John learned to say as a little boy was "telescope." He began observing the visible stars at the age of four. By the time he was ten, he knew the sky like the back of his hand, and by the age of fifteen he could "read" the firmament like a map of the galaxies. When he studied physics, however, he was one of the very few to dream of becoming an astronomer like Copernicus, Kepler, Eddington, and Freundlich.

"After World War II, hardly anyone was looking up at the sky," Beckman said. Entire generations of physicists turned away from astronomy—and from the general theory of relativity. The land of the giants was no longer one of their goals of choice. Everyone wanted to go to the dwarfs, to the subatomic region, to study the composition of matter. In increasingly large and vastly more expensive accelerators, they were chasing after quarks and quanta, measuring decomposition products, and observing the trajectories of particles—and leaving the stars where they were: out there.

As a result, Einstein did not live to enjoy the most beautiful fruits of his theoretical discoveries. Today people's fantasies are ignited by visions of the farthest cosmos and the subjects of John Beckman's work— black holes and brown dwarfs, dark energy and dark matter—but the true significance of these topics was not recognized until long after Einstein's death, even though all of these areas of investigation are based on his findings.

In 1916, Europe was plummeting in World War I. Berlin was experiencing severe supply shortages, the field of science was running on empty, and many researchers had set out for the fronts. But back at home, one man was working at full steam. No sooner was the printer's ink dry on his study of the general theory of relativity than Einstein virtually alone founded the area of research that we now call cosmology. He juggled with his field equations, which John Beckman calls "hellishly difficult"—in fact, unsolvable. According to one anecdote, the English astronomer Sir Arthur Eddington is said to have responded, when asked whether it was true that only three people could understand the theory, with a question of his own: "Who might the third person be?" Yet the field equations make possible something that did not work before Einstein: determining the geometrical form of the entire world.

Einstein was not the first, however, to succeed in applying this theory to spectacular effect. To his surprise, a soldier in the war beat him to the punch. An astronomer named Karl Schwarzschild, director of the Astrophysical Observatory in Potsdam, had been calculating missile trajectories at the Russian front since the beginning of the war, in his capacity as an officer and volunteer. When the news about Einstein's reports to the Prussian Academy reached him in November 1915, he sat right down and applied the new formulas to a minor problem in astronomy: What would the gravitational field look like around a single-point mass in otherwise empty space? In other words: If there were only one celestial body in the universe, how would it act on spacetime without the input of other masses?

According to Einstein, this condition ought not to occur in the first place. He continued to adhere to the view of the Austrian physicist and philosopher Ernst Mach, who contended that gravitation is not inherent in a body, as attraction is in a magnet, but results from interaction with other bodies, or, to put it more universally, with all matter in the cosmos. Einstein later called this interrelation "Mach's principle." Mach did not live to see the ensuing radical revisions of man's image of the universe; he died in February 1916, one day after his seventy-eighth birthday.

Karl Schwarzschild also died that year, in the aftermath of an incurable skin disease while he was still serving in the military, but by then he had assured himself a lasting place of honor in the temple of science by somewhat playfully applying Einstein's field equations to the hypothetical question of how the gravitation inside a celestial body (as opposed to in its environment) might look. His calculations led him to bizarre results. Below a certain posited distance from the center of a celestial body, the formulas reach their limits. If mass and energy were so densely bundled that the spacetime bent back on itself to the smallest volume, not even light could escape. Beyond this horizon, known today as the Schwarzschild radius, everything vanishes, even time, never to be seen again. The astronomer was describing a phenomenon that decades later would be known, and indirectly observed, as a "black hole."

The French mathematician and philosopher Pierre-Simon de Laplace had made similar observations in the late eighteenth century. He noticed that for an object to be able to escape the gravitational field of a celestial body, it had to exceed escape velocity. What would happen, though, if a celestial body combined so much mass in such a narrow space that the escape velocity exceeded even the speed of light? "The largest bodies in the universe could be invisible to us," Laplace concluded.[1] He therefore deserves credit as the first to conceive of black holes. According to Schwarzschild's calculations, this would be the case for the sun if its total mass were concentrated on a diameter of three kilometers, or for the earth if it were not larger than one centimeter. Neither Schwarzschild nor Einstein lived to see this idea go beyond purely mathematical speculation and become a measurable reality.

Einstein was surprised to learn from the astronomer's results that his formulas worked for individual bodies in an otherwise empty universe. "Ultimately, according to my theory, inertia is simply an interaction between masses," he wrote to his colleague on the Eastern Front in dismay. "It can be put jokingly this way. If I allow all things to vanish from the world, then following Newton, the Galilean inertial space remains; following my interpretation, however, *nothing* remains."[2]

Still, he had to accept the fact that, according to his formulas, something like this could well occur: a celestial body that opens out its own spacetime completely on its own, and in the process forms its own reality. But what was it? A universe within the universe? Was the devil trying to put him in his place now that he had just elicited one of the greatest mysteries of God's creation?

At first Einstein continued to work with Mach's principle, presupposing a static cosmos that does not change. To stabilize his model, he took a daring step that seems rather feeble at first glance, but from today's perspective reveals his intuitive creativity. "I have perpetrated something again as well in gravitation theory," he wrote to his friend and colleague Paul Ehrenfest on February 4, 1917, "which exposes me a bit to the danger of being committed to a madhouse."[3]

Four days later the indefatigable Einstein gave a lecture on "Cosmological Reflections on the General Theory of Relativity" at the Prussian Academy. For the first time he was describing the universe as a whole in mathematical-geometric form. In doing so, he was taking a step that he would later characterize as his "greatest blunder." To counterbalance gravitation, he introduced the term *lambda* as a "cosmological constant," and described it as a kind of universal recoil force that keeps the cosmos from collapsing. Goethe might have called it the force that holds the world apart in its core.

Einstein was of course well aware that in the quest to stabilize his theoretical construct, he was introducing a peculiar phenomenon that did not mesh with established physical facts. His colleagues objected. "If I am to believe all of this, your theory will have lost much of its classical beauty for me," Willem de Sitter, director of the observatory in Leiden, Holland, had written to him.[4] Einstein nonetheless bore with the weakness of his construct so that he could get his comprehensive theory of the cosmos off the ground. In his quest to reach new territory, he kept building bridges and serenely watching them collapse once he had crossed them. On the way to a higher truth, he acted like any good theorist, allying himself with the ghost of speculation, which, with any luck at all, dissolves into nothingness in the light of knowledge.

His physical speculation has had considerable consequences for our image of the universe. Einstein's use of lambda built on the foundation of the general theory of relativity the model of a finite but boundless universe in four-dimensional spacetime that is essentially still valid today. Physicists like Einstein and John Beckman can go deeply into such "interpretations of the universe in a closed, quasi-spherical space that is assessed in a hypereuclidean manner" with mathematical formulas the way musicians can use notes for polymorphic compositions. Anyone who does not have these tools has to shift down a gear and first observe the familiar relationships between surface and space, between a two-dimensional world and a three-dimensional one. "According to the general theory of relativity it is probable," he wrote to Heinrich Zangger, "that the universe is not infinite but closed in upon itself, somewhat like the surface of a sphere."[5]

As an example of a finite but boundless geometric figure, Einstein chose a sphere on which a beetle moves along. As long as the beetle keeps crawling, he will never run up against a boundary, even though he is dealing with a finite surface. If the sphere is large enough, the beetle will even consider it flat—the way people long regarded the earth as a disk. Within the range of their experience, it *was* flat. However, because the beetle is cognizant of only two dimensions and is himself as flat as a character cut out of a comic book, he could not in any way perceive the third dimension in which the sphere that is his world curves. He could not even imagine it. Three-dimensional human beings can grasp the concept that our world is a sphere that can be circled, but we have problems imagining the four-dimensionality of spacetime.

The fact that a two-dimensional spherical surface—and, by extension, three-dimensional space in the universe—has no center was an important factor for Einstein. Every point is essentially equal; none takes precedence over the others. A sphere does have a center, but it does not function as such in the beetle universe. It does not lie on the surface and is thus outside the two-dimensional world. The shortest distance from one pole to the other is not the axis, but any geodesic line that describes a semicircle between the two. Einstein applied this idea—in a strictly formal way—to a three-dimensional space that has neither center nor boundaries.

If a light beam is sent on a journey in this world, it does not disappear into the nirvana of spatial infinity, but returns to the starting point at some time on its way through spacetime—just as the beetle on the sphere will eventually cross his own earlier paths. Light enters the telescopes of observers on earth from the greatest conceivable distances, even from distant pasts and from long-extinguished galaxies. Astronomers like John Beckman today use their investigative instruments to look almost to the end of the world and hence to the beginning of time shortly after the big bang. Einstein's stroke of genius did away with the problem of the infinity of space. Once again, this great feat gives us some insight into why people like Beckman call Einstein the "sower" of cosmology today.

Einstein had thus detonated the third phase in his relativity rocket. During the first phase, in 1905, he had unified space and time into spacetime; during the second, in 1915, he molded spacetime to a geometric figure that is defined by masses and in turn defines the movement of masses. And now, in 1917, he was giving the totality of spacetime and matter a structure, thereby paving the way for an additional turbulent development.

Willem de Sitter weighed in once again from Holland with an unmistakable taunt. He showed that Einstein's model did not represent the only possible one. Worse still, de Sitter's universe was empty, and thus contradicted Mach's principle according to which spacetime and matter are inextricably joined. According to de Sitter's calculations, the lambda alone was enough to explain the curvature of spacetime. " 'My' four-dimensional world also has the λ term," he explained with a tacit snicker, "but no 'world matter.' "[6]

The astronomer "emphatically" contested the "assumption that the world is mechanically quasi-stationary," that is, motionless on a cosmic scale. "We have only a snapshot of the world, and we cannot and must *not* conclude from the fact that we do not see any large changes on this photograph that everything will always remain as at that instant when the picture was taken."[7] As a result, he was formulating a central idea that laid the foundations for the work of many of today's astronomers: The universe is variable. It has a history. Presumably it has a beginning and possibly even an end. Penetrating cosmic history with better and better telescopes is therefore at the forefront of astronomical research today.

Einstein vehemently opposed his colleague's contention, and admonished him, "Conviction is a good mainspring, but a bad judge!"[8] De Sitter promptly replied, "The main point in our 'difference in creed' is that you have a specific belief and I am a skeptic."[9] In 1918, Einstein was forced to concede that de Sitter's results on the basis of his equations really did represent a counterexample to Mach's principle. The universe does not have to be at rest, and gravitation is calculable even without the influence of all masses in the universe. Einstein never acknowledged this concession in any published form, however.

Another researcher to make a crucial contribution to the development of cosmology, only to suffer a tragic fate, was the Russian mathematician Alexander Friedmann. Like Schwarzschild and de Sitter, Friedmann made use of Einstein's cosmological equations without respecting Mach's principle, and calculated a nonstatic universe. In September 1925 a typhus infection cut short the life of this highly talented individual, who was only thirty-seven, but by then he had created a model of the universe on the basis of purely theoretical calculations that mesmerizes and energizes astronomers like John Beckman even today. Friedmann was able to use Einstein's formulas to describe a universe that continually expands. In 1929 his prognosis was confirmed in a spectacular manner. Five years earlier, galaxies outside the Milky Way had first been discovered; now the American astronomer Edwin P. Hubble figured out that all galaxies were receding from the Milky Way—the farther away they were, the faster they went. The universe truly is expanding.

Einstein, who was still discounting Friedmann's conclusions in 1930, visited Hubble in California and was persuaded by the data. In 1932 he published an article together with his former adversary de Sitter in which they made lambda zero. They introduced a model of the universe that is still considered essentially valid today. Einstein was happy to be rid of his "greatest blunder" and to see his equations shine in all their former glory with new contents.

The expansion of the cosmos proved not to be an explosion—which would have made the fragments spread out in an existing space, centered around the core of the detonation—but rather an extension of the universe, of space itself, with time. Space expands with the spreading masses. Unlike explosions, there is also no center; the expanding universe looks the same from any standpoint: all other galaxies recede, the speed increasing as the distance from the observer increases, wherever he is.

In order to picture that, we again have to step back and imagine a balloon to which little spots are attached. If the balloon is blown up, making its surface larger, all the spots move away from one another to the same degree. The farther apart the spots were to start with, the greater the increase in distance as the balloon is blown up. The same picture emerges from any vantage point. The spot itself, however, remains the same size—galaxies do not expand.

Einstein had been living in the United States for several years when a man named George Gamow, who was a Russian emigrant and former student of Alexander Friedmann, began in 1942 to develop the idea of the birth of the universe in one gigantic "big bang." This idea holds that the universe as we know it is not eternal, but rather has an age, which appears to be roughly fourteen billion years, based on what we know today. This idea was not entirely new. The Belgian mathematician and theologian Georges Lemaître had formulated a similar suggestion back in 1931 in the journal *Nature*: "We can picture the beginning of the universe in the form of a single atom whose atomic weight equals the total mass of the universe."[10] According to Gamow's theory, space and time first arose with the big bang, from a singularity that Einstein long regarded as physically impossible. Hence, in the lifeline of the universe there is something resembling Newton's absolute time: cosmic time, measured in light-years from the time of the big bang.

Until the late 1950s, the general theory of relativity was virtually consigned to oblivion. Because there was little chance of verifying it, it lay like Sleeping Beauty for half a century. Only a handful of physicists devoted serious attention to it, despite the fact that the preeminent researcher J. Robert Oppenheimer, Einstein's colleague at the Institute for Advanced Study in Princeton and future director of the Manhattan Project to build the first atomic bomb, had developed a crucial presentation back in 1939 with calculations according to which black holes (which were given this name in 1967) would have to be possible under real conditions. They arise when, for instance, a very large star collapses under its own gravity once it is burned out.

The young British mathematician Roger Penrose introduced a new and far simpler method of calculating Einstein's equations in 1960, thereby enabling scientists to conduct systematic investigations of superheavy singular celestial phenomena. Stephen Hawking is the preeminent researcher in this area. Hawking calculated how two black holes merge when they come too close to each other. His *Brief History of Time*, which describes the bizarre theories of cosmologists and carries on Einstein's dreams, has become the biggest science bestseller of all time.

Still, the people making the greatest advances in research are not the astronomers at their desks. "Since the 1950s," says John Beckman, "technology rather than theory has been the driving force behind astronomy." Innovative methods like radio, X-ray, and infrared telescopy enable us to look more and more deeply into space—and time. "In the past we could play only one note on the piano; today we have almost the entire keyboard at our disposal." The Hubble Space Telescope has succeeded in snapping images further back in time than ever before, showing the universe as it was more than 13 billion years ago.

In spite of such spectacular results from the cosmos, says Beckman, astronomy depends more than ever on observations from the earth. A domed building of the Observatorio del Teide houses his 1.5-meter infrared mirror telescope. High above the machine, which weighs several tons and is made of steel girders, struts, cables, tubes, and measuring devices, the roof slides open to reveal the pitch-black sky at the press of a button. The telescope moves automatically until it reaches a programmed position.

This instrument enables Beckman not only to witness the birth of stars—"it is roughly as complex as the birth of a human being"—but also to measure the trail of the big bang. As an alternative to expensive accelerators, the sky is regarded now as the "laboratory of the little man." Shortly after the big bang, the first chemical elements were formed in a "primordial fireball"; these included all the hydrogen in the universe, a large part of the helium, and the first traces of lithium. Beckman and colleagues could infer the density of the conventional matter in the universe from the ratio of hydrogen to helium. Those calculations yielded astonishing results: the known matter made up less than 5 percent of the total mass that would be necessary for a closed universe in Einstein's sense. But where was the remaining more than 95 percent?

The professor explained how interest in Einstein's lambda has reawakened. In the mid-1960s, the term began cropping up in theoretical studies, and in 1995 two American physicists titled their essay "The Cosmological Constant Is Back." Originally introduced in Einstein's failed attempt to coordinate the general theory of relativity with Mach's principle, lambda is celebrating a spectacular comeback.

Astrophysicists can deduce that a portion of the missing matter—about 23 percent—is invisible "dark matter" by measuring "cosmic background radiation," a weak afterglow of the big bang that can still be assessed today in all directions. They attribute the remaining 73 percent to a speculative force field to which they give the name "dark energy." What it is is a matter of speculation, perhaps a "quintessence," a "fifth substance" that fills the universe like a kind of ether. Whatever it is, it acts like lambda and keeps the galaxies apart—or even scatters them. Not bad for a "greatest blunder": Einstein's intuitive discovery describes nearly three quarters of the universe.

Measurements of supernovas—exploding stars—of a particular type made possible an additional sensational discovery in 1998: the universe is not only expanding, as Edwin Hubble observed in 1929, but the expansion is even accelerating. The idea that we are living in an "accelerating universe" is now widely accepted. The index of a book bearing this title, written by Mario Livio and published in 2000, clearly shows Einstein's continued significance to this discussion: his name is followed by twenty citations. Alexander Vilenkin has the second-highest number with ten citations, and he is followed by Hubble with eight.

Because of the acceleration, that is, an increase in speed, the question of course arises whether the cosmological constant is really as constant as Einstein required it to be for his static universe. The matter is still unresolved. To the surprise of many experts, however, the newest data from measurements of a total of forty-two supernovas bear out the concept of a stable constant.

Cosmologists have to rely on astronomers to acquire these data. "Measuring distances," says John Beckman, "is the key to all other astrophysical derivations." For this purpose, astronomers like himself look for "standard candles," which are based on well-defined distances to specific stars and help calculate all other distances using so-called distance ladders. Astronomers work much like surveyors on earth. The Cepheids have served as the primary celestial yardsticks; they are pulsing fluctuating stars and are found everywhere in space.

"To look more deeply into the universe, however, we need stronger standard candles," says Beckman. Using a Type 1a supernova as a "candle" has enabled scientists to measure the ratio between distance and velocity at very great distances and to draw conclusions about the accelerating universe. At present his group is developing a new standard candle based on the "characterization of the collective populations of ionized regions of interstellar gas that surrounds newborn clusters of massive luminous stars."

The machines are humming softly. The neon light is flickering. Outside the twinkling stars are following their course. John Beckman's ideas are ambling along somewhere in this black sky. This is how the natural sciences got started. People look up at the heavens and read the laws of nature. Cosmology has two periods—pre-Einstein and post-Einstein.

"The theory of relativity lives on," says John Beckman, when he is back in his car and riding down to the valley along the narrow road from the Observatorio del Teide. In the light of the moonless night, the shadows of the ghostly lava landscape join together. "Is it a true picture of reality? That is probably more a matter of faith than of proof."

THIRTEEN

SPACETIME QUAKES

THE THEORY OF RELATIVITY
PUT TO THE TEST

Einstein stares down at them in the control room. He has his hat and coat on, one hand in his pocket, the other clutching a book. A fairly well-known picture, taken about 1920 in Berlin, cut out and pasted onto cardboard. They want the master to be watching when the great moment arrives. The general theory of relativity is being put to the ultimate test, the confirmation of gravitational waves.

Einstein had made an incredible claim back in 1916: using nothing but his field equations, he could demonstrate that gravitation must produce waves. Extreme changes in masses and hence in gravitational fields, for example when a star collapses, would have to make spacetime vibrate. Of course he noted that the effect was likely to be so minimal that it would probably never be registered. Since then, scientists have been dreaming of tracking down these cosmic tremors.

Their dream may become a reality about fifteen miles south of Hannover, Germany, near the little town of Ruthe. The sign at the entrance gate says "GEO 600." When Peter Aufmuth of the Albert Einstein Institute of the Max Planck Society rides along the perfectly straight road to his experimental station, a road situated between fruit growing on trellises and fields of grain, no matter how slowly and quietly he drives, his arrival does not go unnoticed. Over the past ten years the physicist and a team of British and German colleagues have constructed one of the world's most sensitive instruments for registering vibrations.

The control room has the typical charm of a construction shack, a look that is nearly universal at the forefront of experimental physics.

There are about ten workstations with flat screens, columns of numbers everywhere, charts, scratch paper filled with formulas, cans of drinks on the big table, empty packaging from that day's take-out food, yesterday's crumpled newspapers, and science posters on the wall. "The general theory of relativity can be summed up in one sentence," Aufmuth explains, and proceeds to state this sentence: "Gravitation is not a force, but rather a characteristic of space." Gravitational waves are consequently nothing but "disturbances in spacetime with finite velocity"—tremors that radiate outward like shock waves at the speed of light. However, the effect proves to be so minute that even seeking to measure it might appear foolhardy.

It has taken researchers a good thirty years to get to this point from the first prototypes in Garching, just outside Munich, and in Glasgow, Scotland. Because they want to confirm Einstein's prediction of gravitational waves in order to show that his fear that they cannot be measured was unfounded, they get into more and more of a predicament with every step they take: the more they boost the sensitivity of their measuring instruments, the more disturbances they register.

"Everything disturbs us," says Aufmuth. "Not just when a car comes around the corner—even clouds passing by. Everything disturbs us." Quite a catchy motto for a research project. His remarks about the gravitational wave detector make it clear how literal he is being. He speaks of it the way a father might describe an oversensitive child who flinches at the slightest sound. "No sooner did we turn it on the first time than it started flailing around." It was immediately clear to the seventy researchers involved in the project that its equipment was not registering cosmic tremors, as they had hoped, but vibrations on earth.

"We have registered a quake of 7.2," Aufmuth recalls. "Not here, but in Vanuatu." Vanuatu is in the South Pacific, halfway around the world. Aufmuth backtracks for a moment to explain how simple the central equation of the general theory of relativity appears at first glance: g equals kappa times t. But when you take a closer look, "there are ten coupled nonlinear differential equations of the second magnitude that cannot be solved; they have to be out-tricked. Only then do you see how rigid spacetime really is." The earth creates an indentation that measures only a one-billionth deviation from the rest of the structure. "That is why I need large masses and rapid events to be able to measure anything at all."

Aufmuth uses extreme numbers to illustrate his point. Even if only cosmic events of the most colossal kind, such as gigantic supernovas or colliding superheavy black holes, come into consideration as sources of measurable gravitational waves, their effect on spacetime is ridiculously small. The distance between the earth and the sun—about 150 million kilometers—is compressed by about the diameter of a hydrogen atom when a wave of this kind passes through. Adapted to the earth, this means that a measuring distance of a kilometer is shortened by a thousandth of the thickness of an atomic nucleus. The research team in Hannover does not even have this kilometer.

GEO 600 stands for two 600-meter-long expanses on the earth. The measuring device consists of two 600-meter-long non-oscillating suspended tubes made of corrugated high-grade steel and pumped to achieve an extreme vacuum. The tubes meet at a right angle. Because spacetime disturbances expand and compress the two "arms" of the apparatus to varying degrees, depending on how it is set up—the variations are in any case quite minute—the differences in length between the two ought to reveal gravitational waves, assuming that the "noise" of the sources of a disturbance can be sufficiently suppressed. The instrument even registers the swell of the distant North Sea: "We practically hear the waves splashing," says Aufmuth.

The function of the equipment is based on the same ingenious fundamental principle as the interferometer with which the Americans Michelson and Morley established proof of the constancy of the speed of light in 1887 and thereby initiated the end of the ether by Einstein. To put it simply, the researchers send light through a semipermeable mirror in a way that distributes it to the two arms. Mirrors on their ends reflect the light back to the starting point, and there the light waves from both arms overlap. The team in Hannover adjusts the "signal" to "destructive interference" to make the light waves extinguish each other to absolute darkness. Even the minutest shifts between the two arms are perceptible by creating luminosity. They can see the interference pattern directly on a black-and-white monitor that appears as a magic eye of their experiment between the modern color screens, like a television from the 1950s.

The signal is produced in the "central house," a simple cube cemented into the ground in which the arms of the interferometer meet.

When Peter Aufmuth enters his "hallowed halls" sporting safety glasses and "clean room" shoes, he can barely curb his enthusiasm. "The great miracle was that the installation functioned as soon as we assembled it." Screwdrivers and strings are lying around, along with soldering irons and boxes with electronic components.

"We have to be better than the available technology in every regard," he says, and floats on silent soles through the narrow passage of the detector, which has already started its test run. Any little speck of dust can interfere with this sensitive device. Because extreme precision is required here, these scientists work with the best special laser in the world. It sends extremely stabilized light of a specific wavelength, by means of beam splitters made of synthetic quartz glass recently designed expressly for this purpose, to a north and an east tube. Like the beam splitter, the "end mirrors" in the tubes are suspended as a triple pendulum. The result is a reduction of any disturbing vibrations to a hundred millionth. All components are custom-made, rigged up on site, and feature vibration insulation, optical resonators, power amplifiers, signal enhancement, noise muting. The equipment is damped, doubly amplified, and polished to perfection.

It takes people like Peter Aufmuth, who have the requisite knowledge of physics coupled with scientific expertise and adroitness—plus a healthy dose of determination—to keep on going if the ministry of research and development suddenly slashes the budget. His team has held on to its dream and built up everything on its own, on a bare-bones equipment budget of 7 million euros, which is a ridiculously small sum compared to the $300 million that has gone into constructing two comparable American LIGO (Laser Interferometer Gravitational-Wave Observatory) projects in the United States. There the antenna arms are four kilometers long, which raises their receptivity to gravitational waves by a factor of nearly ten. "It really is like David versus Goliath," Aufmuth says. "We make up for the others' lead with more sensitive technology."

Is it worth going to such great lengths just to confirm a theory? "Not *a* theory," says the physicist, "but *the* theory about our universe. If we could actually track vibrations of spacetime here on earth, it would be the most direct support of Einstein's equations." Only by confirming the deflection of light he had predicted, which happened with the announcement of the solar eclipse data in 1919, was Einstein able to triumph in such an unparalleled way. "Then I would feel sorry for the good Lord," he once replied to the question of what would have happened had Eddington not confirmed experimentally the deflection of light he predicted. "The theory is correct anyway."[1]

It is far from certain, however, whether it is truly valid for all time or whether it will someday suffer the fate of Newton's theory of gravitation. Even Newton's design came so close to experiential reality that the smallest deviations, like the easily shifted perihelion of Mercury, were long outweighed by the many thousands of pieces of evidence to support it. Could it be that Einstein's theory will eventually be just as outdated as Newton's?

Einstein himself was convinced that it would be. He would not have been surprised to find out that at major physics conferences fifty years after his death, keynote speeches would address the question "Was Einstein Right?" The majority of physicists today work in experimental physics, and a good number of them dedicate their research to confirming—or refuting—the general theory of relativity. Relativity touches on nearly every aspect of physics, which is why more than a few experts believe that sooner or later Einstein's theory will fail. At this point, however, it is considered undisputed.

For researchers who work with very high velocities, Einstein's special theory of relativity is as indispensable as legal texts are to lawyers and judges. Many experiments in particle physics since the 1950s have incidentally served to confirm its predictions time and again. In specially designed experiments, even the bizarre "twin paradox" of the siblings in the spaceship and on earth aging at different rates has been possible to demonstrate as a real phenomenon: In the 1970s, a clock in an airplane that had flown for about sixty hours really did run slower by the predicted length of time—by the amount in spacetime that it had "used up" for its path through space relative to an identical clock on the ground. Experiments of this kind serve as confirmation of the general theory of relativity as well. Because gravity also influences how clocks run, and they run more slowly when gravitation increases (up to a standstill when a black hole is reached), by the same token they run faster as they move away from the gravitational field of the earth. The higher the airplane flies, the more apparent this effect becomes.

Before the advent of space travel, the general theory of relativity was practically forgotten. Only in the 1960s did the issue of supporting its predictions come up again. Experiments with redshifting and slowing down of clocks and atomic oscillation by gravitation are yielding increasingly precise results. Einstein was at least 99.995 percent correct on the issue of time, according to today's measurements. The deflection of light at the sun could be measured more and more precisely during subsequent solar eclipses. Far more exact investigations with radio signals from galaxies and quasars have recently enabled scientists to calculate the value up to four decimal places. In 2003 there was an additional spectacular confirmation when the space probe Cassini, traveling to the planet Saturn, sent radio signals past the sun to the earth. They were deflected by exactly the amount that the general theory of relativity had predicted.

The most widespread technical application of both theories of relativity in one system—and another kind of confirmation of their formulas—has become a virtually commonplace item in industrial nations: the Global Positioning System (GPS). This system, which has been operational since 1995, can be used by anyone seeking orientation in space and time. It would not function without Einstein's equations. The atomic clocks on board the twenty-four satellites are accurate to within a nanosecond per day—that is, to a billionth of a second. They are adjusted to compensate for two factors that cause them to deviate from clocks on the surface of the earth. Relative motion makes them tick slower by about 7,200 nanoseconds a day, but the weaker gravitational field in orbit makes them tick faster by 45,900 nanoseconds per day. Only after these relativistic effects are taken into account can GPS locations be given to the desired degree of precision. If they were not factored in, the entire Global Positioning System would be unusable within a minute and a half. People who use navigation systems in modern cars have Einstein to thank for these systems. The original military aim of GPS has also been realized. Remote-controlled bombs such as those used by the United States in recent wars achieve their accuracy by means of a practical application of Einstein's theoretical achievements, just like the friendly voice of the GPS navigator in cars.

In connection with the bending of light by large masses, Einstein had also predicted the existence of so-called gravitational lenses. If a large quantity of matter were located between the earth and a distant celestial body, the observer on earth would experience a lens effect because of the deflection of light in the gravitational field of this mass. There can even be multiple images. Einstein had discovered this effect in 1912, as a notebook discovered among his papers confirms. However, he waited until 1936 to publish his formulas in the journal *Science*, at the urging of the Czech amateur scientist Rudi W. Mandl.

Afterward he wrote to the editor-in-chief, "I thank you very much for your kindness regarding the little publication that Mr. Mandl talked me into. It is not worth much, but that poor fellow got pleasure from it."[2] This was one of his classic understatements. At the close of the twentieth century, his prediction was confirmed in a sensational manner. Through cosmic lenses, astronomers see crosses, rings, and arcs, which they have named after Einstein. Although the matter that gives rise to the lens effect is itself not visible, images of Einstein rings or crosses are evidence of the dark matter in space that lies between the observer and such objects as quasars.

It would be mistaken to believe, however, that these many confirmations have satisfied Einsteinians. Once the general theory of relativity had been roused from its Sleeping Beauty phase, the toughest test phase was just beginning. The "equivalence principle" of the identity of gravitational and inertial mass was subjected to more and more rigorous trials—which it has been passing with flying colors so far. What appears to be gravity truly does stem from the curvature of spacetime, through which all bodies move on geodesic lines.

In April 2004, NASA sent an experimental satellite named Gravity Probe B into orbit around the earth from Vandenberg Air Force Base in California. After forty-five years of preparation time, in which nearly a hundred dissertations were completed on the planned experiment and $700 million was spent, the satellite was to check on a strange effect: By rotating around its own axis, and thus spinning its mass, does the earth make the surrounding spacetime rotate along with it? The idea stems from Einstein's work with Mach's principle. In 1918 two Austrian physicists had used his field equations to predict that a body floating weightless around the earth would rotate ever so slightly in relation to fixed stars.

A major technical effort with supercooling and clean-room conditions is required for the four gyroscopes, each of which is equipped with quartz balls on board the satellite, which has now been launched. This is because the effect to be measured is so minute, as in the case of gravitational waves. In one year it amounts to only a hundred thousandth of a degree—comparable to the thickness of a human hair seen from a distance of half a kilometer. No one knows if or when the experiment will provide the desired data.

Peter Aufmuth and his colleagues also dream of experiments in space. "We are at the beginning of a new era of astronomy made possible by detecting gravitational waves." The electromagnetic waves that are commonly used in astronomy today—visible light, infrared, radio, and X-rays—reach the earth only when nothing stands in their way. Large portions of the cosmos, however, are hidden by dark clouds. Not until 380,000 years after the big bang—a mere blink of an eye on a cosmic scale—did the universe become permeable to electromagnetic rays such as light. Since the entire cosmos is transparent for gravitational waves, we are able to peer into areas—and eras—that were hitherto invisible.

With the aid of gravitational waves, we may be able to look back to the moment of the big bang. According to cosmologists' calculations, they would have to have been emitted later than 10^{-24} seconds after the zero hour—that is, a millionth of a billionth of a second. And because wherever stars explode, tremors in spacetime precede the visible flash of light by a matter of hours, in the future scientists like Peter Aufmuth will be able to alert their colleagues like John Beckman on Tenerife far enough in advance for them to direct their telescopes to one of the often-overlooked supernovas.

The new professional branch of gravitational-wave astronomy is already gaining in popularity. A belief in Einstein's predictions can sweep away mountains of doubts. Although researchers have yet to achieve success on earth, they are already planning space travel. Beginning in 2013, the European space agency ESA will join with NASA, in the framework of a project called Lisa, to send three satellites into space to act as a gigantic interferometer intended to measure gravitational waves with incomparably higher resolution than is possible on earth. A test of the first prototype is planned for 2008.

"The best part," says Aufmuth, "is that competition makes this whole enterprise possible." Researchers around the world are interdependent. In addition to GEO 600 and the two LIGO detectors, there are stations in Italy and Japan, dubbed "Virgo" and "Tama" respectively. Only when several antennas receive comparable signals simultaneously, and disturbances can be calculated by tremors of the earth, are gravitational waves clearly ascertainable at all.

In 1993 two Americans, Russell A. Hulse and Joseph H. Taylor, discovered just how rewarding it is to confirm predictions of Einstein's theories: they received the Nobel Prize for Physics for establishing proof, albeit "just" indirect proof, of gravitational waves. They made a sensational discovery with the binary pulsar PSR 1913+16, which they discovered in 1974, a pair of astronomical objects orbiting each other within the Milky Way: the orbiting time of the objects is reduced by the exact amount that Einstein had predicted on the basis of the loss of energy due to gravitational radiation.

However, it was only at the beginning of the twenty-first century, ninety years after Einstein's prediction, that physicists saw that the time had come to measure the ominous waves directly as well. Those who succeed in doing so may one day receive the most beautiful mail a natural scientist can dream of getting: a telegram from Stockholm. Three researchers from three countries, who already work quite closely together, would share the well-deserved Nobel Prize for the initial discovery of gravitational waves.

If the project succeeds, Aufmuth's boss, Karsten Danzmann, would receive the distinction on the German end. In the first publications on gravitational waves, his name appears alongside three hundred others. The fact that only three scientists receive the trophy—on behalf of all the researchers—shows how outmoded the Nobel Prize has become since the days of Einstein, the self-described lone wolf. Peter Aufmuth sees no problem with that. Witnessing the confirmation of Einstein's theory would be the triumphant pinnacle of his professional career. But he also knows that his dream may not be fulfilled.

FOURTEEN

HIS BEST FOE

EINSTEIN, GERMANY, AND POLITICS

On May 15, 1915, Einstein gave his wife in Zurich the devastating news. "Mrs. Haber shot herself two weeks ago."[1] No further commentary, no explanation. The previous year, when his family broke up, the wife of his colleague Fritz Haber had devotedly cared for Mileva. Mileva, in turn, had seen Clara as a partner in suffering whose marriage was in equally bad shape and who, like Mileva, felt that she had been cheated in life. "What Fritz gained in these eight years, that—and much more—I have lost."[2]

Clara Haber's reasons for ending her life went beyond a failed marriage and professional dissatisfaction. In contrast to Mileva, Clara had earned her doctorate in chemistry and worked as a scientist until motherhood and domestic duties forced her to give up her career. Her suicide was also prompted by Fritz Haber's role in the use of poison gas in the trench warfare on the Western Front, which Clara had vehemently opposed.

The small Belgian city of Ypres is to chemistry what Hiroshima is to physics. It is the place at which the "science of materials" forever shed its innocence. On April 22, 1915, at 6:00 p.m., when the wind shifted into the westerly direction, the order went out to open the valves of the gas cylinders along six kilometers of the battle front. Within five minutes, 150 tons of chlorine gas were released and blew toward the enemy positions "at the pace of a trotting horse." Estimates put the toll at 5,000 dead and 15,000 wounded in this attack. The historian Fritz Stern calls it "an early instance of science put to satanic service."[3] Thirty years separated chemical warfare from atomic warfare.

Chemist Haber, who was brilliant, energetic, egotistic, and patriotic, meticulously prepared for the use of this gas in Berlin, according to Stern, "in a kind of Manhattan Project before its time."[4] Now he traveled to the front to oversee it himself. The "success" brought the portly little bald man with a pince-nez the promotion he desired, from vice-sergeant to captain. At the ceremony on May 1, Clara is said to have tried again to dissuade him from participating. He admonished her for stabbing him in the back. In the early morning hours, she took his service revolver, went to the garden, and shot herself in the heart.

The very same day, the widower traveled on to supervise new poison gas experiments, this time on the Eastern Front. Although the Hague regulations concerning the laws and customs of war on land had declared these kinds of actions criminal in 1907, Haber was awarded the Iron Cross and other distinctions for his service. Haber embodied the Janus-faced aspect of science. His discoveries both ended and saved human lives. His "manna from heaven," technical nitrogen fixation for synthetic fertilizer, saved millions from hunger. After World War I, he was awarded the Nobel Prize in chemistry for his breakthrough in ammonia synthesis. His institute in Berlin still bears his name today.

There is no record of Einstein's views about chemical warfare. Although he adamantly opposed war and all things military, he fell silent on the subject of military weapons and what they owed to science. "As long as I am interested in working along lines of research . . . the practical aspect, that is, every practical result that is found simultaneously or arises out of it later, is a matter of complete indifference to me," he once said.[5] He drew no distinction between basic theoretical research of the type he conducted and applied science, which uses "pure" knowledge to produce objects and systems—including gas grenades and atomic bombs.

In about 1930 he called Haber a "tragic figure" whose "patriotic convictions had been misused,"[6] yet his friendship with Haber remained intact when Haber mobilized German chemistry for the military, just as he had stayed on friendly terms with Max Planck in 1914 when Planck solemnly sent off his students to military service, and with

Walther Nernst when his enthusiasm for the army led him to practice marching and military salutes on the gravel path in front of his villa. Einstein's own work on the gyrocompass and in the construction of airplane wings contributed to technical developments that were of use to the military. Half of his salary came from an industrialist who also supported military research.

His friendship and collegiality with Haber notwithstanding, Einstein was rankled by Haber's inflated egoism and his patriotic zeal on the heels of his conversion to Christianity. "Haber's picture is to be seen everywhere, unfortunately," he was complaining back in 1913 in a letter to his lover, Elsa. "Unfortunately, I must reconcile myself to the idea that this otherwise so splendid man has succumbed to personal vanity, which, moreover, is not even of the most tasteful kind."[7]

He would have had grounds for disagreement with Haber and his other German colleagues just half a year after his arrival in Berlin. Two months after the war broke out, ninety-three prominent Germans provocatively defended themselves in an internationally publicized "Appeal to the Civilized World" against the condemnation of the Germans after their attack on neutral Belgium and the atrocities committed there. The signers included the painter Max Liebermann and the writer Gerhart Hauptmann, as well as Einstein's colleagues Haber, Nernst, and Planck. The "Manifesto of the Ninety-Three" aroused indignation in many countries. Einstein commented a few years later, "When a group of people is possessed with collective insanity, one should counteract it."[8]

In Germany, by contrast, there was virtually no protest against the manifesto. Einstein himself was not actively involved until someone else took action, at which point he joined the initiative. Georg Nicolai, a cardiologist in Berlin and a good friend of his lover, Elsa (and future lover of her daughter Ilse), presented him with an "Appeal to the Europeans" as a countermanifesto to the "Appeal to the Civilized World." This manifesto states that "the struggle raging today will likely produce no victor; it will leave probably only the vanquished. . . . And if you feel as we do, if you are like-mindedly determined to provide the European will the farthest-reaching possible resonance, then we ask you to please send your (supporting) signature to us."[9] Other than Einstein himself, only two people were prepared to add their signatures. The appeal went virtually unheard, and Nicolai was transferred to the Eastern Front to render medical service. In 1918 he made a spectacular escape to Copenhagen by airplane.

Despite a common perception to the contrary, Einstein was not a political activist at the time of his move to Berlin in 1914. Certainly there was a kind of "political" assessment in a brief note of August 19, 1914, to his friend Paul Ehrenfest in Leiden: "Europe in its madness has now embarked on something incredibly preposterous. At such times one sees to what deplorable breed of brutes we belong."[10] But the antimilitarism he had continued to champion from his youth had a long way to go before it turned into the verbal pacifism that later made him almost more renowned than his physics had. "I am musing serenely along in my peaceful meditations and feel only a mixture of pity and disgust."[11]

Einstein initially retreated into the snail shell of his science. "The decision to isolate myself is proving to be a blessing," he wrote to Ehrenfest in early December 1914.[12] The general theory of relativity required every bit of his strength. In April 1915 he confided to Heinrich Zangger, "I am now starting to feel comfortable in the present-day mad turmoil, in conscious detachment from all things that occupy the deranged public at large. Why is it not acceptable to live enjoyably as a member of the madhouse staff? All madmen are respected as those for whom the building that they are inhabiting is there."[13]

By donning a fool's cap, carrying a Swiss passport, and working on his theory, he gained a degree of independence that his German colleagues could only dream of. One month later he let Zangger know that "my life here is ideally pleasant if you disregard things that actually have nothing in the least to do with me."[14] The main issue that preyed on his mind in the early phase of this "sorry international entanglement"[15] was the threat posed to the global community of his peers—the cosmopolitans of science. "The international catastrophe weighs heavily on me as an international person."[16]

Two issues impelled him to take political action. One was the "Jewish question," and the other was his deep commitment to the global understanding that underlay collaboration and progress in science, which he regarded as his only true home. This commitment spurred him on to co-found the League for a New Fatherland in November 1914, a worthy but largely ineffectual democratic organization (the group was banned in September 1916). His lover, Elsa, joined as well.

While Einstein was wrestling with a particularly stressful aspect of his theory of relativity in the fall of 1915, he was able to come up for air long enough to write an essay for the Goethe League in Berlin, called "My Opinion on the War." This essay made his radical—and utterly modern—political stance known for the first time. His advocacy of a European Union–like solution sounds visionary today. Einstein proposed a "statelike organization in Europe that makes European wars impossible, just as now war between Bavaria and Württemberg is impossible in the German Reich."[17]

In one letter after another, Einstein declared his aversion to the abysses of modernity. "All of our exalted technological progress, civilization for that matter, is comparable to an axe in the hand of a pathological criminal,"[18] he wrote to Zangger—without commenting on the consequences this view might have for his own work. An epistolary exchange with the French writer and pacifist Romain Rolland provides a telling indication of his political evolution. "Through the military success of 1870 and through the successes in the fields of trade and industry, this country has taken up a kind of religion of might," he declared in August 1917. "This religion holds almost all intellectuals in its sway; it has driven out almost completely the ideals of Goethe and Schiller. . . . I am firmly convinced that this straying of minds can only be steered by hard facts."[19] In plain English: Defeat is the only solution.

Einstein's sporadic political engagement did not set him apart from the community of academics and intellectuals, which was starkly politicized by the war anyway. In late 1914 he accepted an invitation to join the board of directors of the German Physics Society and in 1915 even to become its president, despite the group's patriotic stance. He maintained his silence during the sometimes heated political discussions at the sessions of the Prussian Academy (by the end of 1918 he had participated in 135 meetings). According to a report from army headquarters, he subscribed to the liberal newspaper *Berliner Tageblatt* and sympathized with pacifist organizations—but otherwise kept a low profile. By 1918 he had not established any official contact with the Social Democratic Party or any other organization in the democratic reform movement.

Einstein spoke about spacetime while Germany was abuzz with lebensraum; he played Mozart while the Germans blared military marches. The calm and independence he had sought and found in Berlin enabled him to experience a kind of rejuvenation after 1918 that made him appear as though he had spent half of his life dealing with politics. At the beginning of the year his name appeared in the ninth slot on the Berlin police list of the thirty-one leading pacifists and social democrats—not that it took much to get there. On November 9 he remarked dryly in his lecture notes, "cancelled because of revolution."[20]

Shortly after the collapse of the monarchy, he became politically active for the first time, together with his physics colleague Max Born and the psychologist Max Wertheimer. The university in Berlin had been taken over by a students' council, which was modeled on workers' and soldiers' councils. One of its first "official duties" was to depose the rector and other chief administrators and have them locked up. To bring about their release, the three professors met with the students in a conference room of the Reichstag. The young people declared that negotiating a release was not within their purview, but issued them an entrance pass for the new government. Owing to Einstein's renown, the three professors were able to arrange a meeting with Friedrich Ebert, the newly proclaimed president, at the nearby imperial chancellery on Wilhelmstrasse. Ebert wrote a note to the minister in charge, and shortly thereafter the university administrators were released.

Four days after the German Revolution in November 1918, Einstein made his first major political appearance at an official meeting of the *Bund Neues Vaterland* (New Fatherland League), which had been newly constituted shortly before the end of the war. "Comrades!" he exclaimed to the more than one thousand people in the upper hall of the Spichernsäle in Berlin. "Speaking as an old-time believer in democracy, one who is not a recent convert, may I be permitted a few words." With surprising foresight, he warned his audience to be vigilant "lest the old class tyranny of the right be replaced by a new class tyranny of the left. Do not be lured by feelings of vengeance to the fateful view that violence must be fought with violence, that a dictatorship of the proletariat is temporarily needed in order to hammer the concept of freedom into the heads of our fellow countrymen."[21]

This statement confirmed his unwillingness to ally himself with the radical left, which was seeking to establish a soviet republic. At the same time, he admitted to Max Born in January 1920, "I have to confess to you, besides, that the Bolsheviks are not so unappealing to me, as funny as their theories are. It would be damned interesting to take a closer look at the business once."[22] Of course he had never actually set foot in the Soviet Union. By mid-1932, however, he would change his mind about Bolshevism. "At the top there appears to be a personal struggle in which the foulest means are used by power-hungry individuals acting from purely selfish motives. At the bottom there seems to be complete suppression of the individual and of freedom of speech. One wonders what life is worth under such conditions."[23]

Meanwhile, nothing had changed in his leftist viewpoint: "Anyone who sticks his head out of the window yet fails to notice that the time is ripe for socialism is stumbling through this century like a blind man."[24] During the war, literary texts provided sustenance for his social conscience. Einstein spent long evenings at the Bristol on Kurfürstendamm reading Heinrich Mann's novel *Der Untertan* (*The Man of Straw*), which was circulating in insider groups in a private printing. "The differences between social classes do not seem to me justified. I believe them to have been in fact established by force," he said.[25] The aims of the bund, which changed its name to the German League for Human Rights in 1923, sound almost as radical today as *The Communist Manifesto*. The signers included Einstein, Heinrich Mann, Käthe Kollwitz, Max Wertheimer, and Magnus Hirschfeld. He wrote to Besso, "I am enjoying the reputation of an irreproachable Socialist."[26] Today he would most likely be called a Social Democrat.

Einstein's enthusiasm for the coming German Republic with a democratic constitution came across as wide-eyed optimism: "A major thing really has been achieved. The military religion has vanished. I think it will not return anymore."[27] Two days later he gave Arnold Sommerfeld in Munich his assessment of the new state of affairs: "If England and America are sensible enough to agree, wars of some consequence will not be able to occur at all anymore."[28]

There was a distinct note of confidence in Einstein's pronouncements at this point: "I am firmly convinced that culture-loving Germans will soon again be able to be as proud as ever of their fatherland—with more reason than *before* 1914. I do not believe that the current disorganization will leave permanent damage."[29] In June 1919, Max Born received a letter from Einstein with this message: "I am convinced that in the next few years things will be less hard than in those we have recently lived through."[30] A fateful error. Although the swastika had yet to tarnish his optimistic view, it was already making its way through in the back rooms of the up-and-coming fascists as a symbol of the ultraright.

"Without laying any claim to authority, Einstein became for many people the sage and prophet of the age," the philosopher Karl Jaspers later commented. "He had a good eye for the basic conditions of human existence: honest and to the point, but very limited."[31] Whereas for Haber, Planck, and most of his traditionally rather conservative colleagues, a world was falling apart with the capitulation and the end of the empire, Einstein hailed a new beginning. He described to his sister and her husband in Lucerne two days after the November Revolution "the greatest public experience conceivable": "That I could live to see this!! No bankruptcy is too great not to be gladly risked for such magnificent compensation. Where we are, militarism and the privy-councillor stupor have been thoroughly obliterated."[32]

Einstein was reveling in the new era of peace. Just a month later he was making fun of the éminences grises in the academy to Besso, "The Ac[ademy] meetings are amusing now; the old folks are for the most part completely disoriented and dazed. They perceive the new era as a sad carnival and are mourning over the bygone state of affairs, whose disappearance means such liberation to the likes of us."[33] He wrote to Zangger on Christmas Eve, 1919, "Misfortune suits mankind immeasurably better than success."[34]

Even though his name had been well known in Berlin for quite some time, his pronouncements carried far more weight after the noteworthy day in November 1919 when he suddenly attained international renown. After years outside the academic mainstream, followed by a rapid ascent to the position of number-one physicist, his third career as a politically active star of science now commenced—a never-ending roller coaster between high regard and hostility, productive contributions and counterproductive faux pas.

As a Jew, leftist, pacifist, and open-minded thinker, he embodied everything his adversaries hated. Even his appearance indicated that he clearly rejected the dominant order and stood out as an outsider. He lacked the illustriousness of a Max Planck, and his passionate desire for freedom went against the grain of the establishment. He was utterly lacking in the three major forces he saw ruling the world: stupidity, fear, and greed—although he did confess to Heinrich Zangger, "I am only getting more stupid with fame, which is quite a common phenomenon, you know."[35] The aura of mystery surrounding his work, which everyone discussed but no one understood, gave rise to fears and uncertainties.

"Now to the name relativity theory," Einstein said in 1921, "I admit that it is unfortunate, and has given occasion to philosophical misunderstandings."[36] The misunderstandings were not merely philosophical. The very name of this theory encouraged false interpretations, since people understandably confused his theory of relativity with relativism, which emphasized the relativity of all knowledge, discarded universally valid moral norms, and left ethical truths to the discretion of an individual or a nation—and was thus a breeding ground for extreme nationalism in the twentieth century. Einstein's own belief system was the diametrical opposite of relativism.

If everything is relative (which, incidentally, Einstein never said), and relativity is a "general" principle, nothing can be certain and unalterable. This was a hazy sort of theory, which, despite its confirmation during a solar eclipse, could not be verified by laymen. You either believed it or you didn't. It became a familiar topic of conversation, but remained abstruse, and was vulnerable to attacks by its adversaries, especially those who could lay out their arguments in scientific lingo. Moreover, it was considered revolutionary, and its originator was linked to subversion and revolt. To top it all off, Einstein took a political direction that was hardly calculated to win people over.

In mid-December 1919, Einstein officially welcomed the Belgian pacifist Paul Colin—a brave deed in view of the still-charged atmosphere regarding Germany's former enemies. However, he never took the next step to civil disobedience in the style of Gandhi. It would be difficult to picture Einstein on a hunger strike, or heading up a parade of demonstrators. He was more bold than courageous. Even so, his role as prominent advocate of the new social and political order in Germany brought on a barrage of criticism, which, in turn, elevated his prominence.

On the next-to-last day of his fateful year, he spoke out about "Immigration from the East" in an article that attracted a great deal of attention. This lecture was his first public appearance to address the "Jewish question," one of the two major issues that had spurred him on to political activity (the other issue being his concerns about the internationality of science). An outsider shows his solidarity with others—and finds "his" political theme. And the theme finds him. His commitment to the needs and interests of his Jewish "ethnic cohorts," about which we will be speaking later, made him a perfect target for reactionaries and future murderers. Moreover, both leftists and Jews were being held accountable for the German disaster. In the chaotic anti-Semitic climate that ensued, he found that his work had come to be branded categorically as "Jewish physics." He grew accustomed to attacks on "Einstein's fraudulent theory" as long as he continued living in Germany. Even so, according to his early biographer Alexander Moszkowski, Einstein handled this disparagement "not only without anger but with a certain satisfaction. For indeed the series of unbroken ovations became discomfiting, and his feelings took up arms against what seemed to be developing into a star-artist cult."[37] On the other hand, he incurred the envy of colleagues who looked askance at him for having made journalistic sparks fly as the top man in his profession. Not one of them had ever been celebrated like that.

He may have underestimated the danger posed by his opponents, who went up against him under the guise of serious science. He ought to have taken special note of the comments of Ernst Gehrcke, who managed to launch a powerful attack on Einstein and his theory of relativity without revealing an anti-Semitic agenda. Gehrcke, a highly respected experimental physicist and department chairman at the Imperial Institute for Physics and Technology in Berlin had first spoken out against Einstein's theory in 1911, in the form of two essays, and by 1913 he had written seven articles posing objections to the theory of relativity. Gehrcke fixated on his adversary, amassing everything about him, file after file, box after box full of offprints, newspaper articles (which he marked up in several colors), neatly clipped photographs, caricatures, and notes. Later he and a few like-minded colleagues created an international network of Einstein opponents, whose branches extended to the United States.

Einstein met his ferocious adversary shortly after arriving in Berlin. On May 20 and 27, 1914, the topic for the physics colloquium that Professor Heinrich Rubens held every Wednesday at the university in Berlin was relativity. Gehrcke took advantage of the opportunity to confront Einstein with the twin paradox. His argument was essentially this: Since the theory of relativity states that each of the two can claim that the other is moving relative to himself, each of the twins' watches would have to run more slowly, which is logically impossible. Einstein countered (rather feebly) that the acceleration of the one in respect to the other caused the watches to slow down—at which point Gehrcke noted with satisfaction (and rather cleverly), "In the theory of relativity, accelerated motions are absolute."

In 1919, Einstein was still using this line of argumentation instead of emphasizing the geometry of spacetime and the difference between geodesic and nongeodesic lines—as though he were somewhat wary of his own theory. In 1916, Gehrcke attacked the equivalence principle in the respected *Annalen der Physik*, and even accused Einstein of plagiarism. A man named Paul Gerber, Gehrcke claimed, had come up with the same formula back in 1899 that Einstein began using to calculate the deviation of the perihelion of Mercury in 1915. Einstein brusquely informed the secretary of the *Annalen der Physik*, "I am not going to respond to Gehrcke's tasteless and superficial attacks."[38] When Gehrcke did not let up, Einstein countered with a brief newspaper "comment" to "point out that [Gehrcke's] theory is untenable because it is based upon contradictory assumptions."[39]

Gehrcke was no longer fighting alone, however. He now had a prominent comrade-in-arms at his side, Nobel Prize winner Philipp Lenard, with whom Einstein had enjoyed a friendly exchange after 1905. In 1918, Lenard published an essay, "On the Principle of Relativity, Ether, Gravitation," which was later reprinted several times. The essay disputed the theory of relativity and upheld an ether theory of gravity.

In late November 1918, Einstein felt compelled to react to the criticism of his theory of relativity in an article that took the form of a dramatic dialogue between a "Critic" and a "Relativist" in the style of Galileo's "Dialogue." Eventually the critic has to "admit that it is not as simple to disprove your point of view as I previously thought." The imaginary opponent continues to probe, however: "What is the present situation of the sick man of theoretical physics, the ether, whom some of you have declared as finally dead?"[40]

The relativist, namely Einstein, replies, "The sick man has had a fluctuating fate; I don't think at all one could say he is dead now." He closes his statement with the warning, "However, one has to be careful not to attribute to this 'ether' any matterlike properties."[41] By reintroducing the ether, which he had dismissed so emphatically in 1905 as a young unknown, he left himself open to attack. He wrote to Hendrik Lorentz on November 15, 1919, full of remorse, "I would have been more accurate in my earlier publications if I had limited myself to emphasizing the nonreality of the ether's velocity rather than the nonexistence of the ether in general."[42]

Once Einstein had achieved his overnight fame; the crusade against him gained momentum. "The newspaper drivel about me is pathetic," he wrote to Heinrich Zangger in December 1919.[43] He pointed out to Zangger, "Another comical thing is that I myself count everywhere as a Bolshevist, God knows why; perhaps because I do not take all of that slop in the *Berliner Tageblatt* as milk and honey."[44] On February 12, 1920, there was a minor incident, an "Uproar in the Lecture Hall," as the *8-Uhr Abendblatt* reported.[45] Einstein's presentations at the university attracted not only students, but a wide range of auditors. Attending a lecture by Einstein became a must for tourists. The professor did not mind this, since "my popular lectures on the theory of relativity" served the goal of bringing his ideas to a large public. The students protested—with some justification—against the presence of the unenrolled auditors. When Einstein tried to persuade them to yield, there were "unpleasant scenes," as the newspaper reported, in the course of which "remarks of an anti-Semitic nature were made."[46]

In a personal explanation that followed the article, Einstein announced, "I found it necessary to cancel further lectures."[47] Unlike his colleagues, he could afford to do that. Whether it was the clever thing to do is another matter. His friends were distraught once again, and he defended himself. "Your letter, in which you call me a wretched politician, just arrived. Do you trust newspaper gossip more than me?" he asked Zangger. "My conduct toward the students and the university was nothing less than diplomatic."[48] A few weeks later the right-wing Kapp Putsch yet again made the Germans aware of the instability of the political situation. A general strike finally forced the rebels to give up. Einstein commented, "This country is like someone with a badly upset stomach who hasn't yet thrown up enough."[49]

The criticism leveled by the physicists Gehrcke and Lenard, which was at least reasonably scientific in nature, brought the staunchly reactionary freelance journalist and anti-Semitic demagogue Paul Weyland into the fray. He founded the Study Group of German Researchers for the Preservation of Pure Science. This pretentious name was little more than a one-man show with Weyland himself as the major, and nearly the only, active participant. The group's first publication was *The Theory of Relativity: A Scientific Suggestion to the Masses, Presented for General Understanding*—by Ernst Gehrcke.

Weyland, who regarded himself as the "recording secretary of the Einstein opponents," launched a full-fledged campaign. After a series of inflammatory newspaper articles, he planned a succession of large-scale events opposing the theory of relativity. The first one took place on August 24, 1920, at the Philharmonic in Berlin. Einstein could not resist showing up in person. He was accompanied by his stepdaughter and secretary, Ilse, as well as by his colleagues Walther Nernst and Max von Laue.

Weyland accused Einstein of being "publicity hungry" and indulging in "scientific dadaism," and to add insult to injury, he also branded him a "plagiarist." Had Weyland been alone onstage, Einstein would most likely have laughed off the incident with a round of amused applause. The demagogue's true intentions were revealed by the anti-Semitic, rabble-rousing texts and swastikas he arranged to have sold at the entrance. When Weyland was followed by Gehrcke, Einstein recognized the gravity of the situation. His scientific opponents, no matter how unfounded their criticism, were making common cause with the ultraright. As a physicist with an impeccable reputation, Gehrcke gave a straightforward lecture, and skirted any political polemics.

The newspapers reported and commented in great detail—pro Einstein, by and large. On August 27, *Morus* published a parody titled "Einstein Hounding," which mocked the motives of Einstein's adversaries. The reciters are a "chorus of fraternity students" as well as three "rabble-rousing professors." The first of these "professors" draws clear battle lines:

> We Teutons can't escape the views
> Of theories put forth by the Jews.
> Such nonsense! Logic is disputed
> As space mysteriously swallows time,
> The "era of greatness" is refuted,
> He lacks a sense of national pride.[50]

The same day, the *Berliner Tageblatt* printed a text by Einstein called "My Answer," which ought to be considered one of his major "blunders." The subtitle read "On the Anti-Relativity Theory Company." Here he went at his opponents in a text brimming with condescension. First he lumped Gehrcke together with Weyland and declared, "that both speakers are not worthy of an answer from my pen, because I have good reason to believe that motives other than the striving for truth are at the bottom of this business. (If I were a German nationalist with or without a swastika instead of a Jew with liberal international views, then . . .)."[51] He left it to his readers to draw the inference.

Ernst Gehrcke may have been Einstein's major adversary, but, despite his participation in the Philharmonic event, he was never a partisan of the right, unlike Weyland, who later joined the SA, and Lenard, who soon joined the Nazis. Einstein did not respond to a rejoinder by Gehrcke in the *Berliner Tageblatt*—although his antagonist quite reasonably asked him "to offer proof that there is any link between my years of factual exceptions to the theory of relativity and political and personal motives."[52]

With the resoluteness of a rebel, Einstein ignored both justified misgivings and understandable sensitivities. For more than two centuries, people had been able to rely on the cosmos running like a clockwork mechanism according to Newton's rules. Suddenly Einstein came along and claimed that this image was based on an illusion. Like a true revolutionary, Einstein made no apparent effort to deal considerately with those who expressed misgivings and sought to preserve the established view.

When he was finished with Gehrcke, Einstein proceeded to attack Lenard, who had not participated in the meeting at the Philharmonic: "I admire Lenard as a master of experimental physics; but he has not yet produced anything outstanding in theoretical physics, and his objections to the general theory of relativity are of such superficiality that up to now I did not think it necessary to answer them in detail."[53]

This was a recipe for making a lifelong enemy. His decision to attack Lenard of all people, to unmask this most prominent opponent of the theory of relativity, who had once had his eye on the Institute, only to find that Einstein had gotten the top position, and who, like so many of his colleagues, found himself consigned to the shadows by Einstein's appointment in Berlin, was hardly a sign of diplomatic skill. Einstein's inner voice enabled him to speak and write on political matters, but he had no second voice counseling him to exercise caution and reason.

His friends were beside themselves, and scolded him like a child who had been up to no good in an unsupervised moment. Paul Ehrenfest was unable to believe that "*you personally* wrote down at least some of the turns of phrase."[54] Ehrenfest implored his friend, "Dear, dear Einstein—please don't write one more word in reply to these shabby newspapers."[55] And Max Born's wife, Hedwig, deplored "the unfortunately very clumsy reply in the newspaper," adding, "Those who do not know you get a false picture of you."[56] Einstein replied the same day, "Dear Borns! Don't be too hard on me. Everyone has to offer his sacrifice at the altar of stupidity from time to time, for the amusement of God and man. And I did a thorough job of it with my article. This is proven by the exceedingly appreciative letters from all my dear friends."[57] This letter reveals a glimmer of the adult within the child whose understanding of the situation reaches much deeper than the appearance would suggest.

He asked Ehrenfest to bear with him when his friend advised him to exercise tact. "I had to do this if I wanted to stay in Berlin, where every child knows me from photographs."[58] However, when Arnold Sommerfeld, president of the Physical Society, begged him to "retract [his remarks] just as publicly as they were made in the first place," Einstein stood firm. Later events proved that Lenard really was in league with the right-wing anti-relativists, although Einstein and his friends could not have known this at the time. After Weyland visited Lenard in early August in Heidelberg, he wrote that the reactionary journalist was a "supporter of our reforms, and he has singled out Einstein's excessive machinations and the whole style of his behavior to combat systematically—as un-German."

Immediately following the Einstein event within-the-Weyland-event at the Philharmonic, three physicists, Nernst, von Laue, and Rubens, composed a joint declaration of support for their colleague in the *Tägliche Rundschau*. They confirmed that "his influence on the scientific community, not only in Berlin, but throughout Germany, can hardly be overstated."[59] A few days later an additional statement was published in the *Berliner Tageblatt*. Its prominent signatories included the theater impresario Max Reinhardt and the writer Stefan Zweig. They addressed their remarks directly to Einstein, declaring their "outrage at the pan-German propaganda against your outstanding reputation," and assuring him that they were proud "to consider you one of the leaders of the international scientific community."[60]

The impetus for these remarks may well have been Einstein's apparently spontaneous announcement that he planned to leave Germany. This statement was quoted in all the newspapers. Even the German chargé d'affaires in London learned about Einstein's intention from the British media. In a letter to the Foreign Office, he called Einstein a "cultural factor of the first rank" and urged officials to bear in mind, "We should not drive out of Germany a man with whom we could make real cultural propaganda."[61] It is hard to say whether Einstein was seriously considering leaving—even though he groused, "The political mentality of academics is assuming the form of blind obstinacy."[62] On September 9, 1920, two weeks after the noteworthy evening at the Philharmonic, he admitted to Hedwig and Max Born, "During the first moment of onslaught, I probably considered flight. But then I thought better of it and the old phlegm returned."[63]

On September 23, Einstein had further occasion to cross swords with his opponent. At the meeting of the Society of German Natural Scientists and Physicians in Bad Nauheim, his direct confrontation with Philipp Lenard was characterized by "an objectivity and calm that were truly exemplary," as the *Deutsche Allgemeine Zeitung* noted.[64] The *Berliner Tageblatt* printed an abridged version of the discussion the next day:

> LENARD. I am not dealing with formulas, but rather with actual processes in space. That is the chasm between Einstein and me. I have nothing against his theory of special relativity. But his theory of gravitation? If a traveling train brakes, the effects appear in fact only in the train itself, and not outside where all the church steeples remain standing! [He was essentially correct on that score.]
>
> EINSTEIN. The observable effects in the train are caused by a gravitational field, which is induced by the totality of near and distant masses. [He was defending Mach's principle.]
>
> LENARD. Such a gravitational field must certainly also bring about effects in other ways if I wish to visualize its presence.
>
> EINSTEIN. What human beings consider to be *intuitive* is subject to great change, and is *a function of time*. A contemporary of Galileo would also have declared his mechanics to be very unintuitive. These "intuitive" conceptions have their defects, just like the oft-cited "healthy common sense." (*Laughter.*)[65]

The unmistakable note of arrogance in Einstein's remarks undoubtedly served to confirm his opponents' distaste for his views. Lenard's arguments were quite clever. And Einstein again passed up a golden opportunity to have the essence of his theory of gravitation—the geometry of spacetime—speak in his favor. A braking train leaves its geodesic line, but a church steeple does not.

Paul Weyland adopted a shrill tone in the *Deutsche Zeitung*, angrily declaring, "under Lenard's leadership, the desecration of physics by mathematical dogmas is rejected, whereas the Einsteinophiles insist on their position and try to clamber up the Parnassus of their damnable formulas."[66] At the German Physical Society he bluntly announced, "It is high time for this rat's nest of scientific corruption to get a breath of fresh air." Max Planck, who had presided over the meeting in Bad Nauheim, tried to patch up the relationship between Einstein and Lenard. The result, as reported in the *Berliner Tageblatt*, was that Einstein wished "to express his sincere regret for directing barbs in his article . . . at Herr Lenard, whom he greatly respects."[67] But there was no appeasing Lenard. After the meeting he resigned from the Physics Society and posted signs at his Institute in Heidelberg that forbade entrance to any members of this group.

Lenard and Johannes Stark (for whose *Almanac* Einstein had written his visionary essay on the theory of relativity in 1907) launched an "Aryan physics" movement. This movement failed to catch on, even in the Third Reich, despite its leadership by two Nobel Prize winners. Just like Gehrcke, who no longer wished to be affiliated with the reactionary movement, Lenard distanced himself from Weyland. This demagogue had lost any interest in Einstein and his theory anyway, and he canceled his series of presentations after only two evenings.

Einstein's "scientific" adversaries did not let up, however. In April 1922, a ninety-minute film about the theory of relativity premiered at the Frankfurt Fair. No copies of the film, which was the longest educational film to date and featured elaborate special effects, have been preserved. The controversies surrounding this film again highlighted the areas of conflict between fans and foes of Einstein. The fans waxed lyrical about this reasonably comprehensible film, while the foes warned about the "dangers of popularization" and any attempt to "drum this wonder of the world into people's heads." The allegedly "serious" Einstein opponents created a breeding ground for a far more pernicious set of adversaries: ultraright anti-Semitic German Nationals who were out for his blood.

On June 24, 1922, Walther Rathenau, the German foreign minister, was murdered. Rathenau was a Jew, and a personal friend of Einstein. Rumors started flying that Einstein would be the next prominent Jew to suffer this fate. Einstein had known for some time that this rumor was well founded, since he had been receiving a series of death threats. On January 9, 1921, the reactionary *Staatsbürger-Zeitung* stated that for Einstein and others of his ilk "there is evidence of treason. We would consider any German who shoots down these scoundrels a benefactor of the German people."[68]

Einstein kept his distance from Berlin and public life for extended periods of time. He also canceled his participation in the upcoming convention of natural scientists in Leipzig. "Apparently I am part of a group targeted for assassinations by the *völkisch* Germans," he said to Planck to explain his absence. "The only remaining options are patience and getting out of town."[69] Planck's frustration was evident: "So it's come to this; a gang of murders . . . dictates the agenda of a scientific organization."[70]

In Leipzig, Lenard and his cohorts began to bully Einstein even more than they had in Bad Nauheim. Lenard's students distributed flyers protesting the planned plenary meeting on the theory of relativity, on the grounds that its adherents were concealing the fact that "many preeminent scholars . . . not only regard the theory of relativity as an unproven hypothesis, but even reject it altogether as an essentially misguided and logically untenable fiction."[71] The whiff of scientific seriousness that people like Lenard and Stark gave the anti-Einstein league was like oil on the fire of Einstein's political opponents. Even after Weyland left, they still had the scientific credentials they needed for their campaign to "cleanse" German culture and science.

Einstein offered his enemies even more reasons to defame him. His trips abroad enabled reactionaries and revanchists to portray him as a "traitor to his country." He was caught in a tragic dilemma between revenge and reconciliation: as a "cultural factor of the first rank" he was serving his homeland, yet he was rewarded for his efforts with harassment and malice.

His trips abroad offered him a welcome reprieve; they allowed him to stay out of the limelight and away from the threats on his life. Between mid-1920 and mid-1925 he spent about half of his time outside Germany. The Berlin historian Siegfried Grundmann has called these sojourns "a form of emigration."[72] Einstein himself welcomed the "opportunity for an extended absence" and spoke of the "need . . . to spend time away from the tense atmosphere of our homeland, which so often makes life difficult for me."[73]

Einstein's travels offered him two important positive opportunities as well: to disseminate his theory and to work toward restoring the international community of scientists. Weyland had accused Einstein of being "publicity hungry" and claimed that Einstein's efforts to promote the theory of relativity exemplified Jewish pushiness.

Certainly Einstein did welcome the chance to publicize his theory, which was meeting with uninformed resistance abroad as well as at home. Like a founder of a religion on a mission to preach his doctrine and to gather devotees, Einstein gave lectures throughout the world in auditoriums filled to overflowing. Innumerable reports from German embassies around the globe, which the Foreign Office compiled in a file marked "Lectures by Professor Einstein Abroad," attest to his success.

Wilhelm Solf, the German ambassador in Japan, declared that Einstein was serving "the good German cause, which transcends chauvinism" in the interest of the government, the fledgling republic, and the young democracy.[74] Even here, of course, the international league of science, which Einstein considered his only true homeland, was front and center. He put his good name, his reputation, and his unconstrained candor on the line to bring about an end to the boycott of German researchers. As an emissary of German foreign policy, the temporary emigrant helped thaw out the icy shell that had formed around a status-hungry Germany after World War I.

He had a fairly easy time of it during his travels to the neutral countries of Holland and Norway, as well as in Prague and Vienna. Even in the United States, which was trying to establish a good relationship with Germany and even make a separate peace treaty with the defeated enemy, he encountered little rejection. However, Einstein met with standoffishness and even open opposition on his trip back to Europe, when he stopped in England for a visit in the spring of 1921, and even more so one year later, in France.

When Einstein went to the podium to deliver a lecture at Kings College in London, he was greeted with stony silence. To make matters worse, he gave a lecture in the despised German language. But his combination of courage, sincerity, and an unconditional will to mend the rifts of the past, coupled with the magic of his theory, won over his audience. In fact, the intellectual elite of London was overwhelmingly enthusiastic. Even his old nemesis Ernst Gehrcke felt bound in 1924 to cite the anti-German *Daily Mail* for a retrospective text on Einstein: "He is a Jew. . . . He is a revolutionary. And yet he was received in a manner that was not only friendly, but positively enthusiastic. . . . It makes no difference whether his language is understood or not. People know they are under the spell of a compelling personality, a powerful intellect."[75]

During this trip, Einstein went to Westminster Abbey to lay a wreath at the final resting place of his great forebear and role model Isaac Newton. That made a good impression on the British, who maintain a sporting attitude even when negative feelings run deep. No sooner had the gunsmoke of the war settled than they extended a welcoming hand to the researcher from the land of the enemy. One of them, the astronomer Sir Arthur Eddington, refereed the contest between the two Titans Einstein and Newton upon his return from his solar-eclipse expedition to confirm Einstein's prediction concerning the deflection of light. Eddington declared Einstein the winner. (The fact that Eddington's measurements provided less unequivocal results than was claimed, and that he even left aside unsuitable measurements as unwelcome anomalies, was not publicized until much later.)

Einstein fought to restore good working relations between scholars, and in doing so achieved a thaw between hostile nations. His status as a Swiss citizen worked to his advantage, as did the fact that he had spoken out against the war and had not signed the Manifesto of the Ninety-Three. His naïveté in political dealings stood in stark contrast to the wariness of his political rivals—at home and abroad. It was painfully embarrassing for Germany that it took Englishmen to confirm the theory of relativity, and now they had to suffer the indignity of ingratiating themselves with the enemy.

His two-week trip to France in late March 1922 was subjected to more critical scrutiny than any previous venture abroad. The French nationalists rejected his coming outright, and the right-wing press hounded the man from the land of the enemy. Fearing possible violent protests by French students, he left the train station in Paris by way of a side platform, but his fears were unwarranted. The students simply wanted to give the famous physicist a friendly welcome. The man with the little yellow suitcase and the "gray overcoat and legendary artist's hat with an oversized brim"[76] had captured their hearts.

As always, Einstein used the occasion both to introduce his theory— this time in French—and to promote scientific contacts across national borders. In auditoriums that were full to overflowing, debates about his ideas followed the lectures. The *Vossische Zeitung* in Berlin reported, much to Einstein's delight, that "the whole discussion took a turn that was quite favorable to Einstein's theory."[77] The *Berliner Tageblatt*'s coverage of the visit was positively gushy: "This German conquered Paris. All the newspapers printed a picture of him, and an entire literature on Einstein emerged. . . . Einstein is all the rage. Academics, politicians, artists, rednecks, policemen, cabdrivers, waiters, and pickpockets know when Einstein is giving his lectures. The cocottes in the Café de Paris are asking their dandies whether Einstein wears glasses or is chic. All of Paris knows all about him, and anything Parisians don't know for a fact, they say anyway."[78]

Near the end of his trip, his hosts fulfilled his wish to visit the battle-fields of the First World War. His friend and fellow physicist Paul Langevin and his old friend Maurice Solovine accompanied him. This brief visit reinforced Einstein's most significant political conviction—namely, his complete opposition to war. He became an activist for pacifism and sought podiums and audiences to deliver his message.

A French journalist described the stirring hours Einstein spent there: "Right here, where the wheat is sprouting, there are still traces of moldering trenches. This is the first cemetery, not laid out according to rank, but in rows next to each other, the French with their white crosses, the Germans with their black ones. Einstein takes off his hat. He is quite moved. In a soft voice full of melancholy tenderness, he talks about the war, the militarism he despises and has always hated. . . . 'It is imperative,' says Einstein, 'that we bring all the students of Germany here now, students throughout the world, so that they may see how appalling war is.' "[79]

The reaction to the article in Germany was quick in coming. Einstein was labeled a "turncoat." Nobel Prize winner Stark, who was now Einstein's fiercest adversary (apart from Lenard), criticized Einstein's "ingratiation with the French" in print—instead of recognizing that Einstein had preserved the high moral respect of science. This was the thanks Einstein got from Stark. Germans could breathe a sigh of relief that they had such an effective covert foreign minister. "There is no doubt," the German embassy cabled to Berlin from Paris, "that Mr. Einstein, who, after all, has to be regarded as a German, is gaining attention for German intellectual life and German science and has gained new prominence."[80]

Einstein as the representative of Germany, of the "good" Germany to boot, seems quite surprising to us today. Only his deep-seated desire for freedom, which gained a ray of hope in the democratic spring of the Weimar Republic, gave his role a certain plausibility. It even seemed within the realm of possibility that people could get beyond petit-bourgeois conformism in liberal Berlin, which was an issue dear to his heart. If not for the narrow-minded chauvinists, the anti-Semites, the re-vanchists, and the reactionaries hell-bent on revenge—Einstein could have found something like a homeland in the very land of his birth.

Germania was Einstein's femme fatale. If he came too close, he got burned, but he could not pull away either. His *bête blonde* had molded him into the person he was. She spoke his language, praised his dili-gence, shaped his thinking, told the jokes he enjoyed, understood what he said, cooked his favorite dishes, and felt his philosophy. She hated and loved him, and he loved-hated her back as the inner demon he could never exorcise. His self-discipline, coupled with his occasional lack of self-control when dealing with supposedly weaker people, some-times made him more German that he would have liked. On the other hand, he lacked the German traits of self-hatred, an inferiority complex overcompensated for by megalomania, and schadenfreude.

Even before moving to Berlin, he judged the Germans rather harshly, "How coarse and primitive they appear. Vanity without real self-esteem. Civilization (well-brushed teeth, elegant tie, well-groomed mustache, impeccable suit), but no personal culture (coarseness of speech, movement, voice, feeling)."[81] On top of that, he returned to this country during a golden age of militarism, and one of his major reasons for having left Germany earlier had been his determination not to serve in the military. Einstein was revolted by drills and oppression, orders and blind obedience to authority, and masses of men in boots and uniforms all facing forward. Everything about soldiers was inimical to his nature. He could not imagine lusting for power, and detested co-ercion and brutality based on military rank and epaulettes as much as he worshipped the ideal of freedom.

The curse of the war took just over four months to arrive, and Ein-stein, who had once given up his German citizenship, was hoping that for the good of the world, the country of his birth would be defeated. When the Germans were indeed defeated, the utopia of a free and peaceful Germany seemed to have moved within reach. This utopian vision was the basis of his very successful attempts to win over the hearts of people in even the most hostile foreign countries for himself and his work as well as for his suffering fatherland.

In November and December of 1922, Elsa and Albert Einstein spent over six weeks in Japan at the invitation of a publisher. "His trip in Japan was like a triumphal procession," Ambassador Solf reported to Berlin. Solf's report stated that "the entire population of Japan, from the highest dignitary to the rickshaw driver, took part, spontaneously, without preparation and affectation!"[82] Patiently and cordially they followed his lectures, which ran up to five hours. Einstein, in turn, idealized Japan, and he suffered terribly in 1945 when the atomic bombs were dropped on Hiroshima and Nagasaki.

The highlight of the Einsteins' trip was their visit to a chrysanthemum festival with the royal family. "Everyone wanted at least to shake hands with the most famous man of the era," the ambassador reported. "The press was full of Einstein stories, both true and false. . . . There were also caricatures of Einstein, which featured his short pipe and his thick, unruly hair and hinted at the occasional inappropriateness of his clothing." Unruly hair, inappropriate clothing—this is how diplomatic words paint pictures. However, Einstein's charm and air of naïveté made up for his lack of etiquette.

Einstein was using a Swiss passport, which greatly facilitated the logistics of his travels. There is no evidence that he promoted his Helvetian homeland at the same time that he was devoted to the "German cause." But wasn't he a German as well? Hadn't he automatically become a Prussian citizen by entering the Prussian Academy? The unresolved question took on explosive force during one of the journeys he took as a Swiss citizen. When the Nobel Prize was awarded to him in 1922, he was on his trip through the Far East, which meant that he could not appear at the awards ceremony in Stockholm. The question arose whether the German or the Swiss ambassador ought to represent him at the ceremony.

"Einstein is a Reich German," the Berlin Academy telegraphed to Sweden.[83] Just as categorically, Ambassador Nadolny confronted his Swiss colleague, who politely stepped aside. Later the German ambassador came to realize that he had miscalculated the situation but, in a letter to Berlin, requested "that Einstein's Swiss citizenship not be mentioned if at all possible."[84]

A great deal has been written about Einstein's alleged "naturalization" being carried out without his being consulted. He himself told his ex-wife, Mileva, when she asked him about it in July 1938, "In 1919 the Academy urged me to accept a German citizenship in addition to my Swiss one. I was stupid enough to give in."[85] According to the files, he "took an oath for state employees twice, first on July 1, 1920 (to the German national constitution), and then on March 15, 1921 (to the Prussian state constitution)." Hence "the Academy concludes that in doing so Mr. Einstein became a German citizen."[86]

Einstein waited until February 1924 to declare—somewhat defensively—that he had a Prussian citizenship in addition to his Swiss nationality: "I have no objections to this idea."[87] In early 1925 he even applied for a German passport for his upcoming trip to South America. The document was granted to him with the blessing of Foreign Minister Gustav Stresemann. Just like Rathenau, Stresemann respected him not only as an adviser, but also as a kind of special ambassador in places that were off limits to even the best professional diplomats.

Upon his return from the Far East, Einstein saw all over again that a prophet is virtually without honor in his own country. Again he met with malevolence from the Nazis. Again his life was threatened, and two days before Hitler's "march to the Feldherrnhalle" in Munich on November 9, 1923, he left for Holland. Since Einstein had now resolved to turn his back on Germany, Max Planck implored him "not to take any step now that might make your return to Berlin impossible finally and for all time." Planck was well aware that "foreign countries have long envied us for having this precious treasure. But please think also of those here who love and revere you, and don't let them suffer too much for the abject infamy of a vicious pack of dogs whom we must get under control."[88]

Planck needn't have begged and pleaded. Before Christmas, Einstein returned to Berlin, where the tension had largely subsided. The following spring he breathed a sigh of relief to his friend Besso. "The political situation has calmed down, and thank goodness the masses are not concerned with me too much anymore, so my life has become calmer and more peaceful."[89] Meanwhile, Hitler was sitting serenely in his comfortable prison cell churning out his incendiary future bestseller, *Mein Kampf*.

With the introduction of the Rentenmark and subsequent massive investments of foreign capital, especially from America, the German economy got going again after 1924. Particularly in science and technology, the defeated country proved resilient even after the catastrophe of World War I and the chaos of the revolution. The first Atlantic crossing by airship in 1924 provided impressive evidence for all the world to see. Unemployment declined, consumption rose, the culture industry flourished, and the domestic situation stabilized. Within a few years Germany had become the second-largest industrial nation in the world (after the United States) and had resumed a role in concert with the great powers. Berlin in particular took up where it had left off in the great prewar era. With the expansion of Berlin in 1920, the first building boom since the Industrial Revolution, and the extension of the subway and regional train network, the capital city again spread beyond its borders.

From March to June 1925, Einstein accepted a long-standing invitation to visit Argentina, Uruguay, and Brazil. Once again there were well-attended lectures, receptions, meetings with heads of state, an honorary professorship in Montevideo, and an "Einstein prize" in Rio de Janeiro. The German ambassador noted that "his glaringly obvious indifference to grooming was apparently not held against him."[90] Only in Argentina did "the German colony stay away from all the events," because in an interview Einstein embraced views that were deemed too pacifist.[91] After a reception at the home of the German ambassador, Einstein noted in his diary, "A queer bunch, these Germans. I am a stinking flower to them, but they keep on sticking me in their buttonhole all the same."[92]

After this major trip, Einstein was finally able to spend a few years of his life in somewhat the manner he had envisioned. The hostilities had largely ceased—"at least on the outside," as Max Born later noted. Had Germany been spared "Hitlerism," as Einstein called it, he might have lived out the rest of his life as a "German."

The cult of stardom had now extended to quite an array of prominent actors, writers, musicians, athletes, adventurers, and technological pioneers. There were six-day bicycle races, elaborate indoor sports arenas, car races, zeppelins, and airplanes. Josephine Baker, Asta Nielsen, and Marlene Dietrich had been vying for a place in the headlines with

Einstein for quite some time. The new freedom had turned the republic and its capital into a place for artists and entertainers to experiment with new ideas.

Jazz came to Berlin, and cabaret and political satire were all the rage. Architecture and design had brought radical ideas to the unadorned simplicity of modernity. Painting and music rang out the old traditions and rang in the new. The avant-garde reveled in an unprecedented degree of freedom, sexual libertinage, and the emancipation of women. Reports about Marlene Dietrich, Dinah Nelken, Rosa Valetti, and others of this era refer to Einstein as a night owl and regular patron of shady establishments. Einstein and the writer Erich Maria Remarque reportedly celebrated "free love" and the "new woman."

By 1925 the home front was flourishing, and Germany's relationships with foreign countries had also markedly improved. On the whole, the boycott of German science had ceased, and a positive spirit of collaboration prevailed. As the political emissary of Germany, Einstein had done his duty. Colleagues like the patriotic-minded Haber, whom nobody in the "enemy" countries would have received, were able to build up international networks, thanks to Einstein's pioneering achievement. Einstein himself moved out of the spotlight with evident relief.

One of Einstein's notable activities during this period began in 1922, when he agreed to serve as a "German" representative to the Commission for Intellectual Cooperation of the League of Nations (to which Germany was not admitted until 1926). Regarded with skepticism by the Foreign Office, which filled up nine files about his avocation until he resigned in 1932, he signed on in May, signed off again in early July, let himself be talked into it once again shortly thereafter, and resigned yet again just after his return from the Far East in March 1923, justifying his decision with a stinging indictment of the League of Nations: "Recently I have come to realize that the League of Nations has neither the power nor the goodwill to fulfill its great task."[93]

One year later he took back his resignation again. Now Einstein got to know the inside workings of "the most ineffectual enterprise with which I have been associated," the tedious political business behind the scenes that does not make it into speeches and newspaper articles: participation in the "subcommittee for bibliography"; the question of "offensive passages in the history books" of various countries; "efforts on behalf of an international coordination of research on phototelegraphy"; and the squabbling about people that suddenly brought the fascist Italian secretary of education onto the committee.[94] After naming a replacement, Einstein declared that he would resign in 1928, took part in a session again in 1930, and finally resigned for good in 1932, when the committee had swung farther to the right.

Although he was exasperated by the bureaucracy in the various departments, he threw himself into humanist causes, fighting injustice and protecting the weak. During the war he successfully interceded on behalf of his friend Friedrich Adler, who had been condemned to death for assassinating the Austrian prime minister Karl Graf Stürgkh in November. Adler, a socialist, was pardoned after the war. To help the anarchist Erich Mühsam, Einstein even obtained a personal audience with Chancellor Wilhelm Marx and negotiated Mühsam's release from prison in 1924.

Einstein was never wholeheartedly embraced by the Republic. Beginning in 1923, his political activities were monitored by the Reich Commissioner for the Surveillance of Public Order. In 1926 he was classified among "suspicious individuals who draw attention to themselves in political matters." The files linked Einstein to numerous associations considered suspect, including the League for Human Rights ("one of the major organizations to spread pacifist propaganda"), the Society of Friends of the New Russia, and International Workers' Aid, which had close ties to the Communist Party. He was happy to lend his name in the service of a good cause without worrying about possible repercussions. Even though his thinking was moving farther and farther to the left end of the political spectrum while the political climate in Germany was veering toward the right, he never became a radical Marxist or Leninist. "I have my doubts about the productivity of a planned economy as a whole," he wrote to Besso.[95]

Still, he gave both opponents and neutral observers plenty of ammunition to accuse him of communist leanings. In 1929, for instance, under the auspices of "Red Aid," a solidarity organization supporting politically persecuted individuals, he wrote to the children in a reform school, "Make the very best people your guides. Read the letters of Rosa Luxemburg."[96] He gave a speech at MASCH, the Marxist Workers'

School, on the topic "What the Worker Needs to Know About the Theory of Relativity." He also commented about the leader of the Russian Revolution, "I respect Lenin as a man who did everything in his power to achieve social justice, at great cost to himself. I do not consider his method suitable. But one thing is certain: men like Lenin are the guardians and restorers of the conscience of mankind."[97]

Einstein's opening speech to the Seventh German Radio Show on August 22, 1930, in Berlin, which bespoke his touchingly naïve faith in progress, began with the legendary words, "Ladies and gentlemen who are present and who are not! When you hear the radio, think also about how people have come to possess such a wonderful tool of communication." He pointed out "that it is the engineers who make true democracy possible"—as though it were not possible the other way around. He then added, "Up until our era, people of different nations got to know one another almost exclusively by way of the distorting mirror of their own daily press," he went on to say. "Radio shows them to each other in their most vivid form and usually from the most charming angle, and thereby contributes to eradicating the feeling of mutual alienation that can so easily turn to mistrust and hostility."[98] Three years later the Nazis used the same platform to exert totalitarian control over the radio.

Even though Einstein's political activity is often made out to be naïve and dilettantish, one issue remained consistent throughout the Weimar Republic: calling himself a "militant pacifist," he echoed Kurt Tucholsky's still-controversial statement, "Soldiers are murderers." In the august company of Sigmund Freud, Thomas Mann, Romain Rolland, and Stefan Zweig, he signed an appeal that stated baldly, "Military training is the education of the mind and body in the technique of killing."[99]

In July 1930, a year that was sandwiched between Ypres and Hiroshima, Einstein signed a manifesto that underscored the role of science in building up armaments: "Do you know the meaning of a new war that would use the means of destruction science is ceaselessly perfecting?"[100] He pondered the question, "What good is a formula if it does not stop men from killing one another?"[101] His endorsement of conscientious objection to serving in wars became legendary. He now spent a great deal of time on the road—which in a sense foreshadowed his future emigration.

In late 1930, Einstein gave a speech in New York that included a notable proposition, which became the key idea of radical pacifism throughout the world: "If even two percent of those called up declare that they will not serve, and simultaneously demand that all international conflicts be settled in a peaceful manner, governments would be powerless."[102] His call for civil disobedience relied on a risky calculation that nations could not afford to have even 2 percent refuse to serve—a view that his comrade-in-arms Rolland called "criminally naïve" in September 1933.[103]

Einstein began to lose the support of his friends and sympathizers with his often unrestrained, rash, and self-contradictory ideas. Although he was an internationally recognized pacifist, he advocated the death penalty in language that seemed full of contempt for life. He stated baldly that "in principle" he had nothing against "killing individuals who are worthless or dangerous in that sense; I am against it only because I do not trust people, i.e., the courts. What I value in life is quality rather than quantity."[104] In his view, the right to hold such contradictory opinions was an inalienable premise of freedom. "As long as I have a voice, I have not only the right but the obligation to express myself."[105] His naïveté notwithstanding, he invariably demonstrated his remarkable intuition in political matters. His warnings about fascism and war came well before those of his colleagues.

After 1930, Einstein's opponents targeted his political views more than his physics—even in circles that were inclined to endorse his views. He spoke on every topic under the sun. "Abortion up to a certain stage of the pregnancy should be allowed if the woman so desires. Homosexuality should not be subject to prosecution except where necessary to protect young people. In regard to sexual education: No secrecy!"[106] These demands have been realized in modern democratic states.

Einstein exhibited great courage and commitment to what he considered good causes. When other causes were foisted on him, he occasionally allowed supporters of those causes to use his name on their behalf, but he was just as likely to withhold his support: "I have to take precautions to ensure that my voice is not undermined by figuring too often in relatively insignificant matters."[107] He did not sign on to every cause that came along. In 1932, when opponents of the war tried to get him to agree to speak at their conference on the topic of the "inestimable conflict . . . resulting from Japan's openly aggressive stance in regard to the Soviet Union," he declined to participate, explaining that "I would never participate in a conference of that sort, which would seem pathetically ineffectual. It would be like organizing a conference to stop volcanoes from erupting or to increase rain in the Sahara." Nothing whatsoever could be achieved, because "an insatiable appetite for dividends is the powerful motivating factor."[108]

The international economic crisis that began in 1929 made Einstein recognize far more keenly than most the extent of the danger the Nazis posed. He spoke out in the strongest possible terms to denounce the global conflagration they were threatening. "If you allow democracy to be destroyed, you risk your own destruction," he warned.[109] He railed against the secret armament of the Germans; he banded together with Romain Rolland, Maxim Gorki, Upton Sinclair, and Heinrich Mann to call for an international conference to prevent another world war; he took part in the League of Nations' Disarmament Conference in Geneva; he launched an Einstein Fund for conscientious objectors; and in the summer of 1932 he signed, together with Käthe Kollwitz, Erich Kästner, and many others, an "Urgent Appeal" to "establish a unified workers' front" for "an alliance between the Socialist Party and the Communist Party" for the Reichstag election. The resonance was impressive, but the result was disappointing. Even in the face of a looming fascist triumph, the two leftist rivals were unable to forge an alliance. Einstein commented, "It would probably be easier to reconcile Cain and Abel."[110]

His correspondence with Sigmund Freud (published in a tiny print run in spring 1933 bearing the title "Why War?") was virtually ignored by the public. "Is it possible to control man's mental evolution so as to make him proof against the psychoses of hate and destructiveness?" the physicist asked the psychoanalyst.[111] Freud replied that he saw "no likelihood of our being able to suppress humanity's aggressive tendencies," and posed a question of his own, "Why do we, you and I and many another, protest so vehemently against war, instead of just accepting it as another of life's odious importunities?"[112] At the close of his lengthy response, Freud conceded, "Perhaps our hope that these two factors—man's cultural disposition and a well-founded dread of the form that future wars will take—may serve to put an end to war in the near future, is not chimerical."[113] However, Einstein did not want simply to hope; he wanted to act, which is why he was one of the first to rethink his pacifism.

Einstein gave up his pacifist stance as a series of alarming developments unfolded. Hitler was installed as chancellor in January 1933, then the Reichstag burned down, and Social Democrats and Communists were hunted down and arrested and their parties eventually outlawed. The Parliament ceded its sovereignty to the new rulers in March 1933 by the "Enabling Act," and on April 7 the cynical "law to restore civil service with tenure" was passed, as a result of which 15 percent of the faculty at German universities were forced to vacate their positions within two years, and eight Nobel Prize winners in physics had to leave Germany. On April 1, the day of the Nazi boycott of Jewish businesses, the murderous viciousness of National Socialist anti-Semitism was made quite plain, and on May 10, the book burnings exhibited the extent of the war on culture. Einstein declared in July 1933, "Were I a Belgian, I should not, in the present circumstances, refuse military service; rather, I should enter such service cheerfully in the belief that I would thereby be helping to save European civilization."[114]

This statement pitted both pacifists and Nazis against him. Romain Rolland wrote to his colleague, the German writer Stefan Zweig, "Einstein is more dangerous as a proponent of a cause than an enemy of that cause. His genius is limited to science. In other matters he is a fool. . . . He should just stay out of it! He is made only for his equations."[115]

When it came to Hitler's Germany, however, Einstein was not open to compromise. He was more farsighted than most others and anticipated the true catastrophe that would set in with the "seizure of power." Just as prophetically, he warned earlier than most of the now-threatening consequences for world peace—and as a consequence had to endure accusations in his lost homeland of "atrocity-mongering." The *Kölnische Zeitung* quoted Einstein as campaigning to "protect Europe from relapsing into the barbarism of bygone eras. May all friends of our imperiled cause focus all their efforts on eliminating this global malady. I am with them." The newspaper added the comment, "Let him move to wherever he thinks he is safe from the 'poison' of the national cast of mind. A parting of the ways is all for the best."[116]

As early as December 6, 1931, Einstein had noted in his diary while en route to the United States, "Today I decided, in principle, to give up my position in Berlin. And so I'll be a migratory bird for the rest of my life!"[117] In July of that year he asked Planck "to see to it that my German citizenship is rescinded or to let me know whether this change can be reconciled with my remaining in the Academy of Sciences."[118]

Einstein's departure from Germany ultimately came to pass at the same breathtaking pace as the complete nazification of "Barbaria." He happened to be out of the country, in Pasadena, California, when Hitler became the chancellor of Germany, which was probably a great stroke of luck for Einstein. He was one of the Nazis' primary targets; they were out to get him, dead or alive. Supposedly they had even put a bounty on his head.

Five days after the Reichstag elections on March 5, 1933, Einstein declared that he would not set foot on German soil again. He was on a ship back to Europe when he made this statement. A good two weeks later, while he was still aboard the ship, he announced his intention to leave the Academy of Sciences. "Dependence on the Prussian government, entailed by my position, is something that, under the present circumstances, I feel to be intolerable."[119]

The academy adopted the position "that Mr. Einstein's withdrawal renders any further steps on our part unnecessary."[120] In other words, had he not gone of his own accord, they would have dismissed him. Secretary Heymann released a harsh, essentially unauthorized statement to the press: "The Prussian Academy of Sciences is particularly distressed by Einstein's activities as an agitator in foreign countries. . . . It has, therefore, no reason to regret Einstein's withdrawal."[121]

His friend Max von Laue called for a special session, which took place on April 6, 1933. "It was one of the most appalling moments of my life," the physicist later recalled. "I asked the Academy to disavow Heymann. But not a single voice backed me up. . . . The debate came to an end with Schlenk, the chemist, after whispering back and forth with Haber, who was sitting next to him, spoke up to side with Heymann as well."[122]

The situation deteriorated rapidly. On April 1, the day of the boycott of Jewish businesses, the *Deutsche Tageszeitung* printed a caricature of Einstein being kicked in the behind by a giant shoe and hurtling headlong down a staircase. The caption of this gloomy harbinger of things to come read, "A Poor Madman," and the text under the picture explained this cartoon as follows: "An aide at the German embassy in Brussels was asked to disabuse a loitering Asiatic man of the crazy idea that he was a Prussian."[123] Max von Laue recalled after the war, "The Nazis would have been tickled pink that day to have announced that they had thrown Einstein out of the Academy. The fury in the Ministry that he had beaten them to the punch by resigning was indescribable."[124]

Einstein had no illusions about his reputation in Germany: "I am now one of the people they most love to hate."[125] On April 4 he applied from Ostend to have his Prussian citizenship rescinded—an official act with consequences. Instead of simply being dismissed as a "charlatan" and "the most prominent Zionist agitator," he would experience the "tough degrading punishment" of expatriation.[126] It did not take place until March 24, 1934. Einstein's reactions ranged from sadness and indignation when contemplating this administrative travesty to out-and-out sarcasm. As he wrote to his sister, Maja, when he was in Belgium, "I have landed in the German doghouse, where, thank goodness, I have earned the right to be, fair and square . . . but my sailboat and my lady friends are staying there; H. bothered to take possession only of the former, which was an insult to the latter."[127] "H" stood for Hitler.

One of the most painful experiences of Einstein's entire life was watching the behavior of his colleagues from the academy, whose attempt to get him thrown out had been forestalled by his prior resignation. He vented his frustration in verse to his colleagues:

> Whoever writes grim fairy tales
> Will end up in our harshest jails.
> But if he dares the truth to tell
> We'll cast his soul down into hell.[128]

He lamented bitterly to Planck, "I cannot help but remind you that, in all these years, I have only enhanced Germany's prestige and never allowed myself to be alienated by the systematic attacks on me in the rightist press, especially those of recent years when no one took the trouble to stand up for me. Now, however, the war of annihilation against my defenseless fellow Jews compels me to employ, on their behalf, whatever influence I may possess in the eyes of the world. . . . And is not the destruction of the German Jews by starvation the official program of the present German government?"[129]

He later admonished Fritz Haber, who had betrayed him, "Surely there is no future in working for an intelligentsia that lies on its belly before common criminals and even, to some extent, sympathizes with those criminals. Me, they were unable to disappoint, because I never had any respect or sympathy for them—except for a few fine personalities (Planck 60 percent noble and Laue 100 percent)."[130]

In the plenary session of the Prussian Academy of Sciences on May 11, 1933, Planck paid tribute to Einstein's outstanding studies, "whose importance can only be measured against the achievements of Johannes Kepler or Isaac Newton." He felt it was necessary to state this for the record "lest future generations should ever think that Herr Einstein's professional colleagues were unable fully to comprehend his importance to science." Planck felt compelled to add "that Herr Einstein through his political behavior himself rendered his continued membership in the Academy impossible."[131]

It took the dignified, detached doyen of German physics, already a monument in his lifetime, a good many years to learn the lesson that politics is a dirty business. A few days before the session of the Prussian Academy, Max Planck had celebrated his seventy-fifth birthday. The tragic public servant, who had been a faithful servant of the Kaiser, continued to feel committed to the Republic in his heart of hearts, and was unnerved by the thought of what would become of it under the brownshirts who had come into "office" through democratic means. Although filled with misgivings about the "gang of murderers," he, like many others, nonetheless held out in Nazi Germany, considering it his task to protect science from even worse repercussions. In 1938, when he was old and frail, he resigned as president of the Kaiser Wilhelm Society. Shortly before his death in 1947, he declared that he was willing to have the successor to this society bear his name. The Max Planck Society was founded in 1948.

On September 9, 1933, Einstein left the continent of Europe for good. He had spent the spring and summer watching, from Villa La Savoyarde, a little summer home at the Belgian beach resort Le Coq sur Mer, as the land of his birth sank into a brown morass. After the SA had plundered his apartment on Haberlandstrasse and taken his valuables, carpets, and paintings, his stepdaughter Ilse and her husband, Rudolf Kayser, enlisted the aid of the French ambassador to help them transport his papers and most of his furniture to France as diplomatic cargo; these possessions were then shipped to America.

On his way to the United States, Einstein made a stop in England, where he met with Winston Churchill and other luminaries. On October 3, 1933, one week before his final departure from Europe, he addressed an audience of ten thousand in Royal Albert Hall and was given a standing ovation. Without mentioning Germany by name, he soberly warned "in faulty English" about the imminent danger.[132] He beseeched the "leading statesmen" to "devise for Europe the kind of . . . commitments whose meaning is so completely clear that all countries will come to view any attempt at warlike adventures as utterly futile."[133]

He closed his speech by praising the solitude of the scholar; his words sounded like a somber anticipation of what would await him as an aging exile. "Such occupations as the service of lighthouses and lightships come to mind. Would it not be possible to place young people who wish to think about scientific problems, especially of a mathematical or philosophical nature, in such occupations?"[134] He may have sounded naïve, but he was expressing his deep longing for a republic of scholars that enables science to move forward unhampered by politicking and patronizing attitudes.

Was Einstein really, as the Stuttgart historian of science Armin Hermann has suggested, "one of Hitler's major antagonists"?[135] Einstein and Hitler both changed the course of history, one with the power of his ideas, and the other with his idea of diabolical power. How could the physicist hold his own against the politician, whose "cold, barbaric, animalistic resoluteness" was evident to Einstein from the start?[136] When the German people, who owed so much to the researcher, chose a Nazi to lead them, there was no place for Einstein in "Teutonia." Although Einstein's influence extended to all parts of the globe, he became a symbol not only of the politically astute and active scientist, but also of the impotence of science, or at least of the individual scientist, in the political arena.

Fritz Haber, a decorated war hero, the scientific father of chemical warfare and the beneficent inventor of synthetic fertilizer, also began to realize that science and politics were not on a level playing field. Despite his renunciation of the Jewish faith when he was baptized, and despite his unparalleled love for his country, his economic clout, and the support of Max Planck, who appealed to Hitler on his behalf, Haber eventually had to emigrate. In the summer of 1933 he left for England.

In August 1933, Haber received a letter from his fellow exile Einstein, who was staying in Le Coq, Belgium. "I am delighted to know," Einstein quipped, "that your former love for the blond beast has somewhat cooled off."[137] In January 1934, during a stay in Basel, Haber suffered a fatal heart attack.

FIFTEEN

"I AM NOT A TIGER"

EINSTEIN, THE HUMAN SIDE

Erika Britzke opens her eyes wide to reveal the steely blue vigilance behind her oversized eyeglasses. Her hairstyle is boyish, her face frank and friendly. After spending half of her adult life with Einstein, she conjures up ideas, quotations, and pictures, which she paints with words as they rise up behind her closed eyelids: Einstein in a jogging outfit, going to his sailboat, lost in his thoughts, as usual. The friendly neighbors say hello to him over the fence. Einstein can laugh about anything, even when danger looms. He cannot swim, yet he refuses to wear a swim vest. "If I drown, at least I'll really and truly drown."

"Oh, Albert," his wife, Elsa, gripes. In the forest he points out his favorite spots to his friends. He brings back mushrooms for his housekeeper, Herta, to fry with eggs. He tries out a yo-yo with his friend's son; yo-yos were all the rage for schoolchildren in Berlin. A prematurely gray young man wearing a leather jacket, he goes down to the basement and shovels coal into the central furnace.

Erika Britzke has been looking after Einstein's house in Caputh, two train stops from Potsdam, for more than twenty-five years. This house was the only home he ever created for himself. The back offered access to the woods, where he often went walking for hours on end, and the front, beyond the sloping garden, looked out over the village in Brandenburg and over Lake Templin, where he kept his beloved "big fat sailboat," *Tümmler*. The little log cabin, which the Einsteins used as a summer home, has a spacious terrace and a shady porch, fruit trees, flowers, and berries. The house is simple yet elegant, with a uniform style; the window frames, shutters, banisters, and handrails are painted white to contrast with the dark façade.

Einstein was utterly content here, and he would have been pleased to grow old in Caputh. However, his happiness lasted only three summers, from 1930 to 1932. The following year the Nazis drove him out of his little paradise. The house is still there today, renovated for posterity, minus his furniture, books, and clothes. The closets and rooms are bare, apart from a couple of antique furnishings, reconstructed to convey a sense of the original aura, photographs on the walls like documentation to ward off legends, but aside from that they are the empty shell of a bygone dream.

And yet this is where visitors can best get a feeling for Einstein's private life. His study, his sleeping alcove, the bathroom with sink and bathtub, sometimes almost embarrassingly intimate, but preserved in a dignified manner, and devotedly cared for by Erika Britzke, who has been performing her duties in this house at Waldstrasse 7 since Einstein's hundredth birthday, in 1979.

Back then it was still in the German Democratic Republic (GDR). When Einstein's centenary was approaching, West and East Berlin attempted to outshine each other with symposia, ceremonies, and exhibitions. In the West, President Walter Scheel gave a speech in his honor. In East Berlin, Prime Minister Willi Stoph claimed to speak in the name of one revolutionary to another: on behalf of Lenin, who admired Einstein, and Einstein, who admired Lenin. It was up to the East German Academy of Sciences to renovate Einstein's wooden house down to every last detail and to place it in the care of the former art teacher Erika Britzke. "The GDR," she said, "played its trump card then."

A quarter of a century later, on the occasion of the 125th anniversary of the physicist's birth, Mrs. Britzke celebrated "a silver anniversary with Einstein." Back when she first began her job, "I gulped and said to myself, my goodness, where in the world are you?" Then she began to fill herself in on the facts. She got her information from the neighbors in Caputh, from Herta Waldow, who kept house for the Einsteins for six years in Berlin and Caputh, from the photo archives of the *Märkische Volksstimme*, from the National Library in Berlin, and from Einstein researchers around the world. Because she was often alone in the house in Caputh—weeks would go by without a visitor—she began to delve "deeper and deeper." All the available Einstein literature was at her disposal. "I immersed myself in his life," until she evolved from an information seeker to an information disseminator. Self-styled expert, local historian, factotum as expert.

The history of the summer home in Caputh goes back to the early 1920s. The sudden fame and the increasing threat by the wheeling and dealing of his adversaries made Einstein think about getting a place of retreat from intrusive fans and foes. Shortly after the public hostilities and his bitterly angry reply to the anti-relativists, when friends and colleagues already feared that he would leave Germany, he wrote to Max and Hedwig Born, "Now I think only of buying a sailboat and a country cottage near Berlin by the water."[1]

In the meantime he realized a small version of his big dream: in the summer of 1922 he rented a little bungalow, which Berliners call a *Laube*, and a little boat in the bungalow colony in Bocksfelde at the Scharfe Lanke, a cove of the lower Havel River, near Spandau. His son Hans Albert visited him, but his wife, Elsa, begged off. His "little castle" was too primitive for her. The bungalow association reprimanded him for letting the weeds sprout up in his yard, whereupon the contrite gardener assured them he would abide by the rules of the lease. Evidently he continued to use the bungalow until his summer home was completed in 1929.

More than ten years passed between his sudden fame in 1919 and when he moved into his refuge in Caputh. During this phase in his life, he grew accustomed to being "gaped at like a kind of show horse."[2] Einstein, who had immortalized the equivalence of energy and mass in his famous formula $E=mc^2$, became familiar with the energy of the masses as a function of his fame. According to his son-in-law Dimitri Marianoff, he became a "prisoner of the world"—a living person bearing witness to his own immortality. At a banquet in London in 1930, he thanked George Bernard Shaw "for the unforgettable words which you have addressed to my mythical namesake who has made my life so burdensome."[3]

"It is true," the writer Antonina Vallentin observed, "that he can see the attention fixed on him; he is aware of an atmosphere of respect; he hears his name mentioned all around him; and he is even conscious, secretly, of the bores he wishes to avoid."[4] Even the artists who passed

the time in the Romanisches Café gawped at him with such uncon-
cealed curiosity that he beat a hasty retreat.

The object of all this attention complained, "Even Chaplin looks at
me like some kind of exotic creature and doesn't know what to make of
me. In my room he acted as though he were being brought into a temple."[5] Einstein valiantly attempted to remain an ordinary person, to put
his fame as a "fashion piece" in perspective, to feign a normality that he
would never enjoy again. "I have even had the suspicion," Vallentin
wrote, "that he really believes himself to be exactly like everyone else."[6]

The period of blissful anonymity was over for good, however. Na-
talia Saz, a renowned Russian theater director during those years, recalled the public fervor for Einstein: "People are proud of sharing an
era and a city with him!"[7] "Every step into the public eye was an or-
deal," said Konrad Wachsmann, the architect of the summerhouse.
Even in his apartment, Einstein was subjected to a constant barrage,
which his wife, Elsa, did her best to ward off.

"A queue gathered around their door, as though waiting for miracles
to be performed," reported Elsa's friend Antonina.[8] Wachsmann "saw
beggars, cadgers, petitioners, and enthusiastic tourists who just wanted
to catch a glimpse of Einstein."[9] There were also "painters and photographers who claimed with perfectly straight faces that they were the best
in their fields and were therefore duty-bound to make a portrait of Einstein," but if he went to the door, his family would be cleaned out
within a short time, "because he would not hesitate to give a whole
month's salary to professional beggars or the truly needy."[10]

Then there was the matter of the daily mail. "One man writes that
he has finally discovered the essence of sleep; someone else claims to
have found the only proper way to lower the price of coal."[11] Many "experts" (some of whom are still around today) claimed to have refuted
Einstein's theories. When politics and pacifism brought his name into
the headlines, the cast of characters simply changed: "In the place of
inventors suffering from mythomania, in place of the misunderstood
geniuses who used to storm his door, it was now cranks who besieged
him with their panaceas for universal peace."[12] Sometimes, when visitors simply could not be turned away, Einstein had to leave the apartment by way of the service staircase. Far from getting riled up, however,
he was simply amazed that people found him amazing. "Einstein's im-

age was an unfathomable mystery to him," said Thomas Bucky, the son of his doctor.[13]

"People who had lost all faith in higher values during these years of misery, murderous warfare, lies, and slander saw their hopes revived by the appearance of a simple, modest man who calmly contemplated the nature of space, time, and matter in the thick of these hellish circumstances, and helped lift the veil of mystery surrounding creation," Max Born recalled.[14] This side of the story, from the perspective of a friend and scientist, highlights the solace, hope, and miracle of knowledge. "The theory of relativity was celebrated as a new doctrine of salvation, and Albert Einstein was regarded as the Messiah of our century."[15]

There was more, though, to the practically spiritual effect he had on people than the revolution in science. His unforgettable appearance contributed greatly to his charisma. Since his arrival in Berlin he had aged beyond all measure, owing to his lifestyle, Herculean labor, and severe illness. Now the clock was racing until Einstein became the man we know today, the most famous head in the world: his halo-like mane of hair that brought to mind prophets and artists, his long, kind face, his lofty, arched brows, his brown, perpetually moist eyes with the gentle expression of a starry-eyed sage, the prominent nose, the chiseled oval of the cheeks, the mustache, the sarcastic curl of his full lips, the dimple in his chin.

With his voice as bright as a cello, his roaring laughter, his light step despite his bulky frame resulting from his powerful musculature, his leisurely pace, and his avant-garde sartorial disarray—his unbuttoned shirts, his baggy trousers held up by suspenders, the socks he refused to wear—Einstein appeared to have descended from another world.

"He confronted society as though he had been born on another planet," Vallentin wrote.[16] The director Steven Spielberg had good reason to base the trusting look of his alien-child E.T. on Einstein's eyes (for which Michael Jackson is said to have offered millions of dollars). Einstein has become the ideal object for photographers and reporters and all other priests of popularity, with whom he has lived in a strange symbiosis. For them, the man is more spectacular than his theories, and the myth even more spectacular than the man. When he was asked what his profession was, he quipped, "fashion model."

He was an oddball who spoke his mind, as though he lived by a different set of rules. Even his name seemed tailored to work its way into people's memories and take root there. It is a strange, catchy German name pregnant with meaning (it translates as "one stone"), mighty as a monolith. The two syllables of "Einstein" even rhyme.

Virtually overnight, the Einsteins became the leading lights of the Berlin social scene. Einstein preached the ideals of equality and a classless society, and his words reflected deeply held beliefs, yet his dining room table was a meeting place for doctors, bankers, politicians, industrialists, writers, journalists, painters, musicians and, of course, scientific colleagues, including several Nobel Prize winners.

His choice of friends in Berlin sheds a peculiar light on Einstein, who aspired to live a bohemian life in a bourgeois home. He wrote to Mileva, his ex-wife and lover from his gypsy days, "I have noticed that as a rule, the lives of ordinary people with their programmed routines and their daily grind maintain their emotional equilibrium better than the MotRLC (Members of the Rich Leisure Class)."[17] He was just as comfortable rubbing shoulders with the upper crust, with very wealthy high-society ladies and gentlemen, as with leftists who shared his views. He put up with their chauffeured automobiles, ate dinner at their tables, and visited them at their country homes—as if hoping for a foretaste of his own rural bliss. In turn, his munificent acquaintances could indulge in the pleasure of claiming Einstein as a friend, arguably the era's most dazzling human trophy in Berlin. He was even described as a socialite who resembled a child who had put on an Einstein mask just for fun.

His social contact with physicians, most of whom were prominent, well-heeled villa owners, probably had to do with his hypochondriac tendencies. Not only did he tend to exaggerate when he felt sick (and imagined himself at death's door), but he also wanted to be in the proximity of doctors, preferring a social setting to a medical one. "He loved doctors, but not being treated by them," an acquaintance remarked. The surgeon Moritz Katzenstein was the perfect friend to sail and chat with. Gustav Bucky, chief of staff at the Rudolf Virchow Hospital, looked after his stepdaughters Margot and Ilse, and became a close friend of the Einsteins after they had all moved to the United States.

Einstein's friendship with János Plesch, the "prototype of an elegant and snobby socialite who was always a bit too flashy," seemed downright incongruous.[18] Konrad Wachsmann regarded the two "as fire and water."[19] Plesch owned so many cars that he could have used his fleet to run a taxi service. Einstein never owned or drove a car, and he maintained "an ironic reserve"[20] from the flamboyant professor, but relished their stimulating talks. He abhorred stupid and ignorant people: "He maintained an icy silence with people who had nothing interesting to say."[21]

The mere thought of having to attend an evening event cast a pall over Einstein's day. The notion of etiquette was forever alien to him. He hated collars and ties, and was unwilling to succumb to pressures to conform. "One would explain to him the customary formalities, and those who had not known him long would explain patiently, as to a backward child," Vallentin reported.[22]

When the Einsteins threw dinner parties, which Elsa would have liked to do more frequently, and Albert less so, the distinguished guests found themselves in a "somewhat threadbare bourgeois coziness" where Einstein acted like an interloper. "He always looked as though he had wandered into these rooms quite by accident and now had to live there because he couldn't find the exit," said Wachsmann.[23]

The living room was decorated with Persian rugs and a portrait of Frederick the Great. The beautiful grand piano did not mesh with the Biedermeier-style furniture. Elsa kept the decorative porcelain in corner cabinets and the inevitable knickknacks behind polished crystal doors. There was a small goldfish bowl, and in the kitchen a blue parakeet who answered to the name Bibo or to his nickname, "Biebchen."

Elsa, who had once studied acting, could make her guests dissolve in gales of laughter by impersonating celebrities. Albert guffawed when others saw nothing to laugh about. Even on sad occasions he told jokes, and sometimes exploded with laughter when listening to trifling tales. One of his favorite books was *The Hundred Best Jewish Jokes*.

The evenings were jolliest when his friend and colleague Max von Laue came to visit. Von Laue was the only one with whom Einstein indulged in lighthearted fun. But what did it really mean to be his friend? Einstein used the term for dozens of people, even though his life was just as much of an enigma to almost all of them as their lives were to him. A more decorous atmosphere prevailed when Haber or Nernst came to the house, and when Max Planck was there, the tone turned rather formal. As soon as the dinner-table conversation turned to science, Elsa and her daughters silently left the room. As the lady of the house, who never left any doubt that she was Frau Professor, Elsa, with the help of her housekeeper, Herta, felt primarily responsible for the physical well-being of their guests.

Herta followed Elsa's instructions by sticking to an unchanging sequence of foods: "It nearly always began with egg-drop soup. Then there was mayonnaise with salmon, made with canned lox, then pork filet with sweet chestnuts, followed by a strawberry dish with whipped cream, combined into a strawberry meringue. Herr Professor was mad about strawberries, and loved asparagus."[24] Plesch's son Peter recalled, "He gobbled things up the way children do." The meat had to be well done. "I am not a tiger," Einstein informed the cook.[25] He did not touch the wine or Elsa's celery punch. At most he sipped a bit of cognac.

For breakfast, after what was generally a long sleep, rarely under ten hours, he had the cook make him fried or scrambled eggs with toast or rolls that were delivered to his door. His voracious appetite for honey meant that "whole buckets" needed to be purchased. Following his severe illness in 1918, he only drank the decaffeinated Hag brand of coffee, which was brewed especially for him; otherwise he drank large quantities of black tea.

Still, the "perpetually grubby Albert" had come a long way in matters of cleanliness: "Herr Professor took a tub bath every morning."[26] For dinner, which was served between six and seven, he ate cold cuts, cheese, and eggs. The head of the household "always ate two sunny-side-up eggs, at least two." Photos of Einstein, who was described as "stocky," bore an increasing resemblance to a bear with a Buddhalike belly. Under a sketch an artist had made for a portrait of Einstein, the physicist penned a rhyming quip, "This big fat swine is Professor Einstein?"[27]

Although Einstein had invented a refrigerator for which he held several patents, his city apartment had nothing but a simple icebox that kept its contents cold with chunks of ice; oddly, though, this apartment was equipped with an electric vacuum cleaner. The summer home in Caputh was his first residence to boast an electric refrigerator. His sister, Maja, was delighted to report that his "study" in Berlin had "an electric heater that used radiant heat to warm up the individual rather than the room (it costs about 20 marks)."[28]

He spent most of his time lost in thought up in his attic room—usually enveloped in thick blue clouds of smoke. Because Elsa was attempting to enforce a ban on smoking downstairs after his recovery, he puffed away upstairs. His friends were happy to show their support for him. Even Professor Plesch disregarded his own orders and regularly supplied his patient with cigars. Einstein also exploited social events to run riot with smoking in the apartment. His pipe was so much a part of him that he even carried it around with him between his teeth when it was unlit.

Because he had finished his attic (with a storage area for books and a small room) without bothering to get an official permit, he was prohibited by law from using these rooms. He applied to the chief of police for dispensation in 1927, arguing that "the room is used only by me personally, not by others. Any hygienic deficiencies would affect nobody but me."[29] His request was granted.

His attic refuge could not stand in for his dream of the country home indefinitely. In 1929, the year that Marlene Dietrich made her cinematic breakthrough in *The Blue Angel* and the airship *Graf Zeppelin* made its first global flight, the perfect occasion arose for him to realize his dream. Einstein's fiftieth birthday was coming up on March 14. Fearing an onslaught of well-wishers, he left town to take refuge at the Villa Plesch on Lake Tegel so he could celebrate with a small group of friends. Back in his apartment, the presents were piling up. "Tables, chairs, the piano in the living room—every inch had to serve as a place to put down sundry objects," Wachsmann recalled. "Wives of millionaires gave him manuscripts of Brod's novel *Tycho Brahe* after they had read it. These manuscripts would be priceless today."[30] A cigar manufacturer named his new type of cigar "Relativity."

Otto the doorman had to carry letters, telegrams, and birthday cards by the basketful up to the fourth floor. Many senders of telegrams simply left off the address to save money. Mail addressed "To Albert Einstein, Berlin" arrived. Reporters and photographers were stationed in front of the house, but they eventually realized that the bird had flown the coop. Einstein is said to have been tickled pink about his escape.

What do you give a man like Einstein to mark his half-century? His wealthy friends pooled their money to finance a cabin yacht that he could build to his specifications. This boat featured a twenty-square-meter sail, built-in dish cabinets, a folding table, a foldaway stove, and two sleeping areas. Wachsmann claimed that Einstein loved his sail-boat, which he named *Tümmler* (Dolphin), "the way a boy loves his favorite toy."[31] The new boat," he wrote to his ex-wife Mileva, "is wonderful; it is so beautiful that I am a bit afraid of taking on the re-sponsibility for it."[32]

Only one present, from the city of Berlin, afforded no pleasure to the birthday boy. Officials in Berlin had thought up what appeared to be the perfect gift to go with the new boat: a house on the water. This wonderful plan went terribly awry, however, and ended as a pathetic farce. At first everything went swimmingly. The city offered Einstein the Neu Cladow "nobleman's residence," a property it had acquired some time before for several million marks. When Elsa went to have a look at the "gift," she discovered that the previous owner was still living there—and had a right of residence for life. The municipal authorities responded by substituting a lake property near Gatow on the Havel River. Einstein would have to build the house himself. "Modest as he is, my husband has actually accepted that," Elsa complained. "The present is getting smaller and smaller."[33]

Konrad Wachsmann, a young disciple of the master architect Walter Gropius, who was utterly unknown at the time, sensed that this would be his big chance to make a name for himself. Like so many suppli-cants, he simply showed up at Einstein's apartment. When Elsa opened the door, he announced, "I am an architect and would like to build Al-bert Einstein's house." Elsa did not see through the young man's ruse—he had, in fact, completed only a single building project in his life—and asked him in. It did not take long for her to decide to make good use of his expertise and bring him along to inspect the piece of land they had been offered. It turned out that the site was right next to a motorboat club, and there would be no way to achieve the tranquillity they were after. Elsa was nonetheless enthusiastic about a cluster of old trees on the property.

When Elsa invited him to dinner on Haberlandstrasse, Wachsmann saw that the time was ripe to make his pitch. Overnight he drew up drafts in accordance with Einstein's detailed wishes, which Elsa had laid out for him: "It needs to be stained brown and have French windows and a dark red tiled roof."[34] Einstein envisioned something rather modest, and insisted on a wooden house, although Elsa would have preferred a stone structure. He asked for large terraces and his own separate bedroom and study.

Over dinner, Wachsmann outlined the disadvantages of the site to Einstein in front of his dumbfounded wife. "In that case, I might as well move right into a beach resort," Wachsmann's disappointed client realized.[35] The architect seized the opportunity and displayed his drafts to the family. As a former patent expert, Einstein was well versed in reading technical sketches. "I like your sketches and ideas," he remarked.[36] Of course, Wachsmann had to adapt his plans to the sloping site in Caputh, but these drafts provide a good idea of the summerhouse that Erika Britzke has been caring for since Einstein's hundredth birthday.

Once the family had rejected the site in Gatow, the city came up with several new options. When all of these turned out to be unsuitable, the municipal authorities finally offered Einstein the option of selecting a plot of land himself, which the city would purchase for him. As luck would have it, Elsa heard about a plot for sale in the little village of Caputh. On April 24, 1929, the Berlin authorities resolved to buy this piece of land, "which meets all the scholar's requirements in regard to quiet, nice location, access to the Havel River for sailing, and a direct connection to public transportation, with suppliers nearby."[37]

However, the project was derailed in the city council by disagreements between the German Nationalists and the Social Democratic majority. The decision was put off. By this point, the "great scholar" had had quite enough. On May 14, 1929, the Berliner Tageblatt announced, "The Disgrace Is Complete—Einstein Declines." "In view of the slow pace of bureaucracy and the brevity of human life," he opted not to get caught up in a political tug-of-war that would make him the object of gossip and possibly of ridicule.[38]

Still, Einstein could not be dissuaded from realizing his dream. "The house will be built even if I have to starve to make it happen."[39] Evidently at this time he was not plagued by doubts about remaining in

Germany, despite all the threats and the mounting Nazi tyranny. The Einsteins used their own money to buy the land in Caputh and have the wooden house built, using Wachsmann's plans.

Wachsmann could hardly believe his good luck. He was delighted to show Einstein additional sketches that more closely conformed to his own Bauhaus-influenced style, but his suggestions were met with a sharp rebuff: "I don't want a house that looks like a carton with giant display windows."[40] Back in 1922, the physicist had been equally dismissive of the Einstein Tower, one of the most significant works of expressionist architecture, which Erich Mendelsohn had constructed on Telegraph Hill in Potsdam. Einstein found the geometric clarity of Gothic cathedrals more to his liking.

His house straddled the line between tradition and modernity. Still, Einstein again displayed his deep skepticism regarding contemporary design when Wachsmann presented interior sketches by the young Bauhaus artist Marcel Breuer, "I do not want to sit on furniture that reminds me of a machine shop or an operating room."[41] The large living room was decorated to suit Elsa's taste, with furniture from the Haberlandstrasse attic. Erika Britzke's explanation is matter-of-fact: "He wouldn't pay a bundle of money for a designer."

Einstein's ambivalence about modern art is even more remarkable because his thought processes and the views of life they afforded him had a pronounced influence on the arts, particularly on the fine arts—even though he always vehemently denied it. He insisted that one should not confuse physical with everyday relativity. Regarding the alleged relationship between his works and cubism, he declared categorically, "This new artistic 'language' has nothing in common with the theory of relativity."[42]

The Flemish painter Piet Mondrian called space and time two sides of the same coin—whether Einstein liked it or not. And the aim of Pablo Picasso's cubism was, like Einstein's scientific theory of relativity, to explore the geometry of the world in a new, more profound manner.

Ulrich Müller, an art historian in Berlin, has pointed out how intensely the painter Paul Klee experimented with the four-dimensionality of spacetime. Klee was always asking his students at the Bauhaus to bear in mind that the "notion of fluid space with the fourth imaginary dimension of 'time' " imposed limits on painters, who had to stay within two dimensions, whereas architects, who could conceptualize and plan in three dimensions, were liberated by this new dimension. Müller explains that this is how Einstein's idea penetrated into the intellectual cosmos of the Bauhaus artists, and from there Einstein's revolution indirectly shaped the architecture of modernity, the fluid spaces in the structures of Mies van der Rohe and Le Corbusier as well as Frank Lloyd Wright and Walter Gropius.

Although Einstein considered this idea nonsense, he lent his support to the "Society of Friends of the Bauhaus." As a scientist with artistic inclinations, he valued the freedom of art above any misgivings he had about its contents and forms of expression. When the architectural theorist Sigfried Giedion based the argument of his well-regarded book *Space, Time, and Architecture* (1941) on the work of Einstein, the latter emphatically denied that this argument could have originated in his ideas.

Einstein regarded avant-garde artists, even those who claimed to be influenced by him, with a mixture of skepticism and helplessness. The psychologist Howard Gardner believes that "in matters of art, he displayed the conservatism of age, never embracing any of the twentieth-century modernisms that were created in his shadow."[43] Konrad Wachsmann reported, "The man who had revolutionized physics stood helplessly in front of a canvas and could not make heads or tails of the ciphers in color and brushstrokes."[44] Wachsmann remarked that Einstein "could not bring himself to deal with the iconography of a modern painting, and something within him refused to attempt to gain a true sense of the contents and interpretation of these paintings."[45] He stressed Einstein's belief that "viewing modern works of art was a matter only for professional specialists."[46] Einstein agreed to be painted by Max Liebermann, but turned down Marc Chagall when the latter asked him to pose.

His conservative sense of art, which seemed stuck in the nineteenth century, made him a traditionalist, far from ready to embrace the radicalism of avant-gardists like Picasso, Alban Berg, and James Joyce. His favorite writer was Heinrich Heine. Since his own attempts at poetry resulted in lively verses, he loved Wilhelm Busch. He had the utmost regard for the work of George Bernard Shaw. He respected Upton Sinclair mainly because of his commitment to social issues. Literary modernism, by contrast, did not speak to his interests.

Conversely, Einstein's influence on the literature of his era was quite evident in the works of T. S. Eliot, Ezra Pound, Hermann Broch, and Marcel Proust. In Proust's *In Search of Lost Time*, the narrator interprets an aging church as "an edifice occupying a space with, so to speak, four dimensions—the fourth being Time."[47] Thomas Mann's international bestseller *The Magic Mountain* (1924) weaves musings on the theory of relativity into a dialogue between the two main characters: "A minute is as long as . . . it *lasts*, as long as it takes a second hand to complete a circle," the ailing Joachim remarks to his cousin Hans, who is visiting him in Davos. "In point of fact," replies Hans, "that's a matter of motion, of motion in space, correct? . . . And so we measure time with space. But that is the same thing as trying to measure space with time—the way uneducated people do."[48]

Early in his writing career, William Faulkner made an impressive attempt to reveal the nature of relativity in a literary fashion in his shocking novel *The Sound and the Fury* (1929). On the day he will, and must, kill himself, Quentin Compson shatters the glass of his pocket watch, an heirloom from his grandfather, who had handed it down to his father, and breaks off the hands. He carries the watch around with him for the whole day. The watch runs, but does not tell the time. "I took out my watch and listened to it clicking away, not knowing that it couldn't even lie."[49] The whole day he stumbles through spacetime right to the end of his worldline. "Eating the business of eating inside of you space too space and time confused Stomach saying noon brain saying eat oclock All right I wonder what time it is what of it."[50]

That was not Einstein's world. If he did read novels, he preferred the kind with exciting plots, such as Bernhard Kellermann's best-selling utopian novel *The Tunnel*. Although he owned a complete set of Dostoevsky's works and adored *The Brothers Karamazov* and Cervantes' *Don Quixote*, he had an aversion to thick books, especially those by contemporary writers. As someone who was spooked by his own feelings, he steered clear of overblown emotional displays in literature and opera. He dutifully made his way through *Buddenbrooks* for the sole purpose of participating in conversations about it. He never felt any affinity with

Thomas Mann, the author of the novel. He considered men "in silk shirts" peculiar. In his eyes, Mann was "an impressive schoolmaster, always in need of someone to instruct."[51] Einstein quipped that the writer's prim schoolteacher demeanor was so extreme that the physicist feared Mann would try to explain the theory of relativity to *him*.

By contrast, he got along exceedingly well with Thomas Mann's brother Heinrich, who had soared to fame almost overnight (like Einstein) with his successful novel *Der Untertan* (*The Man of Straw*), in part because the two of them shared a political affinity. Einstein's self-characterization as an "irreproachable Socialist" applied equally well to Heinrich Mann. Einstein's son-in-law Rudolf Kayser, who was an editor for *Die Neue Rundschau*, established many contacts with writers. Einstein often got together with Gerhart Hauptmann, a renowned writer who fancied himself the reincarnation of Goethe. According to Wachsmann, Hauptmann disparaged menial laborers as troublemakers who stank of sweat, even though the protagonists of his most celebrated, incendiary play *The Weavers* are exploited workers. Hauptmann's milieu was the sunny island of Hiddensee, not the back alleys of the working class.

Hauptmann expected to be treated like royalty. "Whenever the conversation turned to a famous man," the architect recalled, "he asked, 'And where do I belong?' "[52] That type of behavior was, of course, antithetical to Einstein's. But Einstein "loved characters," even perpetually inebriated ones like Hauptmann, who amused him no end. Moreover, he had admired the new German prince of the writers since his student days. Wachsmann reports that Hauptmann's son, Benvenuto, had his eye on Einstein's stepdaughter Margot. Erika Britzke claims that Margot was head over heels in love with the writer, and sent him love letters. Such are the workings of the rumor mill.

According to Wachsmann, Einstein had "a terribly low" opinion of Bertolt Brecht, who in turn passionately admired him. When Margot pressured Einstein into seeing *The Threepenny Opera*, he was singularly unimpressed. Kurt Weill's music left him cold. Einstein's relationship with music was a delicate subject, especially because he had distinguished himself as a violinist. Once, when he was a college student, he supposedly heard a piano playing from an open window, whereupon he rushed into the house and shouted to the flabbergasted pianist, "Don't stop!" He proceeded to unpack his violin and play a couple of duets with her.

Opinions as to the quality of his playing ran the gamut from loving it to loathing it. His school friend Hans Byland rhapsodized many years later, "For the first time in my life, the true Mozart rose up before me in the full splendor of Hellenic beauty." The more tepid assessments, such as that of Brigitte B. Fischer, daughter of the famous German publisher Samuel Fischer, seem far closer to the truth: "I found that he was very musical, but his technical skill was lacking. He did not really have a fine sound, but rather played more like a good dilettante."[53] The daughter of Erich Mendelsohn put it more bluntly in describing Einstein's "scratching," which was always quite hard on her parents. "Einstein's fiddling was pathetic," said housekeeper Herta Waldow.[54] "His bowing reminded me of a woodcutter."[55] The violinist himself depicted his playing with self-critical irony:

> Still, the dilettante has the right
> To scratch and scrape all through the night;
> But so his neighbors do not mutter
> He must kindly close the shutter.[56]

Although he even performed in public on occasion, generally for charity concerts, Einstein did not play primarily for audiences, but to enjoy himself with friends—or alone, for relaxation and inspiration. He often sat down at the piano and improvised. Music, he felt, was "for the soul, not for the intellect." Wherever he went, his violin "Lina" went with him, as did his pipe.

From the time of his early childhood, when his mother prodded him to learn to play the violin, he had sought out opportunities to play with other musicians. He had musical get-togethers with fellow students and colleagues, with Besso and Born, Planck and Ehrenfest, with private individuals as well as celebrities, with Pauline Winteler, who took care of him when he lived in Aarau, and with Queen Elisabeth of Belgium in Brussels. In the days before phonographs and radio, playing music at home was a matter of course during get-togethers—the way playing the stereo was later on. Einstein was never a great musician, but he was perpetually fascinated by music. However, since he was a brilliant natural scientist who advanced the mathematical rendering of the worldview, quite a bit has been made of his affinity for music. Music, some claim, is mathematics poured into notes, and it may have afforded him insight into the harmony of the world. After all, they would argue, doesn't science discover the unknown in nature, just as music discovers the unknown in the human soul? We would be equally justified in claiming that Einstein loved sailing as an expression of applied physics. He had this to say on the subject: "Music does not *influence* research work, but both are nourished by the same sort of longing."[57]

As a music lover, Einstein found Berlin a fulfilling place to live. Sometimes he went to three concerts a week, often in the company of his artistic stepdaughter Margot, who had made a name for herself as a sculptor not because of her name, but in spite of it. Einstein was at loggerheads with modernity in the realm of music as well as in the visual arts. Someone like Paul Hindemith irritated him no end. He considered a composer like Arnold Schönberg flat-out insane. His early biographer Carl Seelig believed that Einstein regarded romanticism, which was alien to his nature, "as a kind of illegitimate excuse to achieve in the easiest possible way a deeper appreciation."[58]

He felt most comfortable with the music of the eighteenth century, the clear compositional structures of Bach, Vivaldi, and Haydn. When the magazine *Reclams Universum* asked him to write an article about Bach, he scribbled down, "This is what I have to say about Bach's lifework: listen, play, love, revere—and keep your mouth shut."[59] Beethoven was more problematic. He liked chamber music well enough, but could not make any headway with weighty symphonies. He did not understand Johannes Brahms, and he considered Richard Wagner downright repugnant.

He reserved his greatest affection for the perpetual child from Salzburg: "Mozart's music is so pure and beautiful that I see it as a reflection of the inner beauty of the universe."[60] There was not a single superfluous note in Mozart's music, according to Einstein. With its fine balance between harmony and melody, it comes closest to the aesthetic purism of his world of physics formulas. But he was also drawn to Mozart's childlike nature, his humor, and his daring approach to dissonance, which resembled Einstein's preoccupation with paradoxes in physics.

He did not aspire to be a universal genius, but his fans and worshippers made him into one. The secret of his appeal lay in the juxtaposition of his superhuman achievements, destined for eternity, with the all-too-human normality of his everyday routines. The great Albert Einstein, named "Person of the Century" by *Time* magazine, never ceased to be a man of the people, which goes a long way toward explaining his magnificence. When he did not have a car available from one of his rich friends, he simply took the subway (his stop was Bayerischer Platz), or rode in the ten-pfennig bus to the academy. He did not mind traveling third-class in trains. "Possessions weigh you down," he said, and added, "There is nothing I would have trouble giving up at any time."[61] He retained his modesty and Swabian petit-bourgeois mentality in the face of a barrage of temptations and enticements, as was most evident in his relationship to his "little house" in Caputh.

His pride in ownership—he even sent out a postcard with a picture of the "Einstein country home"—has nothing to do with conceit. He wrote a few lines for the guest book, and signed them "The Landlord, on behalf of the Caputh management":

> Men and women of all ages
> Leave your mark upon these pages.
> But not with prose resembling that
> Of people as they stop to chat;
> Arrange your words in flowing lines
> Like lofty poets' neat designs.
> Begin, and push your fears away
> Your writing will not go astray.[62]

Einstein was the only celebrity who could get away with something like this. Everyone knew that he was an unconventional child in a man's body. He had no need for a stately villa, like Thomas Mann, or for a banquet in the Adlon Hotel, like Hauptmann, or for stylish cars, like Bertolt Brecht. He was satisfied with his fragrant little wooden house made of Oregon pine and Galician fir wood, in which he sometimes felt like Robinson Crusoe. Elsa had wanted to build something larger, while he was almost ashamed of the grandeur of the house.

"I think there is enough space here for all the scholars in the world whom I know and who have steered science onto new courses to gather together to talk and eat."[63] What more could he want? Humbly and gratefully he relished the luxury of living the life of a normal man undisturbed by the public whenever he could. "The sailboat, the expansive view, solitary autumn strolls, relative peace and quiet; it is a paradise."[64] Naturally enough, the emphasis was on "relative," even in this context. "This paradise unfortunately lacks only one thing: There are no archangels to drive off curious gawkers and wearisome visitors with a fiery sword."[65]

Compared with Berlin, Caputh offered the Einsteins quite a cloistered existence. The family spent three summers at this house, from the first hint of spring to the autumn storms. They had no telephone. In an emergency, they could be reached via the neighbors, for whom they had bought a horn to send them signals. They even toyed with the idea of giving up their apartment in the city. "Maybe I'll move out to the country altogether," Einstein wrote to Hans Albert.[66]

It goes without saying that Elsa insisted on moving at least some part of her salon to the country. Surprise visitors who had come all that way could not simply be turned away. But since there were few chance visitors, disturbances were kept to a minimum. Einstein was always ill at ease on Haberlandstrasse, but he felt right at home in Caputh. Wachsmann had shielded his bedroom and study from the rest of the house with a double-wall construction with several layers of insulation. As far as furnishings went, Einstein wanted nothing more than a bed, a desk, an easy chair, a lamp, and two open file boxes, one for blank paper and the other for completed notes.

"Einstein was simple and practical," Erika Britzke recalls. "He did not want to show off to the neighbors." Many visitors were apparently disappointed because they had expected a showier house. But that did not bother its owner. "He had the gift of letting material matters roll right off him."

Then she closes her eyes and conjures up the images that a quarter of a century with Einstein has produced in her—fantasy on a factual basis, so to speak. Einstein fiddling away on his violin, Lina, in the hallway, or settling into the easy chair on the terrace. His son riding up on his motorcycle, with grandson Bernhard in the sidecar, draped in thick blankets. Einstein sitting at the big round table with Planck, Nernst, and Haber, eating asparagus. People who know him understand that he cannot stand chitchat. They speak about physics. Herta and Elsa serve the food and disappear without a sound. "Without the others he could not have existed at all," says Erika Britzke. "The others" means both servants and guests.

The Indian poet Rabindranath Tagore came to Caputh with a large entourage. "He is a solitary man," the poet remarked.[67] The two international celebrities talked about causality and music, about the universal truth of existence and the law of goodness. A journalist who had come along took notes. Frau Britzke has her doubts about his record of the day's events. Tagore and Einstein could barely communicate, she says. She does not buy some of the details, such as the account of him watering the flowers or pulling weeds. "I don't believe," she said, that he tampered with anything in nature.

Nor does she believe the famous account of how he left Caputh for the final time. As they were closing the door behind them, Albert supposedly said to Elsa, "Take another look at it. You will never see it again." Frau Britzke considers this story pure myth. She has done her research. If someone thinks he will not be returning, he does not invest 10,000 marks in the adjoining plot, which comes with a little garden house. She found the contract, which was signed on November 15, 1932. "He wanted to come back," she says. Einstein did not think the Nazi takeover would go so quickly.

"I have also had my Einstein crises," she confesses, referring to Einstein's affairs with women. "I do believe that he made advances to young girls." Actually, it went further than that. Einstein's amorous adventures on his sailboat were "an open secret in Caputh." Margot confirmed to her the rumor about an illegitimate child, "conceived here," born in Prague, and given up for adoption.

Shortly before the Berlin Wall was erected, the father of Britzke's daughter went west to the United States. He hated communism, but Erika Britzke put her faith in the German Democratic Republic. "It was a good country for us women," she says, but notes that it was also difficult. Once, when she had questions about restoring the furnishings in Caputh and requested an official meeting with the housekeeper, Herta Waldow, a Communist Party secretary snapped at her, "That is not my responsibility." She empties the sugar packet onto the last drop of coffee. She is a war child who lets nothing go to waste. "Then I knew that the state of workers and farmers was coming to an end."

The GDR wanted Einstein the left-wing pacifist and scientist, but not the Jew. Anything Jewish was seen as a problem. Britzke, who was a teacher, heard the children in her school whispering, "Jews get out!" It sounded like an echo of the 1930s. After Einstein left, a Jewish orphanage used his house. In all the madness, this was almost a rational use for it. However, in the aftermath of Kristallnacht in 1938, the residents were driven out, and the Hitler Youth moved in. The mayor of Caputh was happy to see his village free of Jews. Forty years later, children were again whispering, "Jews get out!" Erika Britzke could not report it, since some of these children were the offspring of party bigwigs. Then she got her new job, and the maliciousness subsided. As the guardian of Einstein's house in Caputh, she has reaped the benefit of authority.

SIXTEEN

A JEW NAMED ALBERT

HIS GOD WAS A PRINCIPLE

Then down to the temple wall (Wailing Wall), where dull-witted fellow clan members pray aloud as they face the wall and bend their bodies forward and back in a rocking motion. A deplorable sight of people with a past and no present."[1] Einstein in Jerusalem, on February 3, 1923, feeling disillusioned and callous, drew a line in his travel diary. On one side of the line were rituals and throwbacks to yesteryear; on the other a vision, the nation without boundaries, model of a global community of like-minded people. And between them the deep discord of Zionism, assimilation, and anti-Semitism, the old and the new rolled into one. The split cuts right through his Jewish soul and the very core of Jewish identity. Judaism as a religious and cultural community—a matter of religion, politics, and culture. Einstein had reached his decision quite some time ago. He opted against the faith and for the tribe.

"Professor Einstein," the president of the Zionist Executive Council called out to him at the ceremonial inauguration of Hebrew University on Mount Scopus in Jerusalem, "please step up on the stage that has been waiting for you these past two thousand years."[2] Einstein was more than a little disconcerted to be celebrated as the "Jewish saint." The previous year he had been honored in England as the "greatest Jew since Jesus." "They are determined to get me to Jerusalem," he commented on the final evening of his visit. "The heart says yes, but reason says no!"[3] It would be his only journey to the homeland of his heart.

His two-week stay made a deep impression on him. "The brothers of our race in Palestine charmed me as farmers, as workers, and as citizens," he wrote to his friend Maurice Solovine. "The country . . . will become an ethical center but cannot accommodate a very large segment of the Jewish people."[4] The idea of Judaism and a Jewish state captivated him for the rest of his life. That had not always been the case. He wrote in a 1921 newspaper article called "How I Became a Zionist" that until he turned thirty-five, "I did not become aware of my Jewishness, and there was nothing in my life that would have stirred my Jewish feeling or stimulated it. This changed as soon as I had taken residency in Berlin. There I saw the predicament of many young Jews."[5]

His Zionist consciousness probably began to awaken sometime after 1918. Until that time, particularly in 1911 and 1912, when he lived in Prague, he had regarded Zionists only as "a small troop of unrealistic people, harking back to the Middle Ages."[6] The German-Jewish intellectuals in Prague, who included Brod, Kafka, Egon Erwin Kisch, and Franz Werfel, were unable to win him over to the Zionist cause. When he was under consideration for an academic post in the Austro-Hungarian Empire, he was required to state his religious affiliation. Identifying himself as "Mosaic" on the form was a matter of indifference to Einstein, and "returning to the bosom of Abraham meant nothing to me. A piece of paper that had to be signed."[7] When he left Prague, he went back to classifying himself as "without religion," and did so for the rest of his life.

Even so, Einstein had developed a sense of the special societal place of Judaism in Germany, where he had spent the first fourteen years of his life. "A liberal spirit, undogmatic in matters of religion, brought by both parents from their respective homes, prevailed within the family. There was no discussion of religious matters or rules," his sister, Maja, recalled.[8] After all, Einstein and his sister were named Albert and Maria, not Abraham and Esther.

When Albert entered elementary school, he was required to enroll in religious instruction classes. The school offered only Catholic lessons, and he attended them. At home, his parents arranged for him to be instructed in Judaism by a distant relative. "As a result," Maja recalled, "a deep religious feeling was awakened in him [and] he observed religious prescriptions in every detail."[9] He stopped eating pork, and demanded the same of his parents. He wrote and set to music songs in praise of God, which he sang at home and on the street. "He remained true to his self-chosen life for years."[10]

The juxtaposition of the two teachings, which were both based on the Old Testament, proved the essential identity of the two religions to him. One day, however, according to his son-in-law Rudolf Kayser, the Catholicism teacher, who liked him, nonetheless brought a big nail to school, and explained that this was what the Jews had used to nail Jesus to the cross. The calculated effect was immediate. Albert's fellow pupils, roughly seventy of them, pointed their fingers at the only Jew in the class. "For the first time, Albert experienced the frightful venom of anti-Semitism," Kayser wrote.[11] Later Einstein reported that on the way home from school he often fell victim to "physical attacks and insults . . . for the most part not too vicious," but sufficient "to establish even in a child an acute feeling of alienation."[12]

A letter of protest drafted by non-Jewish residents of Schwabing on the occasion of the new installation of electric lighting in their town in 1889 by Einstein & Cie. is one indication of the kind of subliminal animosity that prevailed during Einstein's youth. Since only Leopoldstrasse, the main street in Schwabing, had electric lighting, but the side streets did not, the community was said to be creating "two artificial classes of residents: a first class of privileged Semitic citizens of light, and a second class of underprivileged Christian citizens of darkness!"[13] The rich Jew—an inexhaustible topos.

It is difficult to take Einstein at his word that he "did not become aware of [his] Jewishness" until he turned thirty-five, but he evidently did not take his religious affiliation very seriously. During his high school years, he was instructed by various teachers, including two rabbis in Munich, in the Jewish religion and in the Hebrew language—an "unforgettable" experience, according to Kayser. However, his "deep religiosity . . . found an abrupt end at the age of twelve," Einstein later recalled.[14] Ironically, it was the Jewish student Max Talmud who supplied him with writings that brought him away from the faith. "Through the reading of popular scientific books I soon reached the conviction that much in the stories of the Bible could not be true."[15] He replaced his interest in "religion, which is implanted in every child by way of the traditional education machine" with "a positively fanatic orgy of freethinking." "The religious paradise, which was thus lost," made way for a "striving for a mental grasp of things."[16]

In 1901, when he had difficulty launching his career and contemplated seeking a teaching assistantship in Italy, his concern about his Jewish identity was foremost in his mind. "First of all, one of the main obstacles is absent here, i.e., anti-Semitism," he wrote to Mileva from Milan, "which in German countries would be as unpleasant as it is obstructive."[17] In a letter to her friend Helene Savić, Mileva expressed her belief that Albert was unlikely to "get a secure position soon" in a German-speaking country: "My sweetheart has a very wicked tongue and is a Jew into the bargain."[18] In any case, she, a Catholic, then married him, and his children were raised as Christians.

In 1908, when he was asking his friend Grossmann whom to contact about a teaching position, he mused, "Is it not possible that I would make a bad impression on him (not a Swiss German, Semitic looks, etc.)?"[19] That his fear was not unfounded is evidenced by the hiring committee report for his first professorship in Zurich. Since neither the commission nor the faculty wanted to "write 'anti-Semitism' as a principle on its banner," he was appointed to the position despite his Jewish heritage—although "Israelites are credited among scholars with a variety of disagreeable character traits, such as importunateness, impertinence, a shopkeeper's mind in their understanding of their academic position, etc., and in numerous cases with some justification." However, it was not deemed "appropriate to disqualify a man merely because he happens to be a Jew."[20]

When Einstein later failed to hear back from Prague for quite some time in the matter of his candidacy, he immediately assumed, "the ministry has not accepted the proposal because of my Semitic origin."[21] It is hard to fathom a Jew forgetting that he is a Jew during this period in German-speaking Europe.

In Berlin, the situation for Jews was even more muddled. On the one hand, they enjoyed equal civil rights in the empire, but on the other, the anti-Semitic movement was institutionally tolerated. Its ideological basis was the so-called Berlin Anti-Semitism Debate, launched by the academically influential historian Heinrich von Treitschke in 1879, the year of Einstein's birth, with his leaflet "A Word About Our Jews."[22] The key phrase in his pamphlet was "the Jews are our misfortune." The major arguments did not center on race or religion; instead, the Jews, especially Jewish bankers and financiers, were held responsible for the economic lull that had begun in 1873.

In proportion to their numbers, Jews had, in a matter of decades, worked their way up to an excellent position in society as merchants, department store owners, manufacturers, and industrialists, and in the independent professions as doctors, lawyers, and scholars. In cultural life, as writers, composers, musicians, and philosophers, they had assumed a position that extended far beyond their demographic representation in the population. Three German-speaking Jews—Marx, Freud, and Einstein—were among the most influential thinkers of the nineteenth and twentieth centuries.

In Ulm and Munich, the cities in which Einstein lived as a child, the population was approximately 2 percent Jewish after the creation of the German Empire. Berlin records indicate that there were more than 100,000 "Israelites" in the city at the turn of the century. They constituted 4 percent of the population but paid 30 percent of the city taxes. By the beginning of the twentieth century, many were moving to the grand-bourgeois surrounding communities of Charlottenburg, Wilmersdorf, and Schöneberg, where Einstein himself began living in 1915. A modern liberal Jewish urban society sprang up. Only a small group of orthodox Jews continued to support traditional Judaism.

Whether Jews were Germans, whether they regarded themselves as German Jews or Jewish Germans, remained a virulent question. By the end of the nineteenth century it had taken on additional explosive force with the immigration of "foreign Jews" from the East. Zionism, however, had very few adherents in Berlin—and those who were drawn to Zionism were less interested in the political movement directed toward Palestine and more in the phenomenon of "cultural Zionism," as the writer Martin Buber called it.

Until World War I, the Jews in Germany blended into the background and did not stand apart from other Germans. The historian Simon Dubnov writes that they felt it was "their duty to embrace the new order. They were determined not to lag behind the real Germans in hyperpatriotism and idolatry of Germany as a great power."[23] This also applied to prominent individuals, notably Einstein's colleague Fritz Haber and his friend Walther Rathenau, an industrialist who later became foreign minister. Rathenau once said, "I have and know of no blood other than German blood, no other tribe, no other people but the Germans. . . . My father and I have not entertained any thoughts that were not German."[24] There were even attempts to move the traditional Sabbath services to Sundays.

Assimilationist efforts reached their pinnacle with the outbreak of World War I. In early August 1914, a call was sounded to Jewish "co-religionists": "German Jews! This is the hour for us to show once again that we Jews, who are proud of our heritage, are the best patriots our fatherland has to offer."[25] According to the historian Chana S. Schütz, the Burgfriede (domestic peace) between the various communities, proclaimed by the emperor in response to the war, "encouraged the Jews more than any event since the enactment of emancipation in 1869."[26] All quarrels, including those between religions, needed to be set aside.

However, when things began to go badly for the German army, the Jews in Germany also took a turn for the worse. Suddenly they were accused of shirking their duty and were even held accountable for the lack of progress in positional warfare. In the view of the historian Golo Mann, anti-Semitism was more widespread at this time than when the Nazis assumed power in 1933. There is no record of any commentary on this issue by Albert Einstein. The Jewish question was not an especially important one for him during those years.

After World War I, in which more than 12,000 Jews lost their lives as German soldiers, and after his sudden rise to fame in 1919, Einstein explored, in public forums, issues pertaining to Judaism and the fate of his co-religionists. Shortly after the announcement of the solar eclipse data, his bitterly ironic bon mot was printed in the London Times: "By an application of the theory of relativity to the taste of readers, I am called a German of science, and in England I am represented as a Swiss Jew. If I come to be represented as a bête noire, the descriptions will be reversed, and I shall become a Swiss Jew for the Germans and a German man of science for the English!"[27]

"Anti-Semitism is strong here and the political reaction is violent, at least among the 'intelligentsia,' " he wrote on December 4, 1919, to his friend and colleague Paul Ehrenfest.[28] Academic anti-Semitism was preparing the ground for the coming barbarism. Since the press had built Einstein up as an omniscient sage, he used his access to the mass media, raised his voice—and emerged from his isolation for good. On December 30, 1919, he published an essay in the *Berliner Tageblatt* titled "Immigration from the East," which cautioned readers to be wary of demagoguery from the right: "All these arguments call for the most sweeping measures, i.e., herding all immigrants into concentration camps or expelling them." He also made his first—indirect—concession to Zionism by insisting that the eastern European Jews be "given an opportunity to emigrate *elsewhere*. Hopefully, many of them will find a true homeland as free sons of the Jewish people in the newly established Jewish Palestine."[29]

Even before his visit to the land of Israel, Einstein found the Zionist idea of a Jewish homeland in its ancient setting appealing. He was not ready to promote a Jewish nation in Palestine, and therefore opted to support Zionist ideals without approving the ideal of Zionism—the founding of a state. Socialist approaches, such as the kibbutz movement, made sense to him, but he despised Jewish chauvinism.

On November 9, 1919, three days after his "canonization" in the temple of science, he wrote to his friend and colleague Max Born, "Anti-Semitism must ultimately be understood as a real thing, based on true hereditary qualities, even if for us Jews it is often unpleasant."[30] He even showed consideration for the opposing view, which was based on the notion "that Jews exert an influence on the intellectual life of the German people that far exceeds their numbers in the population." He went so far as to assert, "I believe German Jewry owes its continued existence to anti-Semitism."[31]

Einstein's relationship to Judaism at this difficult time was shaped by a social outlook even more than by political and cultural viewpoints. He believed in the equality of all people. The eastern European Jews were profoundly grateful for his instinctive support of the underprivileged, excluded, and underdogs of all religions. Of the 30,000 Jews who had fled to Germany from the east after the war, 20,000 stayed in Berlin. He offered university courses specifically tailored to the students among them. He would have hated the idea that his poor brothers might be barred from getting an education on top of everything else they had suffered. His courageous endorsement of a Hebrew university in Palestine, which he pictured as the intellectual center of an international Jewish diaspora, was also grounded in this conviction.

In his writings about eastern European Jews, Einstein expressed his views on another issue in Judaism, namely assimilation, which perturbed him even more than anti-Semitism. "The undignified mania of adaptive conformity, among many of my social standing, has always been very repulsive to me," he said in late 1919.[32] "One element of this 'integration,'" as architect Wachsmann described the situation, "was, to Einstein's horror, the manner in which Jews grew accustomed to disparagement and even hostilities."[33]

In April 1920, when he received an invitation from the Berlin Central Organization of German Citizens of the Jewish Faith to a session "that is supposed to be dedicated to the fight against anti-Semitism in academic circles," he responded with a caustic statement.[34] His "Confession," which the *Israelitisches Wochenblatt für die Schweiz* later published, reflects Einstein's attitude toward assimilated Jews (such as his friend Fritz Haber) better than any other statement.

> First we must fight with enlightenment the anti-Semitism and submissive sentiments among us Jews. More dignity and more independence in our own ranks! Only when we dare to see ourselves as a nation, only when we respect ourselves, can we earn the respect of others, or rather they arrive at this conclusion themselves. . . . I cannot suppress a pained smile whenever I read "German citizens of the Jewish faith." . . . Can an Aryan respect such pussyfooters? . . . I am a Jew, and I am glad to belong to the Jewish people, even though in no way do I consider them to be the chosen ones. Leave the Aryan to his anti-Semitism; and let us keep the love of our brethren.[35]

He justified his decision not to join the Jewish religious community in a letter dated December 22, 1920: "As much as I feel myself to be a Jew, I stand aloof from the traditional religious rites."[36]

Einstein advised his Jewish students to "adopt an attitude of courteous but consistent reserve to the Gentiles. And let us live after our own fashion there and not ape dueling and drinking customs, which are foreign to our nature. It is possible to be a civilized European and a good citizen and at the same time a faithful Jew."[37] He described his coreligionists as "people of a definite historical type who lack the support of a community to keep them together. The result is a want of solid foundations in the individual, which amounts in its extremer forms to moral instability."[38]

Einstein dreamed of the day that "every Jew in the world should become attached to a living society" and hated seeing "worthy Jews basely caricatured. . . . The sight made my heart bleed."[39] However, even though he had nightmarish visions of "that pathetic creature, the baptized Jewish *Geheimrat* of yesterday and today,"[40] especially in regard to Fritz Haber, who had had himself baptized at the age of twenty-four, he did not back away from his friendship with the chemist. "Haber and Einstein had a sense of science as a call to a special priesthood in a faith only recently established," says the historian Fritz Stern.[41] Their resultant feeling of solidarity enabled them to look past their differences.

"Jewish solidarity is another invention of their enemies," Einstein scoffed. "Anti-Semites often talk of the malice and cunning of the Jews, but has there ever been in history a more striking example of collective stupidity than the blindness of the German Jews?"[42] In retrospect, he came down quite hard on his fellow Jews who closed their eyes to the approaching catastrophe, as Vallentin noted: "Slavery to comfort accentuated the tragic drama that was beginning. Against all evidence, the rich Jewish bourgeoisie was clinging to its fortunes, to its houses, to its furniture."[43]

Einstein did not participate in the debates within the German Jewish community over whether this community was more Jewish or more German. He was concerned about what he was actually seeing right in front of him, and found it just as degrading as the sight of his co-religionists at the Wailing Wall. He made his position clear to both sides: no clergy and no timidity. His idealistic outlook got him nowhere—other than to a belated confirmation of his worst fears. "It is particularly unfortunate," he wrote to Max Born from Princeton in 1934, "that the satiated Jews of the countries which have hitherto been spared cling to the foolish hope that they can safeguard themselves by keeping quiet and making patriotic gestures, just as the German Jews used to do."[44]

After his first visit to the United States in 1921, he told Besso what he thought about his fellow Jews in America: "Most of our ethnic cohorts are more clever than brave; I saw that quite clearly."[45] Before he left for America in 1921, Fritz Haber tried to pressure him not to go: "You will certainly sacrifice the narrow basis upon which the existence of academic teachers and students of the Jewish faith at German universities rests."[46] Einstein had no intention of being derailed by Haber's spineless attitude, although he had no illusions about the true nature of his role: "Naturally, I am needed not for my abilities but solely for my name, from whose publicity a substantial effect is expected among the rich tribal companions in Dollaria."[47] He told his friend Solovine that he would have to "beg for dollars [and] act as high priest and decoy."[48] Together with the Zionist leader Chaim Weizmann, he collected donations for the medical school of the planned Hebrew University. In the end "I'm left with the agreeable knowledge of having achieved something really good and of having battled bravely for the Jewish cause, regardless of all protests by Jews and non-Jews."[49]

When he returned to Europe from Jerusalem, Einstein described his impressions from his journey in the Blüthner auditorium in Berlin. The *Jüdische Rundschau* printed the text of his speech the following day: "My greatest experience was *seeing a Jewish people for the first time in my life*. Ladies and Gentlemen! I've seen a great many Jews [*laughter*], but I have not seen a Jewish people in Berlin or elsewhere in Germany. . . . These people still carry a healthy national feeling within them that has not yet been destroyed by the atomization and splintering of the individual."[50] A reception at City Hall in Manhattan featured both the German national anthem and the Zionist "Hatikvah," which later became the national anthem of Israel. Compared with the situation Jews were facing in Germany, the American Jews appeared to have found something on the order of the Promised Land in the United States.

But what makes a Jew a Jew? As someone with a strong belief in heredity and genetics, Einstein employed a vocabulary that sounds almost racist today, but did not come across as having racist undertones in the early years of the Weimar Republic. "I am now convinced that the Jewish race has kept itself fairly pure in the past 1,500 years," he wrote after his first encounter with the small Jewish community in Hong Kong.[51]

By 1930 the picture had changed dramatically: "I also admit that our cultural tradition does us credit; in a moral sense quite superior, intellectually one-sided, but impressive, from an artistic and ideological point of view mediocre . . . altogether just fine, but characters suffer under it . . . moral nobility . . . our political cosmopolitanism."[52]

On January 29, 1930, at a benefit concert for the welfare office, the Jewish community in Berlin got to hear Einstein playing second violin in the big synagogue on Oranienburger Street. He was wearing the traditional yarmulke. In view of the Nazi insanity and the threat to culture as a whole posed by the Nazis, his attention shifted more and more to social and cultural understanding. "It is of no interest to what extent the Jews are a racial community," he said in July 1932. "I am for Zionism. It is certain that the Jews share a destiny and are in urgent need of mutual support."[53]

Shortly after Hitler came to power, a Nazi pamphlet bearing the title "Jews Are Looking at You" began circulating in Germany. It portrayed the physicist who had gotten away from Germany as an outlaw. "Einstein. Invented a highly controversial 'theory of relativity.' Was celebrated by the Jewish press and the unsuspecting German people, and showed his gratitude by deceitful atrocity-mongering against Adolf Hitler abroad. (Unhanged.)"[54]

One of Einstein's bitterest experiences came when his fellow Jews turned on him in reaction to the new rulers' reprisals. "The greatest tragedy in my husband's life," Elsa wrote to her friend Antonina Vallentin in 1934, "is that the German Jews make him responsible for all the horrors that happen to them over there. They believe he has provoked it all and in their resentment have announced their total dissociation from him. We get more angry letters from the Jews than we do from the Nazis."[55]

Einstein was now drawing a sharp distinction between German Jews who did not see the threat posed by their stubborn love for their fatherland and those who had long since read the writing on the wall. In August 1938 he told Besso an anecdote that speaks for itself. "A German attorney married to a Jewish woman . . . replied to my question of whether he was ever homesick: But I'm not a Jew! That man understood."[56] Einstein did, too. At about this time he wrote a quatrain that reflected his ambivalence toward his fellow Jews:

> The Jews around me that I see
> Little pleasure give to me.
> When the others I then view
> I am glad to be a Jew.[57]

In 1938 he stated, "The bond that has united the Jews for thousands of years and that unites them today is, above all, the democratic ideal of social justice, coupled with the ideal of mutual aid and tolerance among all men."[58] At the same time he continued to oppose the foundation of a state in Palestine: "My awareness of the essential nature of Judaism resists the idea of a Jewish state with borders, an army, and a measure of temporal power. . . . I am afraid of the inner damage Judaism will sustain."[59]

After members of his own family perished in the Holocaust, his tone softened. "Zionism gave the German Jews no great protection against annihilation," he wrote to an opponent of the movement. "But it did give the survivors the inner strength to endure the debacle with dignity and without losing their healthy self-respect."[60] In 1949 he gave the new state his blessing. "The Jews of Palestine did not fight for political independence for its own sake, but they fought to achieve free immigration for the Jews of many countries where their very existence was in danger."[61]

He again warned, as he had a quarter of a century earlier, that Israel would find happiness only if the Israelis worked together with the Arabs, and not if they were pitted against them. Back in 1929 he had written to Besso, "Without mutual understanding and cooperation with the Arabs, nothing will work. . . . Forcing the Arabs to leave their land is out of the question. The country is underpopulated and much more can be made of it with more people."[62] He reiterated his viewpoint like a mantra at every available opportunity. In 1952 the country offered Einstein the office of president after the death of its first president, Chaim Weizmann. He demurred humbly, but did acknowledge that "my relationship with the Jewish people has become my strongest human bond."[63] He told his stepdaughter Margot, "If I were to be president, I would have to say to the Israeli people things they would not like to hear."[64]

His relationship with the Jewish faith took a different path. His "abrupt" departure from his "deep religiosity" added yet another dimension to his lifelong grappling with Judaism. He did not stop being "a religious person," but he put his faith in what he described as "the strongest and noblest motive for scientific research":[65] "The religious geniuses of all ages have been distinguished by this kind of religious feeling, which knows no dogma and no God conceived in man's image."[66]

According to Gerald Holton, an American historian of science, this step signaled Einstein's entrance into his "Third Paradise."[67] The Third Paradise, as Holton defines it, arises out of a fusion of a first and a second paradise, essentially of metaphysics and physics, and is based on Einstein's great "need to generalize," which Holton regards as a wish deeply entrenched in German culture. Einstein sought to unite not only mechanics and electrodynamics, mass and energy, space and time, gravitation and spacetime, and ultimately all theories in a world formula, but also to unite researchers of all countries, the Jews of the world, the nations and religions in a world government. He wanted to put an end to boundaries and barriers, especially between his own religious and scientific views.

"It was the experience of mystery—even if mixed with fear—that engendered religion," Einstein commented in 1931. "A knowledge of the existence of something we cannot penetrate, our perceptions of the profoundest reason and the most radiant beauty, which only in their most primitive forms are accessible to our minds—it is this knowledge and this emotion that constitute true religiosity; in this sense, and in this alone, I am a deeply religious man."[68] Yet he was religious in another sense as well—in his profound belief in the sanctity of life, which he shared with devout Jews. This belief remained in place despite his frivolous statement about the death penalty. He enjoyed quoting a line by Walther Rathenau: "When a Jew says that he's going hunting to amuse himself, he lies."[69]

Although science is considered positively holy in the value system of Jewish traditions, since it reveals the divine in nature, Einstein never found his way back to "faith" in the sense of religious community: "I cannot conceive of a God who rewards and punishes his creatures, or has a will of the kind that we experience in ourselves."[70] He exhorted the faithful to take note that "the Jewish God is simply a negation of superstition, an imaginary result of its elimination. It is also an attempt to base the moral law on fear, a regrettable and discreditable attempt."[71]

He did not accept the authority of God over man, not even conceptually. "Humility is Einstein's religion," said his son-in-law Rudolf Kayser. "It consists of a childlike admiration of a superior mind."[72] There it is again: the perpetual child. Einstein did not believe in a personified God, but he used the divine image with childlike wisdom. This God did not punish or reward, but created nature so systematically that everything in the universe had to follow its system. Everything is predetermined, set, and calculable, with fate an infinite intertwining of regulated events.

"This realistic feeling with respect to the universe into which he was placed was so strong in Einstein that it assumed a form that seems like the complete opposite," his assistant Leopold Infeld remarked in 1938. "When he spoke about God and the divine creation of the world, he meant the logical consistency and simplicity of the laws of nature."[73] His God was a principle—the principle of cause and effect. He believed in the law of causality, which is an expression of divine intervention. And the Jewish saint sought to see eye to eye with his God. He wanted to sneak a look at the cards of the "eternal mastermind of puzzles," understand his craft, "know how God created this world. I am not interested in this or that phenomenon, in the spectrum of this or that element. I want to know his thoughts. The rest are details."[74]

Einstein, a determinist, did not believe in free will. God had done his part, and since then everything worked according to invariable laws. But did he have any chance to exercise free will at least one time? "What really interests me is whether God could have created the world any differently," Einstein mused.[75] Are there alternatives to the laws of nature as we know them? Are they themselves God? Could one give up hope for eternity and finiteness and still face God by deciphering his system?

"Science without religion is lame, religion with science is blind," said Einstein.[76] More than virtually any of his devout colleagues, he acknowledged the religious dimension of his guild. "In every true searcher of Nature there is a kind of religious reverence; for he finds it impossible to imagine that he is the first to have thought out the exceedingly delicate threads that connect his perceptions. The aspect of knowledge that has not yet been laid bare gives the investigator a feeling akin to that experienced by a child who seeks to grasp the masterly way in which elders manipulate things."[77] Perhaps this was Einstein's secret. In his world, God was a living presence, notwithstanding Nietzsche's declaration that God was dead. Einstein roundly rejected Nietzsche's philosophy: "If I am capable of hating anything, it is his writings."[78]

In 1929, when Rabbi Herbert S. Goldstein sent him a telegram from New York asking, "Do you believe in God? Stop. Prepaid reply 50 words,"[79] Einstein replied, "I believe in Spinoza's God who reveals himself in the orderly harmony of what exists, not in a God who concerns himself with the fates and actions of human beings."[80] His reply, in German, had used only half of the prepaid units—a twenty-five-word profession of faith. Einstein was particularly intrigued by Spinoza's twenty-ninth "proposition," which stated, "In nature there is nothing contingent, but all things have been determined from the necessity of the divine nature to exist and produce an effect in a certain way."[81]

His son-in-law Rudolf Kayser summarized the issue as follows: "Image of the world: The transformation of troubled earthly reality into a pure and ideal condition; Leibniz's 'preestablished harmony' and Spinoza's intellectual monism; orderliness as the ultimate triumph over the rowdiness of anarchistic chance and randomness."[82] God does not decide anything that he has not already decided. In spacetime the moment keeps eluding itself. Without matter there is no spacetime, and there is no effect without a cause. The former follows from the latter, not the other way around. Even God cannot play with chance.

"When he was an old man, he often referred to himself as a deeply religious nonbeliever," Rudolf Kayser reported about his father-in-law.[83] That was the origin of Einstein's famous saying in which he wanted to treat the Almighty like a genie in a bottle and forbid him to play around: "God does not play dice."

SEVENTEEN

THE END JUSTIFIES THE DOUBTS

EINSTEIN AND QUANTUM THEORY

The measuring instrument, a detector in the Institute for Experimental Physics at the University of Vienna, goes *click, click-click, click* as it catches the light. Nondescript case with strings, switches, and a digital display whose numbers keep shooting upward. The clicking sound comes from a small speaker. It is not a steady hum, but simply a *click, click-click*, a countable staccato, just as Einstein had prophesied a hundred years earlier. Back then, in his annus mirabilis, he wanted to achieve more than making light triumph over time with his theory of relativity. He wanted to know what light was.

In the spring of 1905 he presented a paper on "energy quanta that are localized in points of space, move without dividing, and can be absorbed or generated only as a whole."[1] This concept, which was one of the most significant ideas of the twentieth century, set in motion a theoretical revolution with technological implications that have had a major impact on mankind today. It has given rise to the high-tech world of microelectronics, cellular phones, digital photography, computers, chips, the Internet, superconductivity, nanotechnology, and modern chemistry.

Starting in 1905, Einstein's theories of relativity had led science into the realm of the giants, into the finite yet limitless cosmos. With his "light quantum hypothesis" in the same year, he pointed the way to the world of the dwarfs, the smallest components of matter down to the subatomic realm. For this achievement alone he belonged on the Mount Olympus of physics—even without relativity and gravitation. "I have thought a hundred times as much about the quantum problems as I have about general relativity theory," he said.[2]

Yet again, Einstein had penetrated one of the world's deepest mysteries: the double nature of light. It can be wavelike, as described in the classical physics of the nineteenth century, or it can be particle-like, quantized as energy in packets. With this early forerunner of "wave-particle duality," the simultaneous presence of mutually exclusive characteristics, Einstein contributed to the foundation of quantum theory. Individual particles of light in quantized energy states, which were dubbed "photons" twenty years later, zoom through space like the tiniest projectiles. If they make contact with a detector, it can render them audible, one by one, *click, click-click.*

Scientists at the institute on Boltzmanngasse in Vienna are using Einstein's work to map out a future with "quantum information processing." The director of the institute, Anton Zeilinger, sprinkles his explanations with terms like "quantum computers," "quantum cryptography," and "quantum teleportation" the way everyone tossed around vocabulary like lightbulbs, generators, and telephones in Einstein's youth. Like Einstein, Zeilinger began as a child prodigy; now, this "Quanton" (as his students call him on the sly) is getting on in years. Zeilinger is a full-bearded, temperamental scientist with a gruff Viennese charm, a master of things no one understands until Zeilinger comes along and explains them.

Zeilinger has written two noteworthy books on Einstein. His latest is tellingly titled *Einstein's Spook.* Two years earlier he published *Einstein's Veil,* a study of "the new world of quantum physics."[3] In 1924, Einstein had used the expression "big veil" to describe the way the quantum world cloaks its mysteries. Zeilinger has been attempting to lift the veil—without succumbing to the illusion that anyone could fully grasp quantum theory. Although it is not understood down to every last detail even today, quantum theory is at the threshold of direct technical application. Individual photons, which were utterly mysterious back in 1905, now have a central role as raw material in the routine operations of quantum physics.

Zeilinger became world-famous in 1997, when he and his colleagues were the first to "beam" with light particles—teleportation à la *Star Trek,* but "merely" of information, not of matter. "That would have shocked Einstein," says Philip Walther, one of several doctoral students on the team in Vienna, as young as the idealist had been back then in the patent office in Bern. Walther admires Einstein's "unbelievable knowledge" and "his bold logic." He calls his own research "cool" and "really exciting." When his friends ask him what he does in his laboratory, he replies, "What I do there is insane."

During the second half of his life, Einstein was no longer able to accept this insane interpretation of the quantum world, which he had launched as a revolutionary in 1905. The quanta (as well as the unified field theory, which will be discussed later) became a noose around his neck, and by 1925 he was left stranded. He would remain a great skeptic until his death, quite possibly the most productive skeptic in the history of science. Einstein himself would have been amazed at what became of his doubts. Or perhaps not.

The members of Zeilinger's group hold the world record in a peculiar discipline: they test the limits of wave-particle duality. Using sophisticated experimental designs, they have been able to demonstrate that even so-called buckyballs, which amount to substantial chunks on the atomic scale with their more than sixty carbon atoms, act like waves. And were the Viennese quanta king to clear some space in his office on the long wall, preferably next to the photo of him standing beside the Dalai Lama, he could someday display a certificate from Stockholm. He would not be the first of Einstein's heirs to win the Nobel Prize.

The master himself received the distinction in 1922 for his work on "the production and transformation of light" in 1905—not for the theory of relativity, which struck the Swedish academy as too speculative even after its spectacular confirmation during the solar eclipse in 1919. Einstein was also not honored for his breakthroughs in wave-particle duality, but for the theory of the "photoelectric effect" he formulated at the same time. This theory explains the emission of electrons from metal by light energy, which is the foundation of today's digital photography. The name of the process behind this effect has become part of our everyday vocabulary: the quantum leap.

Max Planck discovered this phenomenon in 1900 when he realized that light deflection was discrete ("quantized"). When energy is brought to a "blackbody" (such as a hotplate), the blackbody emits electromagnetic rays as it grows hotter—visible at high temperatures as light, from red to white. However, even though the energy increases continuously, the metal (in a vacuum) does not radiate it continuously, but discontinuously, in stages. Between the individual stages the proverbial leaps occur.

Before the advent of quantum theory, virtually all events in nature could be drawn with steady, continuous curves, and after Leibniz and Newton they could also be expressed in the form of equations. According to the laws of classical physics, changes in speed, force, impulse, or energy occur as smoothly and fluidly as those of space and time. This continuity reached its logical conclusion in Maxwell's theory of electromagnetic fields and in Einstein's field equations of gravitation.

When an automobile accelerates, its speed increases continuously. The speedometer registers a steady increase. In the 1970s, speedometers with digital displays came into fashion. On these, the speed jumped from one whole number to the next, although this did not correspond to the functioning of the cars, which accelerated gently and smoothly, without any noticeable jerking forward. The quantum world is quite different, and digital displays of that era represent quantum events accurately. In the microsphere of atoms, events do not occur continuously, but in stages—as though a car could only drive at seventy or seventy-one or seventy-two, but not between those speeds. Qualities as quantities are not manifested solely in whole numbers; they are also quantized. This was the fundamental innovation of quantum theory when it appeared on the horizon of science.

Planck was well aware of the vital significance of his discovery, but this reluctant revolutionary did everything in his power to keep the genie in the bottle. The rather cautious prince of German physics was concerned about the future of his discipline. The ironclad rule states that *natura non facit saltus*: nature does not take leaps. Planck wanted to keep it that way. In 1908 he still believed that "our current understanding of the world is based on . . . certain characteristics that can never be erased by any revolution, neither in nature nor in the human mind."[4] However, the young unknown man in Bern had already let the genie out, never to be recaptured. Einstein made the quantum a reality, and can justifiably be regarded as the intellectual originator of quantum theory, which represents the most fundamental reorientation of physics since its inception.

Einstein's interest in blackbodies and heat radiation had begun when he was still a student. In March 1899 he reported to Mileva, "My broodings about radiation are starting to get on somewhat firmer ground."[5] Two years later he had a much clearer picture of this subject, and wrote to her, "About Max Planck's studies on radiation, misgivings of a fundamental nature have arisen in my mind."[6] Then his writings cease to yield any more insights into the matter until 1904. There is no doubt, however, that he devoted a great deal of thought to Planck's pioneering work, with which the quantum age commenced on December 14, 1900.

Like Einstein's theory of relativity, Planck's derivation of "blackbody radiation" is elegantly based on a brief, fundamental equation—one of the historically most significant formulas in the entire field of physics. Its discoverer wrestled it out of himself in an "act of desperation."[7] It seemed like something that was correct, but ought to have been wrong. To paraphrase the poet Christian Morgenstern, if it shouldn't be, it couldn't be. E equals h times nu is for quantum theory what E equals m times c squared is for the theory of relativity. The Greek letter nu (υ) is used to designate the frequency of light, which decreases as the length of the wave increases. The Latin letter h was the tough part. It stands for Planck's epoch-making discovery—a new natural constant, and one of the most important ones: Planck's quantum of action. Like the c, the speed of light, this h is universally the speed of light. It is valid for the entire universe and essentially describes the boundary between microphysics and macrophysics.

Planck, however, regarded his formula as only "a mathematical hypothesis with no physical reality behind it," says Danish historian of science Helge Kragh.[8] Planck also failed to realize how markedly his new law of radiation diverged from classical physics and gainsaid traditional views. His colleagues also took no notice of the formula's revolutionary potential. "During the first five years of the century, there was almost complete silence about the quantum hypothesis," says Kragh. "Einstein recognized the revolutionary implications of the quantum hypothesis much more clearly and willingly acted as a prophet of the quantum revolution."[9] In the light quanta hypothesis, the only scientific accomplishment that he himself called "revolutionary," Einstein went well beyond Planck's ideas about heat radiation. He described energy and even light itself as granulated or quantized. His astounding suggestion was met with bafflement at first, and it took twenty years for light quanta to be acknowledged as real.

Einstein made his first major presentation of his quanta hypothesis in 1909 at the eighty-first convention of the Society of German Scientists and Physicians in Salzburg. He had just been appointed professor, and was still speaking in the capacity of a guest speaker in the Physics Section. His topic was "On the More Recent Transformations Our Conceptions of the Nature of Light Have Undergone."[10]

Einstein appeared on stage in the city of Mozart in the dual role of "Quantum Einstein" and "Relativity Einstein." The one called the old ether hypothesis "a superseded position."[11] The other made it clear that light could move through empty space without the medium of the ether only in the form of electromagnetic fields, which behaved like "independent structures"—like particles with quantized energy states.[12] His colleagues listened to the up-and-coming star of their guild with a mixture of incredulity and admiration.

"I believe that so far as the development of physics is concerned, we can be very happy to have found such an original young thinker," wrote Walther Nernst in 1910, after visiting Einstein in Zurich. "A great theoretical boldness, which, however, cannot do any harm because the most intimate contact with the experiment is maintained. Einstein's 'quantum hypothesis' is probably one of the most remarkable ever devised. . . . If it is false, well, then it will remain for all time 'a beautiful memory.' "[13]

Einstein was gaining confidence in his hypothesis. "For me, the theory of quanta is a settled matter," he wrote to his former colleague Jakob Laub back in March 1910.[14] But he had lingering doubts that a full-grown theory could ever be developed. "I no longer ask whether these quanta really exist," he wrote to Besso in May 1911. "Nor do I try to construct them any longer, for I now know that my brain cannot get through in this way. But I rummage through the consequences as carefully as possible so as to learn about the range of applicability of this conception."[15]

In the fall of 1911, at the Solvay Conference in Brussels, Einstein again impressed the participants with a call to acknowledge the inevitability of quanta and discreteness in nature. Again he promoted "the so-called quantum theory . . . not a theory in the usual sense of the word, at any rate not a theory that could be developed in a coherent form at the present time."[16] He complained about this meeting to Heinrich Zangger, "The whole story would be a delight to diabolical Jesuit fathers,"[17] and wrote to his colleague Hendrik Lorentz, whom he greatly admired, in regard to Planck's quantum of action: "The *h*-disease looks ever more hopeless."[18] However, the core message of the quantum gradually caught on with his colleagues. Even so, the majority of the participants still failed to recognize the full extent of the break with the classical theory.

"Most of his colleagues were tentatively searching for ways to incorporate some sort of quanta into a theoretical structure of physics," write the commentators of Einstein's *Collected Papers*. "Some were prepared to consider the possibility of fundamental changes in mechanics, but probably no one except Einstein expected such changes to be needed in electrodynamics."[19] Visitors at this time, when Einstein was living and doing research in Prague, were brought to the windows of his big, sunny study, which looked out onto the park of a sanatorium for the mentally ill, where he told them, "Those are the madmen who do not occupy themselves with the quantum theory."[20]

Einstein's light particles, a phenomenon from the field of electrodynamics, did not find favor even with those who were beginning to accept the quantization of energy. In June 1913, Planck, Nernst, and several of their colleagues included this statement in their recommendation for Einstein's membership in the Prussian Academy: "That he might sometimes have overshot the target in his speculations, as for example in his light quantum hypothesis, should not be counted against him too much."[21]

Gradually it dawned on most of them that in order to understand the structure of matter, a new physics that incorporated quanta was needed. "The quantum was in outright conflict with Newton and Maxwell both, and there seemed no way of reconciling the new with the old," Einstein's assistant, Banesh Hoffmann, recalled several decades later. "Science was in deep crisis—deeper than it realized."[22]

Since roughly the beginning of the twentieth century, a micro-macro analogy has prevailed in the popular understanding of atoms. According to this idea, negatively charged electrons orbit like asteroids around a nucleus of positive electricity. At the time people began to think in these terms, electrons were considered the smallest building blocks of matter and electricity. Even after Ernest Rutherford's discovery of an incomparably larger atomic nucleus in 1910, the image of a small-scale solar system remained intact for the time being, with a massive core holding the electrons on their orbits—by means of electrical forces rather than by gravitation. There was just one catch: according to Maxwell's theory, the moving electrons would have to give off energy in the form of electromagnetic rays. This loss of energy would, however, prevent them from remaining in their orbits and they would crash into the nucleus. But they do not do this.

"The electromagnetic theory does not agree with the real conditions in this matter," a young Danish physicist wrote in his doctoral dissertation in 1911, and cleverly sought out "forces in nature of a kind completely different from the usual mechanical sort."[23] The young man was Niels Bohr. Just two years later he proposed an atomic model that enabled the quantum to achieve its triumphant advance. In this model the electrons travel around "allowed" fixed orbits, which correspond to specific energy states of the atom, around its nucleus. There is an orbit 1, an orbit 2, and an orbit 3, but no orbit 2.5. All other orbits apart from whole-numbered ones are not possible. The discrete numbers describe nature "objectively."

According to Bohr, the usual mechanics is valid for the motion of electrons, but electrodynamics is not, because in his model, light is neither emitted nor absorbed during the orbit. Only when there are quantum leaps from one orbit to another do the electrons give off energy in the form of electromagnetic radiation—namely light—or absorb energy in very specific—quantized—quantities. The difference between the energy states, however (and this characterizes Bohr's breakthrough), is derived precisely from the law for hollow-space radiation that Planck had formulated to heat blackbodies. The ratio of the change in energy and the frequency (that is, of color) of the light emitted or absorbed corresponds precisely to Planck's quantum of action h.

This brilliant stroke of intuition suddenly clarified experimental findings that had appeared baffling. Not only did Bohr's model shed light on the differing colors and the design of spectral lines of light when various atoms were heated, but it also made possible the first relatively sensible interpretation of the atomic structure of all chemical elements. "This is the highest form of musicality in the sphere of thought," Einstein announced, with obvious delight.[24]

The idea of quanta as packets of energy gradually took hold, but Einstein's quantum theory of light did not. Even Bohr flatly rejected the idea of light particles. Planck's notion that light is a mass divisible in any number of ways, and that quantization occurs only in interaction with matter, that is, with atoms, still prevailed.

Einstein was undeterred. His need to grasp the essence of light drove him on. No sooner had he completed his greatest work with gravitation theory in late 1915 than he published a very different sort of work in July 1916, which he rightly called one of his best. This study described his successful derivation: "An astonishingly simple derivation, I should say *the* derivation of Planck's formula," he wrote to Besso. "A thoroughly quantized affair."[25]

With this and one additional study in August 1916, Einstein flung open the door to modern quantum theory. He made the nonclassical assumption that with every emission and absorption of light an impulse is transmitted, like the impulse a billiard ball receives and passes along, which means that light does not radiate out in all directions. "There is no emission of radiation in the form of spherical waves," Einstein stated categorically.[26] Instead, light emission entails directed individual processes, like shots from a gun. This realization represented a crucial step in arriving at the definitive concept of photons. "Thus light quanta are as good as established," he rejoiced to Besso.[27]

Einstein distinguished between two types of light emission. While examining "forced emission," he made a profound theoretical discovery that would take thirty-five years to find its first practical application. If an atom is in an excited state, an incoming light particle may eject an additional photon instead of becoming absorbed. In noting this effect of light amplification—one photon goes in, two come out—Einstein was generating the basic equation for an important innovation, "light amplification by stimulated emission of radiation," better known as *laser*, which is now the basis for countless technical systems such as DVD players and medical devices.

By introducing a second kind of light emission, Einstein was again opening a Pandora's box that could never be closed again. He had unintentionally rediscovered chance.

When light is emitted spontaneously, he said, there is no way of predicting when a photon will be released. The chance of this happening can only be described by means of a probability formula as it is used for the radioactive decay of atomic nuclei. Thus the concept of probability acquired a new meaning—and chance did too.

If a coin is tossed, the probability of the result being heads or tails is 50 percent for either outcome. Each individual toss yields only the one or the other. The result is 100 percent heads or 100 percent tails, and nothing in between. Since a coin toss follows the laws of classical physics, its result can also be predicted. If all key figures such as the speed and direction of the throw were all known precisely, one could calculate the throw exactly. In the language of quantum physics, it is just a matter of "subjective probability." If a baseball flies beyond the playing field into the bleachers and a fan catches it there, the fan considers it a lucky coincidence. Subjectively, he is right.

Photon emission and radioactive decay work quite differently. If a substance has a half-life of one hour, a given atom has a fifty-fifty chance of decaying within the hour, and the outcome is decided only by "objective chance." Even if all of the conditions (of which we are aware) are known, it cannot be predicted or calculated when an individual atom will decay. It decides it to a certain extent on its own—or, as Einstein later said, "of its own free will." According to his 1916 study, the same applies to photons: there is no cause whose effect can be calculated to make predictions regarding the emission of any individual one. No one could have suspected that this new principle of randomness would someday become a cornerstone of modern quantum theory—not even the great minds of Einstein and Bohr.

In the spring of 1920, shortly after the sudden onset of Einstein's international fame, Einstein and Bohr met for the first time in Berlin. The two men were a study in contrasts. Bohr had made a name for himself as a soccer star in Denmark. Einstein had never thought much of sports, but proved dynamic, fearless, and powerful in the verbal arena. Bohr was as awkward as a frightened schoolboy; he mumbled incomprehensibly and repeated himself endlessly when he was sunk in thought. He was also a clumsy writer, so much so that he resorted to dictating all his texts—in high school to his mother and later to his wife, and at the institute to his colleagues. Einstein's pleasure in writing is readily apparent in every one of his lines. Bohr had a long, happy marriage as a family man; Einstein turned out to be fairly inept in this regard.

When it came to physics, however, the two had a great deal in common—even though their paths would later diverge. They got right down to discussing the issue that would bring them together for decades to come. They talked about issues in theory, especially as they pertained to atomic structure. "Rarely in my life has a man given me such joy by his mere presence as you have," Einstein wrote to Bohr afterward. "For me one of the greatest experiences ever was to meet you and speak with you," Bohr promptly replied.[28]

Their friendly rapport notwithstanding, Einstein was growing uneasy. He sensed that his discovery of randomness in photon emission would mean that the ultimate fundamental postulate of all natural science had failed, namely the law of cause and effect, in which he believed as fervently as others believe in God. "That business about causality causes me a lot of trouble, too," he wrote to Max Born in early 1920, shortly before Born was appointed professor in Göttingen. "But I would be very unhappy to renounce causality *completely*."[29] This statement is an early indication of the roots of his criticism of quantum physics, which soon followed.

By this point the "quantum crisis" was looming on the horizon, and would reach its pinnacle in about 1924. More and more anomalies were being revealed and experimental findings could not be brought into line with Bohr's model. At the same time, Einstein underscored, as Born observed, his "physical and philosophical creed, the rejection of statistical laws as the ultimate foundation of physics."[30]

In a now-famous letter to Max and Hedwig Born on April 29, 1924, Einstein wrote, "I find the idea quite intolerable that an electron exposed to radiation should choose *of its own free will*, not only its moment to jump off, but also its direction. In that case, I would rather be a cobbler, or even an employee in a gaming-house, than a physicist."[31]

He was convinced that particles had no free will, and could do nothing of their own accord. Things happened to them. There had to be a cause behind every one of their motions.

A year earlier Born had called for a reconstruction of the entire system of physical concepts "from the ground up."[32] A first step in this direction had been undertaken in the years 1922 to 1924 by a very young doctoral student named Louis de Broglie. De Broglie adhered closely to Einstein's way of thinking. The Frenchman proposed a model that would bring together Planck's and Einstein's equations. If energy equals m times c squared, mass times the speed of light squared, and also equals h times nu, and hence equals Planck's quantum of action times light frequency, what does it mean, de Broglie wondered, if the two are juxtaposed? What is behind the equation "m times c squared equals h times nu"? If m is mass and nu the frequency, shouldn't there be a wave for each mass particle?

De Broglie recognized the fundamental connection between wavelength and momentum, and in doing so became the first to give meaning and significance to the "allowed" electron orbits in Bohr's atomic model. He interpreted them as "standing waves." Only when the wave does not rush away like ocean waves, but stays in place like the vibrating string of a violin, has a stable state been attained. The studies that Anton Zeilinger's group in Vienna is conducting today with buckyballs are based on de Broglie's discovery of matter waves. They are seeking to determine the maximum size at which particles act like waves. Zeilinger and his colleague Markus Arndt are thinking of working with much larger molecules, such as the protein insulin, and eventually even with entire virus particles.

How far can this be taken? Might the boundary be concealing an additional mystery of nature in the no-man's-land between the realms of the giants and the dwarfs? "There is no indication," Zeilinger says, "that the quantum world will collapse at the very place where it is ideologically and conceptually desirable, namely at the boundary between microscopic and macroscopic systems."

De Broglie's discovery coincided with an additional brilliant achievement by Einstein, the profound significance of which did not become fully apparent until the end of the twentieth century. In mid-1924 a young Indian physicist, Satyenda Nath Bose, at Dacca University asked him to look over an article in which he derived Planck's radiation formula in an utterly new way—using a process of counting.

Einstein was the first to recognize the significance of Bose's discovery. He generalized the procedure for counting quanta that Bose had proposed to a method that would enter the literature as Bose-Einstein statistics. It allowed scientists to make astonishing predictions about matter in specific conditions—especially about a fifth state of matter in addition to solids, liquids, gases, and plasmas. At temperatures near absolute zero, these Bose-Einstein condensates act like a single giant atom. Carl Wieman and Eric Cornell of the United States and Wolfgang Ketterle of Germany were awarded the Nobel Prize in physics for 2001 for making these superatoms. The heirs of the prophet reaped the fruits of the seeds he had sown.

Because matter in this supercold state has many novel characteristics, the condensate is being treated as one of the hottest discoveries of the past decade. The multitude of possible applications has sparked the fantasies of researchers and engineers. For instance, it could serve to store information in quantum computers. Nanotechnologists have hopes for a previously unattained precision in constructing the smallest structures with the aid of so-called atom lasers. Ketterle and his colleagues first introduced a system of this kind in 1997. These lasers do not emit light, but matter waves of the Bose-Einstein condensate that oscillate in concert. In this way Albert Einstein helped to usher in the twenty-first century.

Back in his own century, in the 1920s, Einstein bore witness to an unparalleled high point in physics, with epoch-making breakthroughs nearly every month. Once Max Born had become a professor at the University of Göttingen in 1921, he and his brilliant assistants, notably Wolfgang Pauli, Pascal Jordan, and Werner Heisenberg, turned the little city in Lower Saxony into an international center of quantum theory. The young people developed utterly new ideas that kept even a grandmaster like Born on his toes. "I find that merely to keep up with their thoughts demands at times a considerable effort on my part," he admitted to Einstein.[33]

In the summer of 1925, Heisenberg wrote an article that would shake up physics "from the ground up," as Born had envisioned. Heisenberg's "latest paper," Born wrote to Einstein, "appears rather mystifying, but is certainly true and profound."[34] Heisenberg suggested introducing "a theoretical quantum mechanics, analogous to classical mechanics." The idea behind it is a discovery that marked a definite turning point. Heisenberg's approach resembled the manner in which Einstein had developed the special theory of relativity in 1905, in that he asked what measurements, such as measurements of time, really mean. Heisenberg similarly introduced a new way of looking at a measurement theory of quantum mechanical processes by pondering what is subject to observation and what is not. Electron orbits in atoms, for example, cannot be observed—but the light emitted when an electron changes its orbit and alters its energy state can.

Heisenberg turned Niels Bohr's atomic theory upside down. Bohr described orbits in 1913, but not transitions, while Heisenberg focused squarely on these transitions. Conditions such as orbits, he said, were defined solely by means of transitions. Heisenberg suggested a method of calculation that was absolutely new for physics. This "matrix mechanics" is one of the tools quantum physicists still use today. Every possible transition from one state to another can be represented as a matrix of individual numbers. These matrices can be used to make computations, in the same way that numbers can.

Heisenberg's "reinterpretation" of mechanics, a highly mathematized model of subatomic reality, was incomprehensible to most of his contemporaries—and even to himself. For a brief time he even thought of tossing his strange formulas into the fire. Einstein also had problems with an abstract mathematical model that had no connection to physical intuition and required great effort to master. Max Born, however, recognized the great value of the new theory, and within a few months Heisenberg, Pascal Jordan, and Born completed their written study in a solid framework.

Major developments on the summit of physical progress were coming from every direction by this point. Independently of the "three-man paper," Paul Dirac, who was only twenty-three years of age, formulated the identical theory in Cambridge, England. And in 1926, at the University of Zurich, the Austrian Erwin Schrödinger, almost an old man at the age of thirty-nine compared to the others, developed his own "wave mechanics." Schrödinger, a lady's man who was always dressed to the nines, described matter, in stark contrast to all concepts of particles, exclusively in the form of waves that moved in extremely abstract multidimensional spaces. Still, his work met with great acclaim, because it tackled the quantum problem with differential equations—that is, with a mathematical approach that was far more familiar to Schrödinger's colleagues than was Heisenberg's matrix mechanics. Amazingly, both theories, applied to simple physical systems, yielded the same results. It soon turned out that the two studies were identical in the abstract mathematical sense.

Just a few weeks after Schrödinger's breakthrough, it was Max Born's turn again. He reinterpreted the Austrian theory—much to Schrödinger's chagrin. Schrödinger even regretted ever having worked with quantum physics. Yes, there were waves, said Born, but probability waves, rather than matter waves. This is one of those concepts that make quantum mechanics seem so bizarre even today. The probability of a particle's location can be calculated using Born's interpretation.

One additional problem remained: at this point, the model was at odds with Einstein's special theory of relativity. In January 1928, Paul Dirac once again established the connection with his "quantum theory of the electron." Accordingly, the special theory of relativity allows for something that the general theory eschews: it is compatible with quantum theory. Dirac's study also predicted the existence of a new particle (and thus of "antimatter"), which was soon found. It was the positron, a positively charged electron. His theory marked the end of the "heroic pioneer" phase of modern quantum physics. Today researchers are busy trying to find out whether antimatter deflects light.

And Einstein? His firm belief that the cosmos existed independently of any observation, that the moon shone even when we did not gaze at it, suddenly relegated him to the sidelines. His friend Born, of all people, was pointing out what was undoubtedly the greatest challenge of his life as a scientist: the world as a probability function. "Quantum mechanics is certainly imposing. But an inner voice tells me that it is not yet the real thing. The theory says a lot, but does not really bring us any closer to the secret of the 'old one.' I, at any rate, am convinced that *He* is not playing dice."[35]

He held to this view for the next few years. The new music of randomness did not appeal to him, although he was the one who had struck up the tune. Einstein, like his predecessor Newton, continued to regard the world as a gigantic, complex machine that ran according to strict rules—with God as the great clockmaker whose work was fulfilled in the perpetual interplay of cause and effect. He clung almost religiously to the deterministic belief that if you had a complete grasp of the present, you could predict the future. However, Einstein's God— the principle of causality—lost his omnipotence. The dice in the casino of the quantum world fall without him. His magnificence lies in the past.

Einstein denounced the new ideas as "boy's physics." Born tried to calm him down by pointing out that "the motion of particles follows probability laws, but probability itself propagates in conformity with the law of causality."[36] That did not go far enough for Einstein. How can you speak of objective nature when no certainties exist, but only probabilities? Even if the particles all follow the laws of the quantum world, how can it be that each individual one acts, as he said, "of its own free will"? Where is the cause for its action; where is the ultimate purpose?

Matters got even worse. In the spring of 1927, Heisenberg formulated a principle that will forever be linked to his name. "Heisenberg's uncertainty principle" states that when man in his macroscopic world takes measurements, he cannot determine all things simultaneously when examining the microcosm. If a given object in the quantum world has two specific characteristics, only one of them can be measured exactly, or "determinately," at a given time, but not the other. If you try to measure them simultaneously, you wind up with uncertainty or indeterminacy. The more determinate the measurement of the one, the less determinate the size of the other. In other words, the more exactly you establish the one, the more you disturb the other. Heisenberg's principle provided a concrete interpretation for the abstract algebra of his matrices. You cannot measure the position of a particle and at the same time determine its precise momentum. By the same token, the momentum of the particle can be determined, but doing so precludes the possibility of determining its precise position simultaneously. From our current perspective, which is oriented to the concept of information, the system has "used up" its information in the process of measurement and will not yield any further data.

If, however, one does not simultaneously know the so-called initial conditions of a particle, namely its position and momentum, one cannot predict where it will be located later. Its future is open. Causality capitulates when faced with the quantum—or does measurement? Without an exact knowledge of the initial conditions, there is no predictable future, at least not for the smallest entities. The quantum theory put determinism in its place. The future is uncertain, as each of us learns as our lives unfold. Hence the theory contains within it an assertion about the freedom of matter—although that is still a subject of controversy in philosophy even today.

At the same time, the world remains calculable in quantum measure. Between the processes, probability waves lurk in predictable determination. A new type of determinism facilitated prognoses in the form of probabilities—certainty based on uncertainty. The predictions of quantum mechanics have come to pass, and continue to do so to the present day.

Strictly speaking, Heisenberg did not establish a principle like that of the conservation of energy or the constancy of the speed of light. Instead he provided an attempt at interpretation. The act of measuring, he said, has a crucial effect on the result. He spoke in terms of the "influence of measuring devices" at the intersection of classical and quantum theory on which ultimately the "statistical character of the interconnection . . . is based." Then he turned to the subject of Einstein's God: "The partition of the world into observing and observed systems prevents a sharp formulation of the law of causality."[37]

According to Heisenberg, then, processes within atoms cannot be discussed in the same way as macroscopically measurable processes. Instead, the values of what is measured emerge from the relationship between two conditions. While classical physics could be regarded as a theory of objects, quantum physics became a pure theory of relations, and its holistic character emerged. In contrast to classical theory, the whole was more than the sum of its parts.

The description of systems became more comprehensive, yet at the same time more ambiguous. When a state is "classical," no other is possible. The vase stands here and in no other place, whereas in the case of a given quantum state there is a certain probability that others can be found. The electron is both here and there. Measurement determines its exact location. Until then, there is only probability. This probability, however, has nothing to do with the kind used to calculate the outcome of throwing dice. It is an integral part, a characteristic feature of quantum systems.

Heisenberg's uncertainty relation marked a turning point in the history of science. It merges physics and philosophy. Not even in principle can the state of the world be known at any given moment. Hence everything that is observed is always just a selection of the totality of all possibilities and thus a restriction of what is possible in the future. Physics has to limit itself to descriptions of relationships between perceptions. The thing-in-itself is not accessible to it, but only relationships between things. The question of whether the uncertainty principle really does exclude causality and determinism, however, is still a hotly debated topic even today. As Einstein remarked, "Isn't all of philosophy like writing in honey? It looks wonderful at first sight, but when you look again it is all gone. Only the smear is left."[38]

Niels Bohr, who by this point was venerated throughout Europe as the pope of quanta, tightened the screws even more in the fall of 1927. He placed physics and the whole of natural science on a new philosophical foundation that went far beyond Heisenberg's uncertainty. In contrast to the classical world, said Bohr, systems in the quantum world could not be observed without altering them in the process. Someone measuring an electron, for example, does so with light—that is, with photons—and thereby affects the measurement. Heisenberg did not exclude the possibility that systems had values prior to being measured, but Bohr went a step further in contending that values arose only by means of measurement. In doing so, he was essentially following an idea of Immanuel Kant, according to whom systems did not have characteristics per se, but rather acquired them in the process of observation.

To understand how it is at all possible to describe the state of a system, Bohr introduced the principle of "complementarity." According to Bohr, physical sizes are complementary when they cannot be simultaneously determined with precision. The best-known example comes from Einstein himself—the duality of wave and particle. As much as the two concepts, which are borrowed from the macroscopic world, seem to be mutually exclusive, they complement each other so as allow for the description of the microscopic world only in combination.

The conflict between the two different approaches arises in the observer, says Bohr. Because we ourselves represent a part of the world that we are investigating, "we are both spectators and actors in the great drama of existence."[39] Einstein's question, "What is light?" must accordingly remain ambiguous—the answer lies in the eye of the observer, who is like a child faced with a choice between blindness and deafness, and then goes to the movies. If the child opts to be deaf, he sees a silent movie; if he is blind, he is aware only of the voices, music, and background sounds. He cannot do both together, but he can nonetheless form an impression about the nature of what he has seen or heard.

The observer has the choice of whether to measure one thing or the other. Since, according to Heisenberg, it is not possible to do both at one time, the act of measuring eliminates the opportunity to comprehend the complementary aspect. The bizarre part is that until the moment when a particle, for example a photon, is measured, it possesses both characteristics and neither of them at the very same time. It is in an extremely strange state between neither/nor and both/and. Quantum physicists like Anton Zeilinger contend "that no individual particle has a well-defined momentum until it is measured; the momentum occurs by means of measurement."

Chance and necessity go hand in hand. At the microscopic level, freedom and caprice appear to reign. The macroscopic obeys—aside from chaotic processes—the law of predictability: Planets can be counted on to follow their orbits. Billiard balls zigzag according to plan. A hunter aims his bullet at the deer's heart. Chance can be surmounted. But no one knows where the boundary runs between the worlds. In any case, it is not well defined. For the transition process, Bohr introduced an additional makeshift solution, the "correspondence principle." It means, roughly speaking, that even in the quantum world certain rules of classical physics have to apply as the limit is approached.

"The Copenhagen interpretation," which is what Heisenberg called Bohr's interpretation of quantum mechanics in 1955, met with less response from Bohr's contemporaries than it often appears in retrospect. This distorted impression was primarily a consequence of Einstein's fierce opposition, which he first expressed in public at the Solvay Conference in Brussels in October 1927. The debate that the two giants of theoretical physics waged there and continued on subsequent occasions is one of the highlights in the history of science, and became a twentieth-century legend. It marks the beginning of Einstein's long road to scientific exile. After more than twenty years at the front line, he suddenly found himself at the sidelines. Scientists of the next generation venerated him as the hero of their past, but he no longer had a place in their future. They had profound respect for him, but shook their heads as they sought new scientific avenues.

Einstein fought back hard and challenged the "boys," which made them hone their weapons, sharpen their arguments, and bolster their theories. Voicing his deep-seated objections and constructive criticism constituted his final "valuable contribution to science," Anton Zeilinger feels. The end justifies the doubts.

Paul Ehrenfest recorded the dialogue in his inimitable way: "Einstein, like a chess player, with ever new examples. A kind of *perpetuum mobile* of the second kind, intent on breaking through uncertainty. Bohr always, out of a cloud of philosophical smoke, seeking the tools for destroying one example after another. Einstein like a jack-in-the-box, popping out fresh every morning. Oh, it was delightful."[40] Whatever objections the older one cheerfully produced at breakfast, however, the younger one had refuted by dinner. "I am almost unreservedly pro Bohr contra Einstein," Ehrenfest summarized. "He now behaves toward Bohr exactly as the champions of absolute simultaneity had behaved toward him."[41]

This confrontation had a bitter aftermath. Within ten years Einstein had gone from championing a new view of the world to uncomprehending critics to adopting their role himself. Exasperated young people branded him a "reactionary." In their eyes, he was acting like an offended old man, reading the riot act to the next generation, and he deserved the mockery being heaped on him in return. His behavior had a parallel in the private sphere: at the same time, he was arguing against his son Hans Albert's choice of a bride in the same way that his mother had come out against his marriage to Mileva many years earlier.

At the next Solvay Conference, in 1930, the last one in which Einstein would participate, this game was repeated. He came well armed, and when Bohr could not invalidate his thought experiment against the uncertainty relation at first, the Dane was so dismayed that he saw "the end of physics" looming. Paul Ehrenfest not only recorded the duel, but photographed it. One of his pictures shows the two combatants, both wearing hats, taking a walk through Brussels in the autumn. Bohr, his coat on his arm, his mouth open to speak, but looking as though he is panting, a look of panic in his eyes, trying to catch up to Einstein with long strides. Einstein is wearing an ironic smile and looks pleased with himself, as though he has just scored a nice hit. It was one of his final ones.

The outcome of the duel brought a reprise of the defeat three years earlier. Bohr and his colleagues emerged stronger than ever. After a sleepless night, Bohr had found the argument to undermine Einstein's thought experiment and turn his own work against him, by demonstrating that this very experiment refuted the general theory of relativity. That was a bitter pill to swallow. Einstein had to admit defeat for a second time. The revolution that he himself had instigated had toppled his religion. Even so, he remained combative.

"You believe in the God who plays dice, and I in complete law and order in a world that objectively exists," he wrote to Max Born in 1944 from Princeton. Born was now living in Edinburgh. "Even the great initial success of the quantum theory does not make me believe in the fundamental dice-game, although I am well aware that our younger colleagues interpret this as a consequence of senility."[42] In 1951, he wrote to his friend Michele Besso, "Fifty long years of purposeful deliberation have not brought me any closer to answering the question, 'What are light quanta?' Even though these days any scoundrel thinks he knows the answer, he's wrong."[43]

Einstein skirted face-to-face discussions with Niels Bohr when the latter came to Princeton in the late 1930s, but he stayed in touch by mail to muse about the question that kept eating away at him, namely "whether God plays dice and whether we ought to cling to a reality that allows for physical description."[44] In April 1949, Bohr sent him a reply that was both stern and cordial: "I would even venture to say that no one—not even the Good Lord himself—can know what a phrase like playing dice means in this context."[45]

The story reads this way in nearly all accounts, but young historians of science tend to see the matter differently; they note that Einstein had stuck his finger into what is still a gaping wound of quantum mechanics today. "I really sympathize with Einstein on this," says Christoph Lehner, who worked with the Einstein Papers Project for five years and is now researching the development of quantum physics at the Max Planck Institute for the History of Science in Berlin. "Bohr did not provide a reasonable explanation; he only described the problem." Lehner claims that Bohr, like most physicists today, was an instrumentalist for lack of an alternative. Quantum mechanics functions wonderfully as an instrument; its predictions are right on target. Einstein never contested that. He did not argue against quantum mechanics, but against its interpretation.

Lehner explains that Bohr failed in his attempt to create a reproduction of reality: "Even today, there is no generally accepted and satisfactory formulation." But this was exactly what Einstein was asking for. His legacy is the essentially metaphysical demand that physics describe the world objectively. Neither Bohr nor his successors lived up to this requirement, however. This is exactly what Einstein was aiming at with his final big triumph, from which quantum physics has yet to recover.

In his final work of lasting significance, written in 1935, Einstein and two colleagues challenged the quantum advocates with a negative argument. The Einstein-Podolsky-Rosen paradox has not been solved even today in the way its authors intended. The three stated that quantum theory provided no more than an incomplete description of the world. As true as that may be, they said, it addressed only a part of the truth. Just because we do not know more does not mean that we cannot know more. Aren't there "hidden parameters," the three skeptics asked, to which every last thing in the universe is oriented—in the spirit of causality?

Einstein demanded physical facts; he insisted on cause and effect. He argued that if a particle did not previously have a characteristic, such as its spin, and now has it, something about it has changed objectively. But how? His argument used physical principles, but Bohr and his comrades-in-arms acted as though it were a purely psychological process. Bohr said that nothing physical was happening to the particle, but only something conceptual—by and large in the head of the observer. Einstein considered it paradoxical to speak of reality at all in regard to this hazy construct. "He had good arguments," says Lehner. "Bohr didn't."

Their contemporaries saw the matter differently. "Einstein has once again come out with a public comment on quantum mechanics," Wolfgang Pauli carped. "As is well known, each time he does that is a disaster."[46] Einstein's argument unsettled them, but did not make them change their minds about the Copenhagen interpretation.

At this point Erwin Schrödinger stepped in to support his friend Einstein with an additional paradox known as "Schrödinger's cat," which has given his name more prominence in lay circles than his "psi functions." A cat sits in a box with a radioactive atom. An apparatus is set up to release toxic hydrocyanic acid if the atom decays. Once the half-life period has elapsed, there is a 50 percent chance that the radioactive material has decayed. In the classical view, the chance that the cat is still alive is 1:1; either it is alive or it is dead, depending on whether the atom has decayed.

According to quantum physics, in the Copenhagen interpretation, the atom exists in a so-called superposition of states. It is only when a measurement is taken to determine whether the atom has decayed that this superposition is destroyed. Mathematically speaking, the wave functions then collapse. The lethal mechanism also makes this apply to the cat. It is in a state somewhere in between, and as long as no one checks, it is neither definitively dead nor definitively alive. The notion that measurement produces reality seems quite absurd, however. Schrödinger wondered whether there ought to be another interpretation.

Einstein congratulated his friend for having portrayed this incongruity so vividly. He described Bohr as "the mystic who forbids, as being unscientific, an inquiry about something that exists independently of whether or not it is observed, i.e., the question of whether or not the cat is alive at a particular instant before an observation is made."[47]

Seventy years later, superposition has become the focus of attention in quantum physics studies. "Superposition is real," says Anton Zeilinger. If light is sent to a double slit in his laboratory, an individual photon has a 50 percent chance of going through one slit or the other—just as the cat is both alive and dead. Only the process of measurement defines the conduct of the photon—or, to put it another way: only in light of the process of measurement is the conduct defined. Does the act of looking into the box determine whether the cat is alive or dead? One cannot deny the truth of an argument by calling it crazy.

Still, what goes on in the laboratories of quantum physicists like Zeilinger in Vienna looks just as normal as high school physics—at first. Light beams imperceptible to humans but registered on their instruments are sent traveling. Their paths diverge in the beam splitters, horizontally for some and vertically for others, but only when they are measured: *click, click-click*. They are united, or "entangled," in their "knowledge" of each other. Once they have been measured, each knows the orientation of the other, no matter how far apart they get. The right hand knows what the left is doing, even if nothing links them any longer. This effect has made beaming possible.

Schrödinger introduced the term "entanglement," which became the basis of the new quantum world, in connection with the Einstein-Podolsky-Rosen paradox. Entanglement occurs by means of the superposition of quantum states—precisely to the state, that is, that Zeilinger's team can now routinely create in the laboratory. Entanglement, along with quantization and uncertainty, is one of the cornerstones of quantum mechanics.

But, Einstein, Podolsky, and Rosen argued in 1935, wouldn't this mean that the photons exchange information across distances instantaneously? They would surely have to do so if quantum mechanics were complete, but Einstein argued that such "spooky actions at a distance" are out of the question. A direct exchange of information contradicts the special theory of relativity, so the particles had to have had their characteristics previously.

Christoph Lehner explains, "Einstein's assertion enabled the Irish physicist John S. Bell in 1964 to derive an inequation that has been empirically refuted." In other words, it is possible to demonstrate with experiments exactly what Einstein deemed impossible. In fact, the Viennese researchers do so daily as a matter of course. The Einstein-Podolsky-Rosen paradox provided Zeilinger's group with the basis for its work. Einstein had to resort to the subjunctive to muse, "If quantum mechanics were complete, it would have to . . . ," but his heirs can confidently assert in the indicative, "It *is* complete, and it *is* happening." Quantum mechanical systems have turned out to be as strange and spooky as Einstein had indicated as an argument against them. "He had not counted on spooky actions at a distance being a reality," says Lehner. "This is a remarkable fact."

Here in Vienna, they play games with the spook. In May 2003, Zeilinger's study made the front page of the renowned scientific journal *Nature*, which was celebrating the creation of "purified entangled photons." In 2003 the Viennese team also succeeded in distributing entanglement over a distance of six hundred meters in open space across the Danube, thereby achieving a world record. In a spectacular experiment conducted the following year, they demonstrated an application of quantum cryptography. With great fanfare, they transmitted a quantum mechanical encryption from City Hall to a bank branch in a manner that made eavesdropping impossible. It was encrypted by entangled photons that went in different directions. It was not the site of the experiment, however, the famous Viennese sewer system, that protected the photons from eavesdroppers, but the principle of entanglement. No unauthorized third party could sneak into the system, since any intruder's measurement of the data would disturb the system at once.

The Copenhagen interpretation, which forms the basis of all these developments, continues to dominate quantum physicists' discussions today. Bizarre new explanations, such as the "many-worlds interpretation," are being considered. The many-worlds interpretation holds that quantum mechanics fully describes reality, Einstein's qualms notwithstanding. This becomes possible by splitting the universe into several universes with every observation, or into several states of a single universe. In the one, Schrödinger's cat is alive; in the other, it is dead.

These new interpretations, which spark science-fiction fantasies, agree with Einstein on one point: they do not rely on "spooky actions from a distance." Still, Anton Zeilinger takes a dim view of them. He considers "Copenhagen" valid, but hastens to quote a statement by Niels Bohr that describes one of the most beautiful holes in the field of physics: "The opposite of a fact is falsehood, but the opposite of one profound truth may well be another profound truth."[48]

EIGHTEEN

OF THE MAGNITUDE OF FAILURE

THE QUEST FOR THE UNIFIED THEORY

It was supposed to be the pinnacle of his life as a researcher. A final triumph to surpass all the others—the realization of his dreams of uniting all theories of physics. Einstein devoted more time to this one great goal than to all his other work combined. For over thirty years he tried to unify everything in the world, from the quantum to the cosmos, from the smallest to the largest, in a single system, a theory of everything.

Initially the world sat up and took notice whenever anything new seemed to be brewing in Einstein's brain. Shortly before his fiftieth birthday in 1929, the hurly-burly reached a high point. The Prussian Academy had a thousand copies of his new work printed, and they sold out in no time. Selfridges, the London upscale department store, displayed the pages in its windows. Throngs of sensation-seekers crowded together to inspect them. The Berlin correspondent of the *New York Herald Tribune* cabled the entire manuscript by telex to its editorial office, which published it in its entirety. Of course nobody understood what was printed and exhibited. About a hundred reporters camped out at the house on Haberlandstrasse, where Einstein lived, to learn the news firsthand.

"The world awaits your explanation," Wythe Williams of *The New York Times* announced to Einstein when entering his apartment for an interview. In November 1928 the paper had gotten wind of his new opus and reported, "Einstein on the Verge of Great Discovery." Einstein had admired *The New York Times* since an editor there had discovered a mathematical error in one of his equations. Williams, who was the only reporter to be admitted to Einstein's home, insisted on an explanation. Einstein just buried his head in his hands and replied, "My God."

Within a decade, two fundamental models had emerged to describe the world, the general theory of relativity and quantum theory. They were mutually exclusive, but each claimed to be valid. The general theory of relativity used field equations to portray the macroworld, whose language was geometry, and quantum mechanics depicted the microworld in the language of algebra. Like siblings who cannot seem to get along, they were two mechanisms of one and the same nature.

Anyone who could succeed in uniting them would be holding the world formula in his hands, which was considered tantamount to the philosophers' stone; hence the never-ending fascination of the masses and the media when the genius in Berlin made a statement. When the *New York Times* photographer asked him to pose for an action photo, Einstein turned to reporter Williams and said, "Maybe he'd like me to stand on my head." The physicist's parting words to the journalist were, "I can't understand why all the fuss is being made about my little manuscript."[1]

A letter Einstein wrote to Michele Besso in early January 1929 reveals how much this "little manuscript" really meant to him: "But the best thing, which I have been pondering and figuring out for days on end and half the night, is now complete before me, condensed into seven pages and titled 'A Unified Field Theory.' "[2]

At the beginning of the 1920s, Einstein had made his first foray into the unknown terrain that would someday bring together the realms of the giants and the dwarfs, the macrocosm and the microcosm. He wanted to merge the two field theories into a single one, electrodynamics according to Maxwell and Lorentz and gravitation as described in his general theory of relativity. He regarded the physical field as the greatest achievement of science, a "point of view," as he later confided to Erwin Schrödinger, "that has driven me into deep solitude."[3] The "unified field theory" became his mania, his mantra, and his martyrdom.

Others of his stature are inclined to found schools, direct institutes, and provide a new foundation of knowledge and wisdom for the next generation. Einstein remained a seeker, supported by only a handful of faithful followers, virtually alone out on a limb. As early as 1925 he optimistically declared that "after an unremitting search during the past two years I now believe I have found the true solution."[4] It was soon evident that he was mistaken in this belief, and that this was just the beginning of an endless series of announcements and publications on an almost annual basis, complete with defeats, confessions, retreats, and renewed confidence.

In January 1929, Einstein described to Besso the idea he had come up with for a new unified field theory. He was exuberant. "It looks old-fashioned," he said, "and our dear colleagues, including you, my dear man, will stick out your tongues as far as you can at first, because Planck's constant (h) doesn't appear in these equations, but when the limits of the statistical mania have clearly been reached, there will be a rueful return to the spatiotemporal interpretation of spacetime, and these equations will form a starting point."[5]

Einstein firmly believed in his new formulas. The success of his general theory of relativity with its "spatiotemporal interpretation of spacetime" and his utter rejection of the "statistical mania" of quantum theory must have made him blind to his own shortcomings. He truly considered a generalization without including Planck's h possible—the universally valid quantum of action. He sought a formula according to which matter was constructed from the convolutions of spacetime, as his gravitation theory described it—a kind of hostile takeover of one theory by means of another. Just as he did not refute Newton, but declared him valid within the boundaries of his own theory, he thought of quantum mechanics as a special case within the unified field theory. Gulliver's dwarfs needed to get along in the realm of the giants, not the other way around. He did not have in mind a country in which both were considered equally at home. The great Einstein remained imprisoned in the gilded cage of his convictions.

Because he did not incorporate the quanta, his pursuit of the world formula, which began in 1925 and spanned three decades, has received only anecdotal mention in the histories on this topic—and only because he was Einstein. In any case, he remained true to his quest. He sought to find laws of nature that epitomized his ideal of science, a complete, unified description of the world. But this time his luck ran out.

For the theory of relativity, he had also picked and chosen among the virtually infinite number of possible theories and tried out one after the other until he happened upon the right one. But now he was lacking something that had previously served him well on two occasions: a principle, along the lines of his relativity principle or equivalence principle, which could give his new structure a foundation. And even more strikingly than for the general theory of relativity, it also lacked any empirical foundation. There were no observations or experimental data, along the lines of the Michelson-Morley experiment on the constancy of the speed of light or the deviation from the perihelion of Mercury, that could somehow serve as a guideline.

"Einstein was unable to find the unified theory for yet another reason," says Thomas Thiemann, a young professor in Waterloo, Canada, near Toronto. "He failed to take all known natural forces into consideration." Einstein wanted not only to unite electromagnetic force and gravitation. In the 1930s, "weak force" (which is important for radioactive decay) and "strong force" were also discovered. The latter holds together the atomic nucleus. "A unified 'theory of everything,'" says Thiemann, "must of course comprise all known forces."

This physicist, on the staff of the Albert Einstein Institute in Golm, near Potsdam, has signed on—as the second scientific pillar—at the Perimeter Institute in Waterloo, Ontario, a new think tank of quantum physics. A local business executive, Mike Lazaridis, who made a fortune with wireless communication devices, has created a paradise for theoretical physicists who care more about flat hierarchies, complete independence, and intellectual intensity than about a permanent appointment. In addition to ample pay, the perks include a bar and canteen, a conference center, and a squash court—a kind of Institute for Advanced Study of the twenty-first century.

Only someone who dreams in the hieroglyphs of theoretical physics and can maintain a clear overview of the universe while zooming in on the tiniest entities and back out to the farthest reaches can contribute to making Einstein's last vision a reality. Thiemann is one of the select few highly talented individuals to match this description, and he has joined the gifted group sitting and ruminating in their cubicles at the Perimeter Institute. His window looks out onto a small city of wood and brick, like hundreds of other cities in this part of the world. His room has the well-organized simplicity of an austere monastery. Desk with computer and telephone, chair and guest chair, bookshelf and the obligatory blackboard on which the syllables of the ultimate truth are obscured in a thicket of ciphers.

The young professor works in the loop quantum gravity division. He is trying out a new approach to the unified theory, and walks as squarely in Einstein's footsteps as possible. His head is like a control center for the various strands of modern physics. He talks about the big bang and about dark energy, supernovas and black holes, gravitational wave detectors, particle accelerators, quantum field theory, and cosmology. A "theory of everything" must ultimately encompass all these models, observations, phenomena and experiments, Einstein's and Heisenberg's formulas, and the measurements taken by Peter Aufmuth in Hannover and John Beckman in Tenerife.

Thiemann and his fellow researchers, who number one hundred worldwide, work in the shadow of the popular string theory, which boasts ten times the number of researchers. Their model could solve a problem that has stymied every attempt at a unified formula up to this point: anyone who starts with infinitely small particles, as described in classical physics, will eventually end up at so-called singularities, which is where the equations will fall apart. "The results are mathematically nonsensical values like infinities and probabilities over 100 percent," Thiemann explains.

If the particles were not singular points, but extended vibrating "strings," the new theory holds, this problem could be elegantly circumvented. Still, the lovely image of a music of the cosmos with vibrating strings has its limitations: no amount of effort will find a geometry in which strings vibrate in spacetime with their four directions. The string equations cannot "function" until six further dimensions of space are added to space and time; the expansion to superstring theory even entails an eleventh dimension.

The idea of the additional dimension is not new. Back in 1921, Theodor Kaluza, a mathematician in Königsberg, in an apparent stroke of genius, superimposed a fifth dimension on Einstein's spacetime in order to unite his gravitational theory with Maxwell's electrodynamics. Once Kaluza's colleague in Göttingen, a Swedish scientist named Oskar Klein, had generalized his work, it seemed for a while as though

Einstein's dream of an all-embracing opus had been fulfilled with the Kaluza-Klein theory.

However, Einstein was soon plagued by doubts concerning whether the fifth dimension really existed or whether it might be nothing more than a purely mathematical makeshift solution. The four dimensions of his spacetime corresponded to physical realities, but the fifth was just hanging in the air, as it were. Ultimately the Kaluza-Klein theory seemed tenable. Thomas Thiemann and his colleagues have similar doubts. "The string lives in ten to eleven dimensions," he says, "but we observe only four." Therefore, he explains, scientists have to make sure that the surplus dimensions are not observable—"but so far nobody has been able to do so." Hence the advocates of loop quantum gravity attempt to achieve their world formula within four-dimensional spacetime and thus to realize Einstein's dream in the way that appears most in keeping with him. They do not picture space as homogeneous. It has a fine-grained structure, comprising innumerable interwoven loops.

Accordingly, spacetime does not represent a continuity, but is quantized like matter. The orders of magnitude in which the quantization occurs are of course astonishingly small: if an atom were as big as a galaxy, a quantum loop would not take up more "space" than another atom. Thiemann mentions "quantum foam" in this context. That sounds just as outlandish as the idea of strings in eleven-dimensional space, but has the advantage of getting by without additional dimensions.

Would Einstein have preferred this route of quantum geometry to that of string theory? Thomas Thiemann does not regard this as an either-or situation. Everything has to fit. "We are on a par with the string theorists," he says. He does not believe there will be a "showdown" between the two. "We see one another as both competitors and cooperative partners." He considers it likely that both will eventually reach their goal, and regards the "what" as more significant than the "who." "If that works out, it will revolutionize physics as sweepingly as the general theory of relativity and quantum mechanics put together."

Although he emphasizes "collaborative efforts," Thiemann, like Einstein, has not given up on the ideal of the individual hero just yet. Everyone has to get the general idea, but ultimately it is always individuals who make the critical headway. "Everyone who plays the game can contribute something new, and then one of them manages to clear the hurdle, and everything falls into place." He believes that this may happen within the next ten years. "We have more or less cleared four big hurdles already and we are facing two additional ones. This bolsters our hope of getting past the last checkpoint as well."

One significant difference between string theory and loop quantum gravity, which the Perimeter Institute is a world leader in researching, is that loop quantum gravity may be subject to verification by means of a particular cosmological experiment. "We are focusing on natural flashes of light that come from very far away," says Thiemann. "Photons that take ten billion years to get to us have a great deal of time to take on something from the granular structure of space." They would have to pass through tiny loops. Energy-rich light particles would be slower than energy-poor particles. They arrive later—and that could be measured. "Differences of up to ten seconds might be expected."

Thiemann's use of the terms "hurdle" and "checkpoint" reflects highly complex investigations of the "mathematical consistency" of the theory. He can easily look up his "operators" and geometric equations on the Internet. Einstein had to ask the mathematicians of his era for help in obtaining the tools he required. Mathematics as the midwife of theory must itself be judged by physical reality, which is the supreme arbiter. Computation has to measure up to reality, and this is the crux of an additional reason that Einstein failed to find the unified theory he was seeking. His assistant, Banesh Hoffmann, emphasizes in retrospect, "The search was not so much a search as a groping in the gloom of a mathematical jungle inadequately lit by physical intuition."[6]

Was Einstein's erroneous belief that mathematical elegance is the only determinant of the value of a theory queering his chance of success? Beauty and symmetry alone are of little value when bridging the huge gap between the microworld and the macroworld. "Einstein took very bold strokes," says Thomas Thiemann. "But he knew it, and had no illusions." After all, hadn't he often been far off the mark with the theory of relativity from 1912 to 1915, only to triumph in the end?

Wolfgang Pauli spoke on behalf of the group of physicists who criticized Einstein's work when he wrote to Einstein, "All that is left to them now is to congratulate you (or had I better say 'express their condolences'?) on your having gone over to the pure mathematicians."[7]

Einstein himself, in his Nobel Prize speech in 1923, gave a clear out-line of the problem inherent in the unified field theory: "Unfortunately we are unable here to base this endeavor on empirical facts as when de-riving the gravitational theory . . . but we are restricted to the criterion of mathematical simplicity, which is not free from arbitrariness."[8]

His belief in his work and in the correctness of his approach kept him going through thick and thin. In an interview with an English daily newspaper, *The Daily Chronicle*, in late January 1929, Einstein explained, "Now, but only now, do we know that the force that moves electrons in their ellipses about the nuclei of atoms is the same force that moves our earth in its annual course about the sun, and is the same force that brings us the light and heat that make life possible on this planet."[9]

This lovely explanation for the question facing researchers in pur-suit of the world formula unfortunately falls short of providing an an-swer. Even today, nobody knows the solution, and perhaps it will never be found. Einstein failed to find it, and it is anyone's guess whether fail-ure is inevitable in this quest for the quintessence of physics. Perhaps this formula does not exist at all? "Whoever undertakes to set himself up as a judge of Truth and Knowledge," Einstein once said, "is ship-wrecked by the laughter of the gods."[10]

His dreams were now zooming way ahead of his reality. The interest of the masses and the media ebbed away, but not Einstein's fervor. He announced the next breakthrough, only to retract it. In 1930 he offered a completely new theory, and this pattern of announcement and retrac-tion was repeated—as it had been shortly before his breakthrough to the general theory of relativity. Couldn't it play out that way a second time?

"Albert is working as he has hardly ever worked before," his wife, Elsa, reported to Hedwig Born. "Has thought up the most wonderful theory. It's getting more beautiful every day. If only it proves to be *true!*"[11] In his usual sarcastic manner, Wolfgang Pauli claimed to find it psychologically interesting that each time Einstein furnished a new the-ory, as he did on an annual basis, he considered the latest one the defin-itive solution.

In the summer of 1938, Einstein announced that "this year, after twenty years of searching in vain [I have found] a promising field theory."[12] Advanced age, exile, and isolation are considered mitigating circumstances to explain these rash statements. If he had been anyone but Einstein, his serious struggles would have been considered a trivial pastime. His hobby meant home and refuge to him—and life, the only life he knew. "I work like a man possessed, i.e., I ride my hobbyhorse like crazy."[13] If he dismounted, he would be left empty-handed.

How long must he have suspected, or known, that he was struggling with a problem that one person alone can probably never solve? He later confided to Besso, "I consider it quite possible that physics cannot be based on the field concept, i.e., on continuous structures. In that case, nothing remains of my entire castle in the air, gravitation theory included."[14]

NINETEEN

FROM BARBARIA TO DOLLARIA

EINSTEIN'S AMERICA

They were screeching "Einstein! Einstein!" and gesticulating as though they wanted to rip his clothes off his body. Several hundred eager girls in short skirts, clutching little flags and red flowers, were lined up at the San Diego shore to welcome the pop star of science. Trumpets, rattles, songs, cheerleaders, the whole nine yards. "They come aboard and hand me all the big red blossoms," the traveler noted in his diary, "so that I am barely able to hold all the stems in my arms. I finally manage to escape to the cabin to have a little breakfast."[1]

Although it is just 7:00 a.m. on December 29, 1930, the Einstein hullabaloo is already well under way. "Now through swarms of people to the car," he wrote. "We eat at the hotel and once again, radio broadcasts to everywhere. Must not forget the morning's ceremonial reception in the columned hall of the city park with organ music ('The Watch on the Rhine'!) and speeches by the officials."[2]

Reporters chased him through the city. One of them handed him a paper with formulas and peered at him as though he were an exotic animal who might take the bait, or an extraterrestrial who would react in some bizarre way. Cissy Patterson of the *Washington Herald* descended on him when he was finally able to get away from the mob. He was sunbathing, wearing nothing but a handkerchief in his typical style, knotted on the four corners to form a curved surface to cover his head. The reporter did not dare to speak to him, but wrote a racy report that ran on the front page of the newspaper. The name "Einstein" sold newspapers in America, and publishers were willing to print anything about him, even a story of how someone *failed* to get an interview with him.

Without this effusive reaction in the United States, Einstein would never have become the superstar the world considers him today. It took the naïve hero-worship of the Americans for the triumphant child to be elevated to the pop icon of science. America admired and celebrated him—sometimes more than he could bear. He had discovered new continents, which was bound to impress Americans more than any other accomplishment. Europe was hailing him as the new Copernicus, but the Americans regarded him as the "new Columbus of science voyaging through the strange seas of thought."[3] The president of Princeton University had greeted him—in German—with those words when he came to be awarded one of his many honorary doctorates. That was in 1921, nearly a decade before the overwhelming reception in San Diego.

With this first trip to the United States, Einstein's American adventure had begun. (He had been invited to Columbia University back in January 1912, but had rejected the offer.) If he had had any doubts in those days as to the dizzying heights his sudden fame would propel him to after the announcement of the solar-eclipse data in November 1919, he now saw that he had no choice but to capitulate to the unrelenting applause. When he arrived in New York harbor on April 1, 1921, a mob of reporters stormed the steamer and cornered their astonished victim.

According to a report in *The New York Times*, "Einstein timidly faced a battery of camera men. In one hand he clutched a shining briar pipe and with the other clung to a precious violin."[4] He had never experienced anything of this sort. Cameras aflash, questions pouring in his direction, pressure to respond to the loudest questioner. All of them were clamoring for his explanation of the theory of relativity.

Einstein, who did not speak English, could only communicate with the aid of his wife, Elsa, who had learned English in school. Nonetheless, he is said to have declared on this occasion, "It was formerly believed that if all material things disappeared out of the universe, time and space would be left. According to the relativity theory, however, time and space disappear together with the things."[5] After he came ashore, the atmosphere became even more frantic. An energized sea of howling humanity awaited the guests at the harbor. A concert of ear-splitting honking poured forth from the endless series of cars flying

American and Zionist flags. Instead of driving straight to the hotel, the open car with the Einsteins and the Zionist leader Chaim Weizmann followed the procession across Manhattan to the Lower East Side. "The sidewalks were lined nearly all the way uptown with thousands who waved," *The New York Times* reported.[6] They stretched out their hands to touch him. Einstein was literally overwhelmed.

A few days later, reporters followed Albert and Elsa into their suite in the Waldorf-Astoria. Einstein still reacted to this unending onslaught with a mixture of astonishment and amusement. "The ladies of New York want to have a new style every year. This year the fashion is relativity."[7] His humorous style was a crowd pleaser. He was a clown entertaining his audience.

However, the fun was soon over, and duty called, as it would for the rest of his life. On April 8, the city awarded Einstein and Weizmann honorary citizenships. On April 12, 1921, Einstein gave the shortest lecture of his life to some eight thousand people in the packed auditorium of the sixty-ninth Regiment Armory in Manhattan: "Your leader, Dr. Weizmann, has spoken, and he has spoken very well for all of us. Follow him and you will do well. That is all I have to say."[8] Thunderous applause greeted this abbreviated speech. Einstein could hardly fathom his effect on the crowd. He gave Adolf Ochs, the owner of *The New York Times*, to understand that he considered the public's interest in him "psycho-pathological."

The meeting with President Harding at the White House on April 25 was, as Elsa informed the *Berliner Tageblatt* after their return, "a pantomime." Einstein spoke next to no English, and Harding spoke nothing *but* English. The two of them restricted their exchanges to a series of friendly handshakes until the cameras had captured their fill. On May 13, the *Chicago Daily Tribune* reported, "Einstein here, not relatively, but in flesh and blood—says he needs five days to explain his theory." At that point the reporter decided "to restrict the interview to other matters." The caption to Einstein's picture with Elsa read, "The smartest man in the world?"[9]

In Boston, one of the waiting reporters handed him a questionnaire that had been arousing controversy there for quite some time. Thomas Alva Edison was testing every applicant for a job in his company by posing a total of 150 questions along the lines of "Who invented loga-

rithms?" and "What are leucocytes?" and "How great is the distance between the earth and the sun?" This was Einstein's big chance to show
them how clever he really was. "What is the speed of sound?" one of
the reporters translated into German. "I don't know offhand," Einstein
replied. "I don't carry information in my mind that's readily available in
books."[10] Score one for Einstein, who had often praised Edison as a
great man.

Later that month, the *Cleveland Press* described the celebrated
guest, whose appearance at a reception nearly caused riots to break out
among the three thousand bystanders, as a "typical professor. He has
gray hair that he wears rather long."[11] Whether by accident or design,
Einstein fit the cliché of the avant-garde artist of science to a T.

While Einstein was visiting Princeton, a rumor started circulating to
the effect that the experiments on the constancy of the speed of light
had been refuted, thus casting doubt on his theory of relativity. In response to this rumor, Einstein made the now-legendary remark: "The
Lord God is subtle, but malicious he is not."[12] This statement was in-
scribed in stone, and can be seen on the mantelpiece of the faculty
lounge of the mathematics department at Princeton University.

Einstein's first trip to the United States had an unpleasant sequel.
On July 4, 1921, he granted an interview to a reporter for the *Nieuwe
Rotterdamsche Courant*. Einstein, who was inexperienced in dealing
with the press, made a statement that caused a commotion. On July 8
The New York Times printed this statement as a series of headlines:
"Einstein Declares Women Rule Here—Scientist Says He Found
American Men the Toy Dogs of the Other Sex—People Colossally
Bored."[13]

There was a storm of angry letters to the editor, and the newspaper
added fuel to the fire with a lead story about Einstein's alleged miscon-
duct. "His tirades are received by most Americans now, as showing the
errors of judgment natural in the circumstances to a man of his sort,
and particularly to one who, though not German by blood or birth"—
this was incorrect—"is a product of training exclusively German." (The
last part of this sentence was more or less true.) "Dr. Einstein will not
be forgiven and should not be, for his boorish ridicule of hospitable
hosts."[14]

Einstein was filled with remorse and published a statement to the effect that he was "horrified" about the Dutch report. He assured the Americans of his "feeling of gratitude."[15] This would not be the last exasperating incident. Brutally honest as he was, he could not conceal the ambivalence of his views. He was still unaware of the tricky nature of American friendliness, which is often confused with friendship by outsiders. He became acquainted with the country and its people, but he did not actually know them. The America of his era was neither openly nationalistic nor offensively patriotic. But it *was* defensive. If someone attacked Americans, they banded together. That is still true today. On the other hand, America is bighearted, often positively guileless, and forgiving, which is one of its most appealing characteristics. When he traveled through California ten years later, no cloud was hanging over his head from the previous visit.

This was only his second stay in the country. Einstein was still flirting with America, and his cheerleaders were exuberant. Two additional visits followed, and then, in 1933, exile. The flirtation lost its romantic edge when it evolved into a long-term relationship, and wound up a marriage of convenience when he became a citizen in 1940. It was never a love match. "He was never happy in the United States," the Israeli ambassador in Washington, Abba Eban, later recalled.[16]

Expelled from Germany, with Europe lost to him, Einstein was a man without a country for the final third of his life. To the end, a helpless homesickness bound him to the continent of his upbringing. However, his secret passion for Germania was transformed into manifest, irreconcilable hatred. This perspective is the only way to understand Einstein's life in the United States after 1933, his political involvement, his commitment to helping refugees, his (temporary) relinquishment of pacifism, his contributions to military research, his role in the development of the atom bomb, and his compromises on the subject of Israel. With all of these issues, he was somehow always struggling in gloomy indignation against the German demon in his heart.

Organs played the German song "The Watch on the Rhine" to ring in the New Year in 1931—an unwelcome reminder of his original homeland here in "paradise," as he frequently called California. California, a recreated Garden of Eden with palm trees, parks, and palaces as a backdrop to a futuristic workshop where packaging takes precedence over product. The name of the game was simulation—with wide smiles in wide limousines on wide boulevards. "There is a car for every two inhabitants," the guest remarked.[17] He returned to this idea a week later: "Rarely does anyone walk anywhere."[18]

Einstein's friends back in Germany saw and heard Einstein in the *Tönende Wochenschau* newsreels. In reaction to a picture of the "floral float containing lovely sea-nymphs," Hedwig Born wrote, "However crazy such things may look from the outside, I always have the feeling that the dear Lord knows very well what He is up to."[19] Einstein raved about his first trip to a supermarket: "The stores are organized in a wonderful way. Everyone helps himself and pays for the things he has put in his basket. The packaging is ingenious, particularly the egg cartons. Everyone recognizes me on the street and grins at me."[20]

He was offered film roles, and Elsa went to great lengths to fend off these offers. A millionaire donated $10,000 to the California Institute of Technology (Caltech in Pasadena, where the Einstein Papers Project is housed today) in the hope that he would allow her to visit him. He declined. They asked him to contribute his old shoes to an exhibition of worn-out shoes of film stars and presidential candidates. They served him Alaskan salmon, baskets of tropical fruits, and prize-winning hams. "Food is delivered to your door by people of all races, who have nothing in common but their American smiles," Einstein noted.[21]

He did not always make things easy for his conservative Californian hosts. He granted an interview for a Socialist weekly to Upton Sinclair, the writer and social critic, with whom he shared a bond of friendship. And he announced to the students at Caltech, "Concern for man himself must always constitute the chief objective of all technological effort, concern for the big, unsolved problems of how to organize human work and the distribution of commodities in such a manner as to assure that the results of our scientific thinking may be a blessing to mankind, and not a curse. Never forget this when you are pondering over your diagrams and equations!"[22] These words were not exactly geared to warm the hearts of the two hundred wealthy members of the philanthropic group with whom he ate dinner in the research center's stylish faculty club.

In Pasadena, Einstein also met the legendary Albert Michelson, who had proved the constancy of the speed of light with Edward Morley in 1887 and thus signaled the end of the ether. Without his work,

Einstein assured Michelson, who was now eighty-nine years old, the theory of relativity would have remained speculative. No mention was made of the fact that he had not cited the two Americans and their epoch-making experiment with their interferometer in his 1905 publication.

At the famous observatory on nearby Mount Wilson, he visited Edwin Hubble, who had discovered the redshifting in the light of distant galaxies and thus the expansion of the universe. Einstein was so impressed by the results that he dropped his "cosmological constant," deeming it superfluous—the lambda that would make a triumphal return to cosmology at the close of the twentieth century. Camera teams always accompanied him on his excursions—Einstein at the world's largest telescope, Einstein in the desert, Einstein in Palm Springs, Einstein at a picnic with beautiful women.

Then came Hollywood, where Einstein was driven "to the film czar Laemmle, a shrewd, hunchbacked Jewish chap with a paltry, wretched family, who lets the film stars dance and is a master of humbug."[23] Laemmle screened for him the film *All Quiet on the Western Front*, adapted from Erich Maria Remarque's novel, which graphically depicts the horrors of World War I. "A good piece, which the Nazis succeeded in banning in Germany," Einstein noted in his diary, "a serious weakness vis-à-vis the street mobs."[24]

In fulfillment of his wish, he also met the British king of the silent-film mimes, who had been adopted by Hollywood. His common bond with Charlie Chaplin was reflected in his similarity to Chaplin's cinematic characters. Einstein embodied what Chaplin played. No man of his standing so closely approximated the image of the tramp, and certainly no scientist. In real life, however, the film tramp lived like a prince in spacious rooms, where the guest could coax Mozart sonatas out of his violin. When Chaplin later came to visit the Einsteins' roomy apartment on Haberlandstrasse in Berlin, he found it cramped and modest compared to his own home.

Charlie invited Albert to the premiere of *City Lights*. Einstein and Chaplin came in tuxedos. The festive clothing looked dazzling on Einstein. Elsa, in her ample evening gown with the inevitable lorgnette in her hand, looked like a gussied-up country bumpkin. At the end of the performance, there was a standing ovation for the heroes of film and physics. The theorist asked the practitioner what the applause meant. "Nothing," said Chaplin. How to explain this hero worship? "People cheer me because they all understand me, and they cheer you because nobody understands you."[25]

Einstein's sometimes Chaplinesque grasp of etiquette was revealed in a little scene that he recorded later the same year in his diary during a trip to England, "When I changed clothes, the fresh shirt could not be buttoned shut and my manly hairy chest peeped out occasionally when I moved. I search my bachelor's apartment, find needle and thread, and sew the thing shut in such a way that I can still slip into it, although my freedom of motion is limited in a decent fashion. Then, at dinner, I cannot find my spoon for the ice cream (following a typically English meal with conventional conversation in a foreign tongue); and so, with a serious mien, I eat it using the little spoon from the salt dish. Whereupon the benevolent waiter shows me the small spoon that had slipped under my plate."[26] Everyday life as slapstick. If only he had always kept a diary.

Einstein allowed people to parade him around. Even though he was "gaped at like an orangutan in the zoo," he played along.[27] During the return trip by railroad across the American continent, the couple made a stop at the Grand Canyon. The chief of a Hopi tribe humorously dubbed Einstein the "Great Relative" to reflect the scientist's familial role and his role in the invention of the theory of relativity. Einstein agreed to pose in a feather headdress. Fodder for the photographers' cameras, but he did not have to be prodded.

In Chicago he stood at the side of the architect Frank Lloyd Wright on a train platform and spoke to a large crowd of pacifists, urging them to refuse to serve in the military. In New York he agreed to attend another fund-raiser for the Zionist cause. The sixteen hours in Manhattan again became an Einstein frenzy. The German consul general even claimed that there were "outbreaks of hysteria."[28] Einstein's image is carved in stone at the Riverside Church; he was the only luminary of world history to be honored in this way while still alive. When he and Elsa finally went aboard the steamer *Deutschland* to return to Europe, they breathed a sigh of relief that this exhausting tour was finally over. Now the first hints of criticism began to crop up, which set the tone for his future relationship with the United States.

"One notices that one feels more attached to the Old Europe, with its heartaches and hardships, and is glad to return there," he confessed to his friend Queen Elisabeth of Belgium.[29] "To enjoy Europe, one has to visit America," he wrote to Besso. "Although the people there are freer of prejudice, they are mostly shallow and uninteresting, more so than back home."[30] To Ehrenfest he depicted "a boring and barren society that would soon make you shiver."[31]

Is this the sentiment of a man speaking about the country of his choosing? Or is it a despairing premonition by someone who had no choice left because of the situation in Germany? In retrospect it appears as though Einstein had already detached himself from Germany quite a while before his actual forced departure. During his second visit to California, in 1932, he met a man who would have a profound influence on his future.

Abraham Flexner, who had been secretary of the Rockefeller Foundation for many years, came up with the idea of founding a small, select think tank. He envisioned this plan as the pinnacle of his career as a coordinator of research. Einstein did not need to be convinced of the merits of this idea. Since an educated populace appeared to be out of the question, there might as well be a community of scholars—an oasis of blissful intellectuals, such as he had been picturing in his mind for quite some time. Like Flexner, Einstein had envisioned an institute for world-renowned scientists who could enjoy the privileged status he had in Berlin: free to pursue their research, free of obligations, *sine cure*.

The screenplay of human fortunes sometimes seems to be written by a higher hand. Flexner's dream materialized at the very moment when the Nazis overwhelmed Germany and within a few months eroded the international center of natural science research. In 1933 the Institute for Advanced Study opened its doors in Princeton. Albert Einstein became its most prominent showpiece. He would receive $10,000 a year plus travel expenses for himself and his wife. At this point he was still planning to spend half the year in America and the other half in Germany.

A little incident preceded Einstein's unplanned exodus into exile, and provided a preview of the conflicted relationship between Einstein and America in the two decades to follow. His pacifist and socialist statements during the previous visits had made him less than welcome in the eyes of conservative groups. The National Patriotic Council publicly denounced the "German Bolshevist," and the feminist conservative Woman Patriot Corporation even campaigned to bar him from entering the country. In a sixteen-page document, this organization warned about Einstein's communist activities—"*not even Stalin himself is affiliated with so many anarcho-communist international groups*"—and insinuated that the theory of relativity was eroding the authority of state, church, and science.[32]

Einstein was summoned to the American consulate in Berlin to respond to these charges. After the standard procedures, he was asked a question that would become routine in the Cold War era: "What party do you belong to or sympathize with? For instance, are you a Communist or an anarchist?" At this point his patience ran out.[33] Elsa gave a blow-by-blow account to the Associated Press correspondent in Berlin of how her husband told off the consulate official. "What's this, an inquisition? Is this an attempt at chicanery? I don't propose to answer such silly questions."[34]

For the first time, Einstein was looking at the other face of America, and seeing paranoia, intolerance, and administrative callousness. America, in turn, was getting to know another Einstein who differed markedly from the nice, approachable professor whom Americans had come to expect. This Einstein was cold, calculating, and discomfiting. "I didn't ask to go to America," he proclaimed to the assistant. "Your countrymen invited me; yes, begged me. If I am to enter your country as a suspect, I don't want to go at all. If you don't want to give me a visa, please say so. Then I'll know where I stand." Neither he nor the official could sense the irony of this exchange, since this trip would signal Einstein's definitive change of residence. He grabbed his hat and coat, asked, "Are you doing this to please yourselves, or are you acting upon orders from above?" and left the consulate with Elsa without waiting for a reply.[35]

When he got home, he called up to announce that if he did not have his visa in hand within twenty-four hours, he would cancel his trip. Elsa went further. She told the Associated Press correspondent that she had packed six trunks, and if they were not sent to Bremen the next day, it would be too late. "That will be the end of our going to America."[36] She told *The New York Times* that her husband was commenting,

"Wouldn't it be funny if they won't let me in? The whole world would be laughing at America."[37]

On top of that, Einstein wrote a furious philippic against the Woman Patriot Corporation, which appeared in *The New York Times* on December 4:

> Never yet have I experienced from the fair sex such energetic re-jection of all advances; or if I have, never from so many at once.
>
> But are they not quite right, these watchful citizenesses? Why should one open one's doors to a person who devours hardboiled capitalists with as much appetite and gusto as the Cretan Mino-taur in days gone by devoured luscious Greek maidens, and on top of that is low-down enough to reject every sort of war, except the unavoidable war with one's own wife? Therefore give heed to your clever and patriotic womenfolk and remember that the Capitol of mighty Rome was once saved by the cackling of its faithful geese.[38]

Pressure and threats brought about the desired result. The Einsteins got their visas in time and without undue difficulty. The peculiar evalu-ation by "the fair sex" did not, however, go where it belonged, in the wastebasket of history. It had a strange future ahead as the basis for Ein-stein's file at the Federal Bureau of Investigation.

In February 1933, the famous photograph of the laughing physicist rounding a curve on his bicycle was shot in California. Today it seems like a symbol of his situation at the time after the forced exodus from Germany. "People are like bicycles," he wrote to his son Eduard in 1930. "They can keep their balance only as long as they keep moving."[39]

When Albert and Elsa, accompanied by his secretary, Helen Dukas, and his assistant, Walther Mayer (known as "the calculator"), arrived in Princeton after their stay in Pasadena and their final sojourn in Europe in the fall of 1933, the media once again started keeping tabs on their new fellow resident. Whether he enjoyed a vanilla ice cream with chocolate sprinkles or bought a comb ("Einstein's first act in Prince-ton"),[40] the most banal items seemed to merit headline coverage. The anecdote with the comb was printed in newspapers around the world. "People are like the ocean," Einstein wrote to Eduard in late 1933, "sometimes unruffled and friendly, sometimes stormy and treacherous—but essentially just water."[41]

The press, however, was not the only group hot on his trail. Under the lame suspicion of leftist activities, he was under observation from virtually the first day of his life in the United States. In a public declaration in early March 1933 that he would not be returning to Germany, he had said, "As long as I have any choice, I will only stay in a country where political liberty, tolerance, and equality of all citizens before the law prevail."[42] But his new home turned out to be not quite as tolerant, equal, and free as he had imagined. Surveillance, racism, and anti-Semitism—covert and overt—were a societal reality.

In 1932, Einstein had published an article on this subject in the NAACP monthly *The Crisis*. He lamented that "minorities, especially when their individuals are recognizable because of physical differences, are treated by the majorities among whom they live as inferiors." Only by "closer union" and "educational enlightenment" could the "emancipation of the soul of the minority . . . be attained. The determined effort of the American Negroes in this direction deserves every recognition and assistance."[43] The similarities to his deep commitment to promoting Jewish self-awareness in the early 1920s are compelling.

He published that article in 1932, before settling in Princeton, but once he was there, he confronted the conditions in his own new neighborhood. For Einstein, Princeton was not just "a wondrous little spot, a quaint and ceremonious village of puny demigods on stilts."[44] The town had a population that was 20 percent African American, living segregated from the white majority, usually under poor conditions. Black Princetonians had to sit in separate seats at the movies and send their children to their own elementary school. Princeton University did not accept black students—and very few Jewish students, whose number was kept down by means of quotas. Although the institute was then located on the Princeton University campus, Albert Einstein was not a member of the Princeton faculty. His friend and future trustee Otto Nathan did hold a university appointment, but for only two years. Einstein wrote to his sister, Maja, about the "powerful anti-Semitism that flourishes in all parts of the economy even in this free country."[45]

In 1937, when the Nassau Inn refused to give the famous African American opera diva Marian Anderson a room, Einstein put her up in his own home. The two became friends, and from then on the singer always stayed with him when she came to Princeton—which was not to the liking of some white Anglo-Saxon Protestants in the neighborhood. Einstein never let his behavior be dictated by such conventions. Fearless, obstinate, and virtually oblivious when it came to unwritten rules, he lived out his freedom in the framework of the written law.

It did not take long for Abraham Flexner to find out how testy this free spirit became if he sensed any attempt to clip his wings. Flexner, the proud director of the new institute, took his responsibility for his new protégé so seriously that he opened Einstein's mail, and even turned down invitations without consulting Einstein. In walling off Einstein from the outside world, Flexner feigned concern about ensuring his peace and quiet, but in reality he feared the consequences of Einstein's political escapades. When he heard through the grapevine that Flexner had turned down an invitation for him to visit President and Mrs. Roosevelt in Washington, Einstein put his foot down about this "interference . . . of a kind that no self-respecting person can tolerate." He openly threatened the director that he would quit, and proposed discussing "ways and means of severing my relations with your Institute in a dignified manner."[46]

Flexner had no choice but to give in. Elsa and Albert dined with Eleanor and Franklin Roosevelt in the White House in January 1934, and spent the night there. Einstein strongly supported the Democratic president's social and economic reforms. By contrast, his relationship with his director had been irreparably damaged.

Flexner wondered whether he had done himself such a great favor by bringing the prominent professor aboard permanently. Einstein resignedly wrote to his friend Maurice Solovine that he felt "highly esteemed, like an old museum piece or curiosity."[47] Unlike many of his colleagues, such as the mathematicians Kurt Gödel and Johann (John) von Neumann, he was not adding much to the scientific renown of the institute beyond lending it his name.

"This sophisticated Princeton cultivates scholarship in a greenhouse," he groused in a letter to Maja in 1935, "but I am consoled by the fact that the main things I have achieved have become mainstream science."[48] A year later he wrote, "But can it be any other way for someone who is obsessed? Just as when I was young, I keep sitting and thinking and calculating, hoping to get to the bottom of deep mysteries."[49] He had about twenty years still ahead of him at this point.

Einstein's daily life in America proceeded according to a charmingly straightforward routine. Every morning, after his bubble bath and a big breakfast, with a minimum of two sunny-side-up eggs, he went to work. He generally walked the twenty to thirty minutes from his house on Mercer Street to Fine Hall, where he delved into his equations, filled up countless pages with formulas, or stood at the blackboard with his assistant. (The first assistant was Walther Mayer; he was followed by Leopold Infeld, then more than a half-dozen others.) Einstein twirled his hair around his fingers, stroked his mustache, and discussed the progress and setbacks of the unified field theory. Although he did not achieve any results that posterity would consider noteworthy, science was more than ever the core and anchor of his existence.

According to Infeld, "working with him was no easy task," since he "always forced [his staff] to live in a constant state of frenzied activity."[50] Infeld found that the "overdose of science began to exhaust me."[51] He described a scene with the Italian mathematician Tullio Levi-Civita, who could barely contain his laughter when he looked at "Einstein, who kept pulling up his baggy pants every few seconds." He recalled how the two "spoke in a language they considered English."[52]

Einstein never really learned the language of his country of refuge—which only intensified his isolation and outsider status. He allowed only speakers of German to work with him, and he was perpetually quoting passages from Goethe's *Faust*. "Einstein's English was quite simple and consisted of about three hundred words, which he pronounced very weirdly," Infeld recalled.[53] Infeld's successor Banesh Hoffmann reported that Einstein could not pronounce the English *th*, and he used German syntax to form his odd pronouncements. The droll result included Infeld's favorite, "I will a little tink."[54] Anyone who listens to tape recordings of his Swabian-tinged English can readily imagine sentences like one a Princeton student recalled, "Oh, he is a very good formula."[55] When he gave a speech in English on television protesting the use of the hydrogen bomb, he spoke in such an incomprehensible hodgepodge that English subtitles had to be added to enable viewers to make sense out of his words.

According to legend, while Einstein was proctoring an examination in 1940, a student raised his hand and announced, "Professor, these are exactly the same questions as last year." "Yes," Einstein replied, "but the answers are different." There is no documentation to substantiate this incident, but it nonetheless provides a fitting illustration of the way he interacted with students.

When lunchtime came, he always headed home, where Helen Dukas served him European dishes. Friends from Europe regularly sent bouillon cubes and packaged soups so that he could continue to enjoy the foods he had grown up with. The furniture had been shipped from his apartment on Haberlandstrasse in Berlin. He often walked home with his colleagues or with Gödel, with whom he talked about God and the world. The German expression "talking about God and the world" means "talking about everything under the sun," but this is one of those rare instances in which the German turn of phrase can be taken literally.

Princeton and the media gradually got accustomed to his presence and left him alone for the most part, although people could not refrain from "staring at Einstein with hungry, astonished eyes," as Infeld has reported.[56] He learned to ignore their stares. When a car stopped and someone asked for permission to take his picture, he cheerfully struck a pose and promptly forgot the incident. Perhaps that was the best way to deal with it. Instead of griping about these intrusions, he acknowledged that curiosity-seekers were part of his daily routine, much like pesky flies. In fact he did care about his public image; it is reported that whenever he saw photographers approaching, he mussed up his hair with both hands to freshen up his typical Einstein look.

After lunch, he went up to his study on the second floor, which stood in as a place of retreat for his attic room in Berlin and his refuge in Caputh. Gazing through a large custom-built window that looked out over the garden, he watched the seasons change. He took a nap, and afterward he and Helen Dukas dove into the stack of mail that arrived every day. Einstein bemoaned his epistolary fate in a poem:

> The postman brings me every day
> Piles of mail to my dismay.
> Oh, why does no one ever reason
> That he is one while we are legion.[57]

There was always a substantial load of fan mail, which included countless eccentric letters, such as one addressed "to the beloved Albert Einstein, professor and messenger of God and servant of mankind" and another to "Mr. Oberwirrkopf [Chief Muddle-Head] Albert Einstein." These letters are housed in the Einstein Archives in Jerusalem, in the "curiosity file." A few choice snippets: "It would be an honor to wash your feet." "A secret voice in my innermost soul (which rarely deceives me) tells me that I must devote my life to you." "My brother, who is now sixteen, refuses to get haircuts. He is an admirer of yours and replies to urging that maybe he will grow up to be an Einstein." "I must speak to you alone. I am the successor to Jesus Christ. Please hurry." "When a day, a week, or any unit of time passes, where does it go?" "Please inform me whether it is necessary to study physics to prolong life."[58]

In the afternoon he sat back down to work, often with his assistant. It is perhaps pointless to speculate whether he really believed he would find the formulas he was seeking; for Einstein, the process was everything and the product was just a product. He supposedly once said that he loved to travel, but hated arriving. "I would no longer wish to live if I did not have my work," he wrote to Besso in October 1938.[59] In the same year he confided to his old friend from the "Olympia Academy," Maurice Solovine, "I can still think, but my capacity for work has slackened. And then: to be dead is not so bad after all."[60] He still had seventeen years ahead of him.

Although his fame had spread to all parts of the globe, his sphere of activity kept shrinking. Here and there he went on excursions, generally no farther than New York. He rarely took long trips. In the summer he fled the muggy weather in Princeton and spent a few weeks on vacation, whenever possible at the water, often on Long Island. The successor to the stately cabin yacht *Tümmler* was *Tinnef*, a little dinghy devoid of creature comforts. His companions even had to beg for seat cushions. Einstein, who still did not know how to swim, refused to wear a life vest.

Apart from these trips, the former inveterate traveler barely left his small town. Still, in the course of twenty years, he appears to have led an eventful life there as well. There were visits from and to the great men of his era, from Niels Bohr and Wilhelm Reich, Nehru and Ben-

Gurion to Thomas Mann, who lived just a few blocks away from him for several years. Einstein attended concerts, lectures, and ceremonies, and the media kept making a beeline to him. These were two decades of pre-retirement, which did not end with his retirement. He also enjoyed a symbiosis with a bevy of women as dedicated to seeing to his daily needs as he was to ensuring their livelihood.

Helen Dukas, who devoted herself to Einstein like a nun to the Son of God, took care of him as a girl Friday, from cooking to correspondence. His stepdaughter Margot had separated from her husband, Dimitri Marianoff. Like Dukas, she spent her mature years living in the house on Mercer Street, and stayed on after Einstein's death. Margot introduced cats to the household. Einstein loved them like a child, but by the time there were more than thirty of them, he cheerfully decided that enough was enough. Once, during a seemingly endless rainstorm that had the cats meowing incessantly, he supposedly said, "I know what's wrong, but I don't know how to put a stop to it."

Elsa lived only two years after going into exile. When her daughter Ilse died in Paris in 1934, she lost her joie de vivre, and, in late 1936, her life. Einstein described their situation in a letter to Max Born, the punctuation of which is notable in that a mere semicolon separates his sorrow from his joy, "My wife is unfortunately very seriously ill; I personally feel very happy here, and find it indescribably enjoyable really to be able to lead a quiet life."[61]

Einstein retreated even deeper into the protective shell of his refuge in work. "I have settled down splendidly here," he wrote to Born. "I hibernate like a bear in its cave, and really feel more at home than ever before in all my varied existence. This bearishness has been accentuated still further by the death of my mate, who was more attached to human beings than I."[62] For him, "at home" really meant "undisturbed."

As long as Elsa remained in good health, she kept their social life in Princeton on track with dinners and receptions, to the dismay of her ensnared husband. Once her strength had ebbed, Einstein declined all invitations with the excuse that *he* was ill.

Elsa's place in the female household was taken by his sister, Maja, in 1939. She too passed away after a devastating illness, in 1951. Death kept burning Einstein's bridges to Europe as well: Ehrenfest died in

1933, Grossmann in 1936, and Mileva in 1948. "The grief that the death of Paul Langevin has left me with," Einstein said in 1946 after the death of the French physicist, "is particularly poignant because I have the feeling now of being quite alone and deserted."[63] He had little empathy for the sufferings of others, but keenly felt his own suffering. His quest for solitude was also an effort to escape human contact: "Whatever I do in the human arena," he told Carl Seelig in 1949, "always seems to degenerate into monkey business."[64]

Einstein delighted in his "splendid isolation," but made sure to keep up with world news during his years in the United States. He listened to the radio and read *The New York Times*—at least sporadically. The economic crisis that followed Black Thursday in October 1929 set America back by decades. The gross national product, private incomes, and foreign trade sank by 50 percent after 1929. Construction came to a virtual halt. In New York, Rockefeller Center, which was built in 1932, was the final skyscraper project to be realized for more than twenty years. Approximately 15 million Americans were unemployed in 1933, roughly a quarter of the potential workforce.

At roughly the same time that Einstein announced he would be going into exile, Franklin D. Roosevelt was inaugurated as the thirty-second president of the United States, on March 4, 1933. He introduced a series of social and economic reforms that became the symbol of his years in office, and were collectively known as the New Deal. With state-controlled capitalism he was able gradually to stem the decline. However, Einstein regarded even the toned-down capitalism in America with the critical eyes of the enlightened welfare-state European. "Everything here is rushed and ruthless," he wrote to Eduard in May 1937, "a real dance around the golden calf, a wild and ugly dance."[65] He told his sister, Maja, in the same year, "The truth is that nothing but money counts here."[66]

When Hans Albert announced in 1937 that he would be coming, his father was delighted with the news, but warned him, "I am just afraid that you will not really be able to handle this pack of swindlers."[67] In 1938 he impressed upon Besso, "Money rules everything here too, as well as the fear of Bolshevism or simply the fear that rich people have regarding their privileges."[68] He explained to his friend, "The business-

man is the national hero, which is to say that a new type of garter matters more than a new type of philosophical theory."[69]

These problems paled in comparison to the news from Germany,
however, where the persecution of Jews and people with dissenting
views had assumed more and more alarming proportions. Einstein undertook a project that may have been the greatest accomplishment of
his life, but is often overshadowed by his scientific and political
achievements. "Miss Dukas and I run a kind of immigration office," he
wrote to Maja in 1938.[70] This was far more than just a way to keep busy.
In the past, a series of angels had reached out to him when he needed
help; now he became an angel himself. He was living proof that
humanism and humanity have a common root. No one has put a number to how many people's lives Einstein saved. It is probably in the
hundreds.

Einstein wrote countless testimonials and affidavits—almost too
many to achieve the desired result in individual cases. "Everyone had a
testimonial from Einstein!" Leopold Infeld reported.[71] He also paid a
small fortune to enable imperiled Jewish artists and scientists to leave
Germany and come to the United States. Entering the United States
was quite difficult owing to the restrictive American immigration policies in the 1930s. In addition to a birth certificate and a job offer (or a
financial guarantee, which Einstein issued by the hundreds), adults had
to present a police certificate stating that the holder had no criminal
record for the last five years. People who had been persecuted by the
Nazi regime found it virtually impossible to secure this paper. The
American State Department also refused entry to anyone whose name
the Gestapo listed—there were quite a few of those. Until the United
States entered the war in late 1941, the authorities on both sides of the
Atlantic were effectively working together. Any Gestapo comment
about "communist sympathies" was enough to turn down people wishing to enter the country.

Thus began the saddest chapter in the relationship between Einstein and America. The greatest democratic nation began to compile
an extensive dossier about his actual and alleged political activities.
The culprit was a cool hothead: J. Edgar Hoover, the archconservative,
professed anti-Semitic head of the FBI. He fought, as the American

journalist Fred Jerome phrased it in the subtitle of his book *The Einstein File*, a "secret war against the world's most famous scientist." No sooner had Einstein escaped the claws of Hitler's henchmen than Hoover's office started keeping tabs on him.

Since the FBI has opened the Einstein file, the extent of snooping, slander, and simple ineptitude with which the immigrant was denounced as an "extreme radical" is now plain to see. The more than 1,800 pages resemble the scheming of the Stasi in the German Democratic Republic. The man who fought for nothing more than peace and subscribed to the ideals of the French Revolution, who did not have the slightest talent for subversion or posing a threat to the security of the state, appeared like a subversive agent of an imminent communist coup in these files.

One of the motives to spy on Einstein covertly was his support of refugees to whom the Gestapo attributed "communist sympathies." The fact that their lives often hung in the balance, which was why Einstein had lent his support to them in the first place, did not matter to his detractors. An additional factor that made Einstein an object of suspicion was his support of the Spanish democrats in the civil war against Franco's fascists. Even at that time the right-wing press decried his political activities.

"Professor Einstein . . . was given sanctuary in this land," said an editorial in the archconservative Catholic newspaper *The Tablet* on May 14, 1938. Now he was telling the government how to run its business. "What is worse is to have him endorsing a move to shoot down and continue the persecution of Christians in Spain." The article closed with the recommendation that "Einstein be sent back to Germany where he may fully realize how to mind his own business."[72]

TWENTY

"PEOPLE ARE
A BAD INVENTION"

EINSTEIN, THE ATOMIC BOMB,

MCCARTHY, AND THE END

July 1939. The Einsteins were spending their vacation in the little town of Nassau Point at Great Peconic Bay on Long Island. Anyone who could afford to flee the muggy heat in Princeton did so. Up here, not far from New York City, it was easier to cope with the heat. Beach, gentle wind, sailing weather. A car stopped in front of the vacation home, and two men got out. Einstein welcomed his old friend and colleague Leo Szilard, with whom he had been awarded patents for new refrigerators in the 1920s in Berlin, and the young physicist Eugene Wigner. When the three emigrants sat down at the terrace, the guests told the sixty-year-old vacationer some news that once again gave his life a new direction: the Germans might be building an atomic bomb.

Legend has it that Einstein now recognized for the first time the potential of atomic chain reactions. Supposedly he exclaimed, "I never thought of that!"[1] Whether or not this was so, what followed can best be understood against the backdrop of his deep distrust of the homeland he had once loved. He imagined the country "Barbaria" capable of anything. A "uranium bomb" in the hands of Germans would be like an "axe in the hands of a pathological criminal." He had not forgotten how consistently the Germans had adapted scientific achievements in employing poison gas for military purposes in World War I under the leadership of his friend Fritz Haber. He declared on the spot that he was prepared to go to the top level of the administration to warn of the danger.

The second part of the legend stemmed from Einstein himself:

"They brought me a letter and all I had to do was sign it."[2] This statement was a barefaced lie. In reality, he dictated the draft of the letter that President Roosevelt received on October 3, 1939, during a second meeting with Szilard and his fellow Hungarian Edward Teller (the future "father" of the hydrogen bomb) in late July, and graced it with his signature: "A. Einstein":

> Some recent work by E. Fermi and L. Szilard . . . leads me to expect that the element uranium may be turned into a new and important source of energy in the immediate future. . . . The new phenomenon would also lead to the construction of bombs. . . . A single bomb of this type, carried by boat or exploded in a port, might very well destroy the whole port together with some of the surrounding territory. However, such bombs might very well prove to be too heavy for transportation by air.[3]

This letter set the stage for a didactic drama of guilt and atonement, the likes of which have rarely been seen. The protagonist in this drama was Albert Einstein. His memory diminished the role he had played while his conscience boosted the measure of responsibility that he took upon himself. On March 7, 1940, he followed up this first letter to Roosevelt with a second, more urgent one. "Since the outbreak of the war, interest in uranium has intensified in Germany," he contended. "I have now learned that research there is carried out in great secrecy and that it has been extended to another of the Kaiser Wilhelm Institutes, the Institute of Physics."[4]

Even though he had long harbored doubts about the technical feasibility of using atomic energy, Einstein had no illusions about its devastating potential. Asked by his first biographer, Alexander Moszkowski, about the possible consequences of Rutherford's experiments in splitting the atom, Einstein conceded in 1919 "that we are now entering on a new stage of development, which may perhaps disclose fresh openings for technical science."[5]

Moszkowski's speculations went further still, perhaps inspired by H. G. Wells's *The World Set Free* (1914), which portrayed the unimaginable destruction wreaked by atomic bombs falling on Paris. "When such a measure of power is set free, it does not serve a useful purpose, but leads to destruction," the Einstein biographer wrote in 1920.[6] Moszkowski added, "All the bombardments that have taken place ever since firearms were invented would be mere child's play compared with the destruction that could be caused. . . . Heaven preserve us from such explosive forces ever being let loose on mankind!"[7] The atomic bomb was a frequent topic in intellectual discourse a full quarter-century before Hiroshima. The author closed with the prophetic statement, "Einstein's wonderful 'Open Sesame,' mass times the square of the velocity of light, is thundering at the portals."[8]

Still, it would be wrong to draw a direct line from the theory of relativity in 1905 to the atomic bomb in 1945. The equation $E=mc^2$ was not used in constructing the atomic bomb, although Hiroshima was a gruesome confirmation of it. The mega-explosion was instead the product of nuclear physics, in which Einstein had only a passing interest. The development of the bomb would have happened even without the existence of the "Open Sesame." It had begun back in 1896 with Antoine-Henri Becquerel's discovery of radioactivity.

In 1932, when the Briton James Chadwick found the neutron, which theorists had been predicting for quite some time, atomic research entered a crucial phase. Just two years later the Italian physicist Enrico Fermi and his fellow researchers announced that they had bombarded uranium nuclei with neutrons and produced the first "transuranic" elements—so called because their nuclei have swallowed neutrons and thus ought to be heavier than those of uranium. Prompted by this news, the chemists Lise Meitner and Otto Hahn carried out similar experiments in Berlin.

The German chemist Ida Noddack, who was overlooked for a long time in the history of science, soon provided a new interpretation of the results. "It is conceivable that in the bombardment of heavy nuclei with neutrons, these nuclei break up into several large fragments."[9] On December 18, 1938, her far-reaching prognosis was experimentally confirmed by Otto Hahn and his colleague Fritz Strassmann in Berlin.

It was Lise Meitner, however, who took the decisive step. In July of that year, she had fled from Berlin to Stockholm. Together with her nephew Otto Frisch, who worked with Niels Bohr in Copenhagen, she was the first to interpret the process correctly after Ida Noddack: if a uranium nucleus is bombarded with neutrons, "fission" can result. The name comes from the field of biology, where the nucleus splits in the process of cell division. This arguably most momentous discovery in nuclear physics triggered hectic activity in laboratories throughout the world. By late 1939, more than a hundred publications on nuclear fission had been printed.

Since 1934, Leo Szilard had been pondering the possibility of neutron chain reactions and the massive explosions that would result. While Einstein purportedly still doubted whether atomic bombs could be built at all, the press was shouting the news from the rooftops. "Physicists Here Debate Whether Experiment Will Blow Up Two Miles of the Landscape," *The Washington Post* reported on April 29, 1939.

Against this backdrop, Einstein, who was being pressured by Szilard, took the step he would rue more than any other in his life—writing his legendary letter to the president of the United States. His role in the development of the atomic bomb ended at this point, however. Roosevelt appointed a commission, but it had very limited means at its disposal and only an advisory function. Even if Einstein had wanted to collaborate on the project, they would not have let him do so. As we know from a dossier on which FBI director J. Edgar Hoover based his decision not to allow Einstein's participation in 1940, he was thought to pose a heightened security risk.

In this dossier, Einstein's summerhouse in Caputh was portrayed as "the Einstein villa at Wannsee" and "the hiding place of Moscow envoys," and his support of the November 1937 conference of the American League Against War and Fascism was cited as evidence of subversive activity.[10] American policies of the time were happy to turn a blind eye to right-wing movements, because they needed Germany and other fascist countries as a bulwark against Bolshevism. Their left eye was wide open, however, and any involvement as a pacifist was enough to get someone classified as a leftist.

At first, Einstein took this suspicion of communism lightly. "It is easy to see that E. is in Princeton," he quipped in 1937 in a letter to Maja, "recently a P. University publication featured a red cover."[11] He soon lost his taste for jokes of this kind.

"In view of his radical background," the FBI document stated in summary, "this office would not recommend the employment of Dr.

Einstein on matters of a secret nature."[12] The pivotal primary source the FBI used to support this position was the sixteen-page pamphlet with which the Woman Patriot Corporation tried to block his entry into the United States in 1932. Fred Jerome has established that this pamphlet amounted to 80 percent of the dossier.

Even though he probably never learned of his official exclusion from work on the atomic bomb, Einstein must have noticed that a gigantic government project was being launched. Most of the well-known physicists, including many of his fellow exiles, suddenly seemed to disappear. Although even his friends by and large kept their new project under wraps, he surely knew what they were toiling away at.

Why participation was denied specifically to Einstein remains an FBI secret. The only document from his file to disappear without a trace was the letter that spelled out the reasoning behind this decision. Hoover's blatant hostility to Einstein's alleged communist affiliations, or his more subtle aversion to Einstein's Judaism stemming from Hoover's own anti-Semitism, could not have been the sole deciding factors. Quite a few researchers on the secret Manhattan Project had far stronger ties to communist groups, and without the involvement of many Jewish immigrants, the undertaking would have failed in any case.

The marines had no qualms about allowing Einstein to participate in the development of torpedoes for the sum of twenty-five dollars a day, and he had no illusions about his role in this effort: "I am curious," he wrote to his son Hans Albert, "whether I will be able to accomplish something for the Navy before the war is over; as a recluse, I am somewhat cut off."[13] His more significant contribution to the military effort was monetary. Einstein copied out the manuscript of his theory of relativity and had it auctioned off for $6.5 million; the proceeds went to the war chest. This valuable document is housed in the Library of Congress today.

The FBI man and the physicist were poles apart on the issue of the role of the United States in World War II. Hoover, a secret Nazi sympathizer with ties to the head of the Gestapo, Heinrich Himmler, stood on the side of the isolationists until late 1941 and hence against an American entry into the war. Einstein, by contrast, once he was farsighted enough to predict the German war of conquest and put aside his pacifism, knew that America was the only nation powerful enough to combat sweeping Nazi victories, a conviction he shared with Thomas Mann, who regarded Roosevelt as the "born opponent . . . of the creature that must be toppled."[14]

After the Japanese attack on the United States Pacific Fleet in Pearl Harbor and the American entry into the war in late 1941, the Americans stepped up their work on the atomic bomb. The fact that it would later hit Japan and not Germany, which is the only target Einstein would have condoned, was in his view the greatest catastrophe in these times of great catastrophes. The key decision to build the bomb ultimately had nothing to do with the Japanese attack. It was made on December 6, 1941 — one day *before* Pearl Harbor.

"We must strike hard and leave the breaking to the other sides," Einstein told *The New York Times* in late 1941.[15] Now that he had become a citizen of the United States (on October 1, 1940), he felt "particularly fortunate to be an American."[16] In 1935 he had declared to his pacifist friends, "Different times require different means, although the final goal remains unchanged."[17]

In early 1943 the American and British efforts were combined in a military organization with the code name Manhattan Engineering District. In the spring, the atomic bomb laboratory at Los Alamos National Laboratory in New Mexico took up its work under the leadership of the American physicist J. Robert Oppenheimer of Princeton, whom Einstein knew personally. Branches throughout the country worked together with the head office in strictest secrecy. Most of the key physicists involved in the project were from Europe; they included Szilard, Fermi, Bohr, Hans Bethe, and James Franck.

Any degree of force was acceptable to Einstein when it came to fighting Hitler — even war to achieve peace, even the atomic bomb. When the first rumors about the Holocaust started leaking out, Einstein more or less equated the Nazis and the Germans. In 1944 he wrote, in an essay addressed "To the Heroes of the Battle of the Warsaw Ghetto," "The Germans as an entire people are responsible for these mass murders and must be punished as a people."[18]

By mid-1944, when there was little doubt that Germany would not be able to produce a bomb on its own, a debate ensued among the scientists in the Manhattan Project as to the ethical consequences of their cooperation. For the majority of the physicists, the most important reason behind their involvement no longer applied. However, the enormous undertaking had progressed too far to let it drop.

Now many physicists were faced with the burning question of the consequences of their clandestine arms race for the postwar period. Although Einstein had no official knowledge of these activities, he heard about the discussion and talked it over with Bohr. He, like others, shrewdly predicted a "secret technological arms race"; a "supranational government seems the only alternative," the cosmopolitan and humanitarian said, feeling a deep-seated mistrust of the abysses of human weakness.[19] He told Bohr that the leading international scientists ought to support this idea. Bohr was horrified. Once again, an Einstein escapade was threatening to cause trouble. Bohr and his colleagues could face devastating consequences if they violated their oath of secrecy. He hastened to Princeton to dissuade Einstein from pursuing this plan.

Einstein was persuaded. Still, in the spring of 1945, when the atomic bomb would clearly not be used against Germany, but against Japan, he sprang into action. He was not supposed to know anything, but he knew enough to feel compelled to write another letter to Roosevelt on March 25. "I am writing to introduce Dr. L. Szilard, who proposes to submit to you certain considerations and recommendations."[20] With Einstein's help, Szilard would attempt to prevent the use of the weapon in cities and where civilians would be targets. However, Roosevelt died on April 12. The letter was found unopened on his desk a day later.

His successor, Harry Truman, took an unequivocally anti-Soviet line. An appeal by James Franck and six of his colleagues fell on deaf ears. In June 1945 they wrote a report to the secretary of war urging that the American bomb be "first revealed to the world by a demonstration in an appropriately selected uninhabited area."[21]

But no one was paying attention to the scientists by this point. In the largest research project of all time, the cost of which has been estimated at $2 billion, and that entailed the strictest supervision, they were able to create the most terrible weapon that ever fell into human hands, but these hands belonged to people in uniform, and they were pursuing their own goals. The first atomic bomb explosion, on July 16, 1945, did in fact occur "in an appropriately selected uninhabited area"—in the Alamogordo Desert in the American Southwest. Instead of demonstrating "success" to the world for the purpose of deterrence, however, the military conducted the so-called Trinity Test in strictest secrecy to prepare for actual use.

On August 6, 1945, at 8:16 a.m., a four-ton uranium bomb called Little Boy exploded over the city of Hiroshima. Little Boy wiped out the lives of more than 100,000 people on the spot; 75,000 more were wounded, and tens of thousands died in the aftermath. A surface with a diameter of four kilometers was totally destroyed.

At the time the Einsteins were spending their vacation at the idyllic Saranac Lake in the Adirondack Mountains. Helen Dukas was the first to hear the awful news on the radio. When she told Einstein what had happened, he let out a sigh and murmured, "Oh no." What might he have said if the bomb had exploded over Hamburg or Frankfurt?

The Americans came out strongly in favor of the new weapon. A survey revealed that even a month after Hiroshima, nearly 70 percent of the American population was convinced that dropping the atomic bomb had been "a good thing."[22] "The Jap Must Choose Between Surrender and Annihilation," ran the headline in *Newsweek*.[23] On August 9, at 11:02 a.m. local time, the four-and-a-half-ton plutonium bomb Fat Man, with nearly twice as much explosive force as Little Boy, wiped out the city of Nagasaki. Five days later, Japan capitulated. It was clearer than ever that science could determine the outcome of wars and change the course of history. But at what price?

That the first atomic strikes shortened the war, and thus saved the lives of American soldiers, is undeniable. That their number was disproportionate to the number of civilians who were killed is equally indisputable. The political rationale behind the nuclear insanity from the perspective of the United States, especially today, lay in its self-appointed destiny: Little Boy and Fat Man symbolize the beginning of America's inexorable rise to become a—and now *the*—world power.

At the end of the Third Reich, one of the most conflict-ridden power triangles of history collapsed. It had held together the Soviet Union, the United States, and the German Empire (and each of their allies) in shifting alliances for a while, like card sharks of international politics. At first, America and Germany regarded Communist Russia as their enemy, but then the Hitler-Stalin pact suddenly made the United States and the rest of the world the common enemy of Germany and Russia. The pact ended with the German attack on the Soviet Union, which in turn led to a cooperative effort between Russia and America against the Third Reich. Right after the two great powers defeated their common adversary as comrades-in-arms, each became the enemy of the other as a result of their deep-seated differences—with a portion of the defeated enemy territory as a buffer between them.

Calculations about power politics were not Einstein's forte. His temporary support of the war was directed solely against Germany. With the defeat of Germany, which he did not regard as a liberation, but as a necessary conquest of the Germans collectively embroiled in guilt, the deployment of the atomic bomb was, as he saw it, no longer an option—for purposes of a show of power vis-à-vis the new enemy. He had trouble acknowledging that the bond between the two adversarial partners, the Soviet Union and the United States, came unglued in what seemed like the blink of an eye once their common goal had been attained.

At a dinner to honor Alfred Nobel in December 1945, Einstein warned that "the war is won but the peace is not."[24] Oddly, he never explicitly criticized the dropping of the bombs on Japanese cities. In view of the millions upon millions of dead in World War II, he said that the atomic bomb had "affected us quantitatively, not qualitatively"—a misconception, as the second half of the twentieth century demonstrated.[25] The new quality of this destructive force had raised the level of inhibition quite significantly, and an atomic war has yet to occur. Einstein hoped the Allies would be able to remain in an alliance. But what does that really mean? Idealism may appear naïve, but that is what makes it refreshing and liberating in the face of ideology and practical constraints.

In an interview with the *Sunday Express* in London eight days after Nagasaki, Einstein expressed the reasonably realistic view that the atomic bombing was "probably carried out to end the Pacific War before Russia could participate"—that is, to keep the Soviet power bloc away from this part of the world.[26] Still, he was "sure President Roosevelt would have forbidden the atom bombing of Hiroshima had he been alive."[27] There is no way of telling whether he was right on this last point, and whether his third letter to the White House would have remained as ineffectual as the first two if Roosevelt had read it.

Nonetheless, Einstein believed that what he had written had had some impact. Although the FBI had barred him from any involvement in the Manhattan Project, the shadow of partial guilt would hang over him for the rest of his life. "Had I known that the Germans would not succeed in producing an atomic bomb, I would not have lifted a finger," he said after the war.[28] A few months before his death, he admitted to the Nobel Prize–winning chemist Linus Pauling, "I think I have made one mistake in my life, to have signed that letter."[29]

By any strict moral standard, Einstein was far less guilty than he considered himself. In 1939 and 1940, when he wrote his letters to Roosevelt, he did not have nearly enough factual information on hand to call for an American atomic program. Had he not written these letters, history would not have turned out differently, but his recognition of this subjunctive-cum-double-negative did little to resolve his dilemma. Nonetheless, he repeatedly claimed that he was not really to blame. In early September 1945, he wrote to his son, "Dear Albert! My scientific work has no more than a very indirect connection to the atomic bomb."[30] In November he told *The Atlantic Monthly*, "I do not consider myself the father of the release of atomic energy."[31]

These protestations did nothing to blunt the force of public opinion. On July 1, 1946, Einstein's distinctive head graced the cover of *Time* magazine, along with a mushroom cloud and the formula $E=mc^2$. The caption reads, "Cosmoclast Einstein. All matter is speed and flame." The cover story stated, "Through the incomparable blast and flame that will follow, there will be dimly discernible, to those who are interested in cause and effect in history, the features of a shy, almost saintly, childlike little man with the soft brown eyes, the drooping facial lines of a world-weary hound, and hair like an aurora borealis."[32]

Einstein parried attacks of this kind with unassuming equanimity. Either he paid no attention or he simply set the record straight, as he did on March 10, 1947, when *Newsweek* also ran a cover story: "Einstein, The Man Who Started It All." However, his conscience went far beyond the horizon of his actual involvement. Although he had been excluded from the Manhattan Project, he sided with those who did participate. The researchers had handed over their creation, but not their responsibility. He agreed to serve as chairman of the Emergency Committee of Atomic Scientists, as the only member who had not worked on building the bomb. He considered the problem of physics a collective burden, like original sin, with which all people who believe in the Bible have to live. "We have helped to create this weapon to prevent the enemies of the human race from doing so before us," he said.[33] Now the fire needed to be contained.

Einstein's conscience told him that since the only justifiable war—the one against the Nazis—had come to an end, he was now more obliged than ever to voice his advocacy of world peace. With his return to strict pacifism, his criticism of America awoke to new life. There were also dramatic shifts in his internal power triangle after 1945. The strength of the United States, which he had welcomed against Germany, now struck him as progressively more dangerous. "Success has gone to people's heads here," he wrote in 1947 to Hans Albert, who was living in America by that time.[34] " 'God's own country' is getting cockier and more power-crazed by the moment."[35] Antonina Vallentin reported that Einstein "feared . . . the growth of nationalism in the U.S.A. more than in the U.S.S.R. because he observes among us a kind of mob hysteria unbecoming to a nation otherwise so great."[36]

He strongly opposed the militarization of the United States, which reminded him of the "military religion" in the German Empire. When he accepted the One World Award at a ceremony in Carnegie Hall in New York in the spring of 1948, he again shrewdly predicted the coming dismal years that have since entered the annals of American history as the McCarthy years. "The proposed militarization of the nation not only immediately threatens us with war; it will also slowly but surely undermine the democratic spirit and the dignity of the individual in our land."[37]

Disenchanted with the United Nations—which was founded in June 1945 in San Francisco—he insisted, "The secret of the bomb should be committed to a world government."[38] His lifelong theme of unification now focused on the political arena. However, communism and capitalism proved just as resistant to harmonization as quantum mechanics and the general theory of relativity. He criticized both Moscow and Washington for resisting the idea of a jointly governed world. The nationalism of earlier eras was, he emphasized, no longer in keeping with the times.

In 1945 he had warned, "As long as there are sovereign states with their separate armaments and armament secrets, new world wars cannot be avoided."[39] With unerring instinct he predicted that the United States would soon lose its monopoly of the atomic bomb. In his capacity as chairman of the Emergency Committee of Atomic Scientists, he called for cooperation rather than confrontation. "We delivered this weapon into the hands of the American and the British people as trustees of the whole of mankind, as fighters for peace and liberty."[40]

These were all reasonable ideas, but to many people in the postwar era they seemed rather naïve. And dangerous. The FBI began keeping a close watch on the Emergency Committee just after it was founded in 1946. With Einstein at its head, Hoover's staff reasoned, it was virtually inevitable for official secrets to be divulged. The FBI did not go so far as to accuse him of spying, but his non-anti-Soviet attitude was enough to classify him as anti-American and to place him under continuous surveillance.

Although the FBI veiled its activities in secrecy, Einstein had no illusions about his status in America. At a dinner in July 1948, he told the Polish ambassador to the United States, "I suppose you must realize by now that the U.S. is no longer a free country, that undoubtedly our conversation is being recorded. This room is wired, and my house is closely watched."[41]

His anxiety even took a heavy toll on his physical well-being, and he showed marked signs of aging. "The years had weighed heavily on Einstein. He came to meet me with a step in which I did not find the usual elasticity and silent movement," Elsa's friend Antonina Vallentin reported after visiting him in the spring of 1948 in Princeton. "The most moving change, however, was in his eyes. The burning glance seemed to have singed the skin underneath, the mauve and brown shadows descended on the sallow cheeks, forming a strange pattern crossed by the bluish line of the lips. . . . The eyeballs had become hidden in the deep sockets, but the glance, the inextinguishable black fire, was just as glowing. The face, paler, might be consumed from within. . . . When he spoke of some particularly disappointing conversation or absurd incident he burst out laughing, as he used to do when something incongruous tickled his sense of humor. But his laughter was harsh and abrupt. It did not come from the heart any longer—it was a laugh that did not extend farther than the lips."[42]

The case of Einstein was handled innocuously enough at first, and was just one of many in the swamp of files of the Hoover bureaucracy. After 1949 it became an out-and-out witch hunt, although the surveillance continued to be carried out covertly. The stated goal of Hoover's henchmen was Einstein's expatriation and deportation as an undesirable alien. Although Einstein was surely not aware of the extent and details of the harassment directed against him, he was unpleasantly reminded of the situation twenty years earlier in Germany. It was like experiencing déjà vu. Einstein had now shed any illusions about a freedom-loving America, and the last five years of his life may well have been the saddest chapter of his life.

In March 1946, Winston Churchill began to use the term "Iron Curtain." The Western European countries were given American support to stem Soviet expansionism, especially by means of money from the Marshall Plan. Moscow's response was quick in coming: In February 1948 there was a communist coup in Czechoslovakia, and in June 1948 a blockade of West Berlin as a response to the Western Allies' currency reforms. In October of that year, the German Democratic Republic was founded, and in the same year, communists proclaimed the People's Republic of China. By mid-1950 the United States was involved in another war, this time to support South Korea against the communist North.

The most significant incident to step up America's fear of communism to a state of sheer hysteria took place in September 1949. The Soviet Union tested its first atomic bomb, thus ending the American monopoly on nuclear weapons. To make matters worse, the Soviet "success" was evidently made possible by espionage. On January 13, 1950, the German physicist Klaus Fuchs admitted to the British security agency that he had been passing along top-secret information to Moscow for years while enjoying the trust of the British and Americans. Now virtually everyone was suspect—particularly every scientist. Stories about "Red spies" dominated the headlines even in reputable newspapers.

A series of events now unfolded in rapid succession. On January 31, 1950, President Truman announced a program to push through the development of the hydrogen bomb, an incomparably more destructive atomic weapon than the bombs dropped on Japan. Edward Teller, an exile from Hungary, would direct the project. The arms race was escalating. At the same time, a dark chapter in American political history was commencing. On February 9, the Republican Senator Joseph McCarthy of Wisconsin gave the first of his infamous anticommunist speeches. That same year he became the chairman of a Senate committee to investigate "un-American activities." The McCarthy era had begun.

Suspicion, slander, and false convictions poisoned the social climate in the United States for half a decade. Arrests, ruined careers, and suicides were the order of the day. An estimated 10,000 people lost their jobs solely because they did not cooperate with the communist-hunters.

Einstein was predictably one of the first to speak up shortly after McCarthy's address: on February 12, 1950, he warned about the consequences of the hydrogen bomb in a nationally televised broadcast. For the premiere of NBC's *Today with Mrs. Roosevelt*, millions heard and saw him explain that the fate of mankind as a whole was hanging in the balance. The next day's headline in the New York *Daily News* read, "Disarm or Die Says Einstein." Even the widow of the former president, hostess of this early political talk show, now appeared in a suspicious light.

The very next day, FBI Director J. Edgar Hoover requested a detailed dossier about this man who had fallen from grace. Hoover must have exulted when he read that Einstein might have entered the United States illegally, since he had been a member of several subversive organizations, including International Workers' Aid. Moreover, the physicist had allegedly functioned as a personal courier for the central committee of the Communist Party in 1930, transmitting letters, telegrams, and telephone messages. And he was on record as stating, in December 1947, "I made a mistake in selecting America as a land of freedom, a mistake I cannot repair in the balance of my life."[43]

The first comprehensive FBI report on Einstein was completed on March 13, 1950. It contained a long list of his alleged communist ties and activities during his years in Germany—including a great deal of information that Einstein would have been happy to volunteer at any time. He had never made a secret of having lent his name to all kinds of organizations, including communist ones, in their battle against war and injustice.

Einstein had already made himself suspect in the eyes of the FBI by voicing his outrage about the lynching murders of blacks in the United States, the number of which soared after the war ended. He spoke out publicly against racism. He protested the arrest of American communists. He interceded on behalf of criminals who had been sentenced to death. In 1948 he supported the independent left-wing candidate Henry Wallace in his bid for the White House. To top it off, he called for a world government (including Russians) to guard nuclear secrets. Hoover hoped that since the Soviets were after the hydrogen bomb and the American fear of enemy agents had reached an all-time high, Einstein would be found guilty of espionage.

The FBI had been collecting mail pertaining to Einstein since 1934—even suspicions raised by false witnesses and certifiably insane individuals, who seemed to be drawn to Einstein as to a magnet, as he himself once put it. Some claims were quite weird, such as one report by a woman named Lucy Apostolina from Jersey City, who contended that Einstein had invented an electric robot capable of reading and manipulating the human brain. An anonymous informant claimed that Einstein was experimenting with rays to destroy aircraft and tanks.

"I've become disreputable," Einstein told Besso. He complained that the Americans "are well on the way to outdo even the Germans' militaristic attitude," adding that he had never felt so alienated from his fellow men, and there was nothing he could identify with.[44] Everywhere he looked, he found nothing but brutality and lies. He had never been more isolated. "Most of our dear friends," he pointed out to Max Born in May 1952, "are already gone."[45]

Einstein was also experiencing the isolation of the cosmopolitan who no longer has a place to call home. The only spot that had felt like home to him, namely the summerhouse in Caputh, had been his for a mere three summers. If there were such a thing as a passport for "citizens of the world," Einstein would have been one of the first in line to get one. Now that he had shed his illusions about America and was even detecting fascist tendencies, there was no country left in which he could see his ideals realized—not even in Switzerland, which he had once loved so much and with which he maintained his ties as a dual national to the end of his life.

"I like this country as much as it dislikes me," he said of Switzerland three years before his death.[46] His vexation was understandable. After his departure from Germany in 1933, he had asked Swiss authorities to protect the property that had been confiscated by the Nazis, which legally belonged to him as a dual citizen, but the government in Bern was indifferent to his plight. The Swiss reasoned that since "Professor Einstein traveled with a German diplomatic passport," it was "obvious that the conditions have been met to withhold Swiss diplomatic protection from Einstein."[47]

The argument ran that since he had been celebrated as a German, he should channel his request to the Germans. "Moreover," the Swiss officials reasoned, "the measures against him were enacted by the German authorities specifically to deal with his citizenship and the alleged violations of his duties as a citizen." Switzerland was essentially claiming that Einstein had turned his back on the German government with his "atrocity-mongering."

Einstein refused to accept this argument, and pointed out "that I cannot be equated with people whom the Swiss give the telling designation 'Swiss on paper only.'" In any case, the Swiss stood idly by while the Gestapo seized his assets in the amount of 58,000 marks. He complained to Hans Albert that "Switzerland didn't lift a finger to prevent the confiscation of my savings in Germany,"[48] and told Mileva in Zurich, "The Swiss did not make the slightest effort to help me."[49]

Isolated and disillusioned, now nearly seventy-four years old, he once again decided to go on the offensive when Julius and Ethel Rosenberg were facing execution. The couple had been arrested at the height of the Cold War spy hysteria in 1950, accused of passing on atomic secrets to Moscow, and in April 1951 sentenced to death. The indictment was based in large part on a statement by Ethel's brother, who sold out his sister and brother-in-law to save his own skin. In late 2001 he finally admitted that he had lied under oath.

As the day of the execution approached, strong doubts about whether this sentence was justified resulted in massive demonstrations around the globe. Hundreds of thousands of people marched through the streets of Paris, London, Rome, Moscow, Prague, Warsaw, New

York, Washington, and Toronto. Einstein initially tried to make headway with a personal appeal, and wrote the outgoing President Truman
a letter. When he saw that this approach was getting him nowhere, he
joined the fight to pardon the Rosenbergs, which was supported by
many prominent people including Pablo Picasso, Jean-Paul Sartre, and
even Pope Pius XII. Notwithstanding these protests, the Rosenbergs'
lives came to an end in the electric chair.

Einstein's plea on behalf of the Rosenbergs gave Hoover and Mc-
Carthy new grist for their persecutory mill. The file swelled, and grew
still bulkier when Einstein courageously took on the self-righteous
sleuths—and exposed the harm they were causing their victims. Comparing the situation in America with that of Germany shortly before
Hitler's assumption of power, he called upon the American people to
break the law.

"I can only see the revolutionary way of noncooperation in the sense
of Gandhi's," he wrote on May 16, 1953, in an open letter to a persecuted teacher who had asked him for help. "Every intellectual who is
called before the committees ought to refuse to testify, i.e., must be prepared for jail and economic ruin, in short, for the sacrifice of his personal welfare in the interest of the cultural welfare of this country. . . . If
enough people are ready to take this grave step, they will be successful.
If not, then the intellectuals deserve nothing better than the slavery
which is intended for them."[50]

These lines alone would have sufficed to summon him before the
committee, but McCarthy, like Hoover, shied away from taking this ultimate step. Although he called Einstein "an enemy of America,"[51] he
must have sensed that the physicist would turn the hearings inside out
and prosecute the prosecutors.

In the same year, 1953, FBI Director Hoover had new reason to
hope that this time he would be able to nail Einstein. On September 4,
a well-dressed man in his sixties walked into the FBI office in Miami to
make a statement about Einstein's communist past. He identified himself as Paul Weyland—the very same right-wing demagogue who, along
with Ernst Gehrcke and the Study Group of German Researchers for
the Preservation of Pure Science, had launched a public attack against
Einstein and his theory of relativity in the Berlin Philharmonic in August 1920 before dropping out of the picture.

Now he stated for the FBI record that Einstein had declared himself a communist in a 1920 newspaper article. It is unclear why Weyland was making this statement at this time, apart from his old hatred of Einstein. Possibly he was hoping to speed up the citizenship process for himself and his wife. Hoover did not know that he was dealing with a shady character whom even the Nazis had incarcerated. After the war, Weyland worked for the Americans in Berlin, first as a translator, then in the documentation center, where he had access to all National Socialist Party files. He blackmailed former party members with the information he gathered there, until he was caught. Weyland was able to clear out in the nick of time—and move to the United States.

Hoover's hopes were soon dashed. The alleged newspaper article never surfaced, and the matter fell apart like nearly everything else in Weyland's life. By contrast, Einstein's call for civil disobedience caused a sensation around the world, and the ensuing publicity surrounding McCarthyism stirred up anti-American resentment everywhere. Even in the United States, he received considerable support—but also a great deal of opposition, in pronouncements like this one: "You should be shipped back to your homeland and a camp!"[52] Einstein said he wished he were a plumber, since he would then be able to say what was on his mind without all these repercussions, whereupon the Plumbers Association offered him an honorary membership.

The general tenor of the criticism directed at him in the last years of his life focused on his alleged ingratitude. A chorus of voices was eager to remind him that he had "received far more from the United States than he has given."[53] The most influential newspapers came out against Einstein. The "forces of civil disobedience," wrote *The New York Times* on June 13, 1953, were not only illegal, but also "unnatural."

Even so, Einstein had accomplished a heroic deed once again at his advanced age with his open letter to the persecuted teacher. His ingenious hyperbolic rhetoric, which went beyond criticism to proposing illegal resistance, allowed him to achieve the great resonance he was after, and helped hasten the end of the McCarthy era, which survived him by only a few months.

This matter could have had a very different outcome for Einstein. In the opinion of his vilifiers, he was a "leftist" who worshipped Lenin and had belonged to numerous communist-affiliated organizations. In the hysteria of the McCarthy era, this would have been grounds for an in-dictment—if J. Edgar Hoover had had access to all the information we have today.

In 1945 the FBI began gathering documents about Einstein's secre-tary, Helen Dukas, as well. A few weeks before his death, two officials paid a visit to his closest collaborator. Although their questioning led nowhere, her pursuers were essentially on the right track. The "com-mies" under siege by McCarthy really had been conducting business in Dukas's apartment. From spring 1931 to mid-1933, the Communist Party used a sublet to run an illegal office in which espionage reports were written and possibly news was encoded and decoded. It is impos-sible to tell how much Helen Dukas knew about these wheelings and dealings. Still, it appears unlikely that she was completely in the dark.

However, Einstein's perilous proximity to secret service activities did not stop there. According to research conducted by Siegfried Grund-mann, a historian in Berlin, it is even possible that Einstein's own apart-ment served as a communication center for Soviet spies. A Moscow agent had free access to Einstein's apartment on Haberlandstrasse: his son-in-law Dimitri Marianoff. Especially during Einstein's extended absences— when he spent summers in Caputh or winters in the United States—Mar-ianoff and his wife, Margot, had the city apartment all to themselves; sometimes he spent extended periods of time there by himself. It is not known what he was doing there, Grundmann writes, but "everything we do know suggests that Einstein's apartment was used for espionage."[54]

If Hoover and McCarthy had also known about Margarita Ko-nenkova, Einstein would most likely have wound up in their clutches. He met the forty-one-year-old Russian woman in 1935, when her hus-band, the sculptor Sergei Konenkov, was making a bust of him. The two fell in love and wrote each other passionate letters (which turned up in 1998). Their affair lasted until 1945, when the Russian couple re-turned to Moscow. Einstein got the brilliant idea of calling everything the lovers shared "Almar"—a combination of the beginning syllables of Albert and Margarita.

"Everything here reminds me of you," he once wrote to her. "Al-mar's blanket, the dictionaries, the wonderful pipe which we thought was lost, and all the other little things in my hermit's cell; and my lonely nest."[55] The physicist was most likely unaware that Konenkova probably worked for the Soviet secret police. It is uncertain whether she was actually spying on him and trying to coax military secrets out of him—which he didn't know anyway. J. Edgar Hoover, however, turned out to be right in a way for having refused Einstein clearance to work on the Manhattan Project.

The Russian blonde's beauty was not her only captivating characteristic. For Einstein, she also embodied the old Europe, for which he longed in vain until the end of his days in the New World. According to his Princeton friend Gillet Griffin, Einstein also saw in Johanna Fantova, with whom he had a relationship from the late 1940s, a "part of the old world." "She read aloud to him from Goethe's works. She was a link to the things he was missing. . . ."[56]

"Hanne," as he called her, was twenty-two years younger than he. She had emigrated from Prague to the United States in 1939 at his insistence. He had known her husband's family since living in Prague in 1911. Her husband's mother, Bertha Fanta, had organized the salon in which Einstein had met Max Brod and Franz Kafka. Now, in his advanced age, he was carrying on an affair with the daughter-in-law of his former hosts. In his usual way, he wrote her letters and poems, which he occasionally signed as "Your Elephant." He took her sailing (dressing in the sloppy clothes he favored, while she wore an elegant dress) and to concerts. They spoke on the telephone several times a week. She visited him regularly at his home, and was even allowed to clip his hair. He also dedicated several poems to her, including this one:

> Dear Hanne!
> Exhausted from a silence long
> This is to show you clear how strong
> The thoughts of you will always sit
> Up in my little brain's attic
> Holy Saturday is close
> Anxiously looks for his rose
> The elephant—and apropos
> Nothing better did he know.
> > Happy reunion with sausages and a
> > (smacking) kiss.
> > Your A.E.[57]

Fantova, who died in 1981, recorded her experiences with the celebrated senior citizen in a diary. In February 2004 the sixty-two typed
pages were accidentally discovered in her personal file in the Princeton
University Library, and caused a minor sensation. This diary is the only
detailed record of the final phase of Einstein's life by someone who was
so close to him, and reflects the verve with which the self-appointed
"old revolutionary" was combative right to the end. "Politically I'm still
a fire-spewing Vesuvius."[58]

On the subject of McCarthy's supporters, he told her, "The rule of
the dimwitted is unparalleled because there are so many of them and
their votes count just as much as ours." That is the dilemma of every
democracy: to choose between demagogy and populism. "God's own
country keeps getting stranger," he wrote to Hans Albert in December 1954. "Everything, even idiocies, are produced en masse."[59] He remarked to Fantova, "Politically things are looking quite ugly here." One
of the most prominent victims of the "witch hunt" was his close friend
in Princeton J. Robert Oppenheimer, director of the Manhattan Project and now director of the Institute for Advanced Study. Oppenheimer lost his access to the Atomic Energy Commission in 1954
because of his—admitted—sympathy for left-wing groups.

According to Fantova, Einstein was upset that his friend had "not
dropped out long ago" instead of taking the matter as a personal defeat.
"Oppenheimer is not a gypsy like me; I was born with the skin of an
elephant," Einstein said. "There is no one who can hurt me; it washes
over me like water off a crocodile." His emotional shield, which he
used to keep the "merely personal" at a distance, functioned well into
his old age.

He once wrote to Hans Albert, "I have experienced jealousy in all its
permutations. It is no different now, but I no longer depend on those
who hate me."[60] He did admit, however, that he also enjoyed a certain
freedom to play the fool: "Sticking out my tongue reflects my political
views," he said to Fantova. A picture of him sticking out his tongue,
which became his ultimate trademark and a pop motif for posters, but-
tons, and T-shirts, was taken on his seventy-second birthday. The complete photograph shows him between Frank Aydelotte, the ex-director
of the Institute for Advanced Study, and Aydelotte's wife in the rear seat
of an automobile. But only the cropped photograph displaying his head

became an icon. Einstein himself was the disseminator of this iconic image: he had many copies made of this photograph and sent it to friends, acquaintances, and colleagues. But by this point things were not washing over him quite as easily as his well-known picture suggests. When his institute colleagues could not agree on a joint statement on behalf of their director Oppenheimer, he turned away in disgust. "It was abominable; people are a bad invention."

He was also quite outspoken about the issue of German rearmament: "Instead of coming to an agreement with Russia, the United States is helping Germany to rearm." He added that since "all the terrible things the Germans did . . . have been forgotten . . . a great danger of war" was looming once again. When Werner Heisenberg visited Princeton in October 1954, Einstein remarked that the inventor of the uncertainty principle was a "big Nazi" and "a great physicist but not a pleasant person."[61] Even today, opinions diverge about Heisenberg's affiliation with the Nazi regime. In any case, Heisenberg was not an adversary of the brownshirts.

Einstein was finished with the Germans, for good. He curtly refused any overtures to reestablish old ties. He turned down an offer to become a Foreign Scientific Member of the new Max Planck Society ("simply out of a need for cleanliness"),[62] and had no desire to return to any other academic affiliations. As he wrote to Arnold Sommerfeld in Munich, "I do not wish to have anything more to do with Germans, not even with a relatively harmless Academy."[63] He told West German president Theodor Heuss that it was "evident that a proud Jew no longer wishes to be connected with any kind of German official event or institution."[64] Only to German youth was he willing to make concessions, and he agreed to have schools named after him. Graduates of the Albert Einstein Gymnasium in Berlin-Neukölln are given a copy of Einstein's letter together with their diploma even today.

His last girlfriend recorded the decline of Einstein's health. He frequently complained about the pain he was experiencing, especially in his liver. "I ate only a very small amount of butter because my skin is getting so dry, and I had to suffer for it; I'm still feeling pretty bad." When Fantova asked him why he no longer played the violin, he replied, "The piano is far better suited to improvisation and playing solo; after all, I play the piano every day. It would also be too much of a physical strain for me to play the violin." When the world-famous Juilliard String Quartet came to visit him, however, he took Lina out of her case. The musicians had to slow down so that he could play along, but reliable sources reported that they left the house with tears in their eyes.

During his final months, in the long, cold Princeton winter, he sought solace in his formulas. "Wrestling with these problems makes one independent from the human sphere, and this is an invaluable blessing," he confided to the object of his affections.[65] Since he had conceded in his final letter to Besso that quite possibly "nothing remains" of his "castle in the air," his reflections on the nature of the world started to seem like the reflexes of a hermit who is still writing down his final formulas on his deathbed.

"I am a heretic in physics," he said. "It will take a long time, and I will be dead for quite some time before my current work is appreciated." Although he had been exceptionally successful in forging links in the past, he now found that he was in limbo. "The physicists say that I am a mathematician, and the mathematicians say that I am a physicist," Fantova quoted him as saying. "I am a completely isolated man and though everyone knows me, there are very few people who really know me."[66] Einstein mastered the art of calculating, thinking, and protecting himself from dying alive—right to the end.

"The devil counts out the years conscientiously, we must admit," he wrote to Maurice Solovine shortly before his death.[67] "To the immortal Olympia Academy," he addressed a spirited obituary in 1953: "In your short active existence you took a childish delight in all that was clear and reasonable. . . . We three members, all of us, at least remained steadfast. Though somewhat decrepit, we still follow the solitary path of our life by your pure and inspiring light; for you did not grow old and shapeless along with your members, like a plant that goes to seed. To you we swear fidelity and devotion until our last learned breath!"[68]

In the early 1950s Michele Besso adopted a tone of hero worship. When he addressed Einstein as a "dear, brilliant [*geistesgewaltig*] friend," Einstein responded with a Yiddish pun on *gewalt*, "I may be an old friend, but I can only say 'nebbich' to being addressed with the word 'gewalt,' if you know this expressive word of our fathers [*oy gevalt*], which is a blend of sympathy and disparagement."[69] Simply enjoying an old friendship was most difficult for Einstein to achieve.

When Besso died on March 15, 1955, in Geneva, Einstein lost his mainstay in life—a sincere companion who also understood his physics. "Now he has preceded me by a little bit in his departure from this strange world as well," Einstein wrote to Besso's son and sister. "This means nothing. For those of us who believe in physics, the distinction between past, present, and future is only an illusion, however tenacious this illusion may be."[70]

He had drafted his last will in March 1950. One week before his death, he placed his signature under his political testament, in which he and the philosopher Bertrand Russell used strong language to warn the governments and peoples of the world about the mega-catastrophe that would result from an atomic war. This document asserted that "a war with H-bombs might quite possibly put an end to the human race." The use of many hydrogen bombs, Einstein and Russell wrote, could bring death to the entire planet—"sudden only for a minority, but for the majority a slow torture of disease and disintegration."[71]

The Russell-Einstein Manifesto laid the foundation for the most important peace initiative of the past fifty years: the Pugwash Movement, named after the village in which it was founded. The tireless efforts of this group contributed greatly to the fact that the Cold War stayed cold. For his forty years of service as the leading figure in the Pugwash Movement, the physicist Joseph Rotblat, one of the few to leave the Manhattan Project (in December 1944), was awarded the Nobel Peace Prize in 1995. He may have been the noblest of Einstein's noble heirs.

Einstein said that he wished to die "with grace." "I want to go when I want," he supposedly said. "It is tasteless to prolong life artificially."[72] He felt certain that "one can die without the help of a doctor." On the evening before his death, when his friend Gustav Bucky was leaving him, Einstein asked him why he was going. "You should sleep," the doctor replied. "Your presence would not stop me in the least," the invalid retorted as mischievously as ever.

One of the last people to see him alive was his stepdaughter Margot. As chance would have it, she was a patient in the Princeton Hospital when he was brought in on April 15, 1955, and occupied the room next to his. "I did not recognize him at first—he was so changed by the pain and blood deficiency," she described the encounter. "But his personality was the same as ever. He . . . joked with me and was completely in command of himself with regard to his condition; he spoke with profound serenity—even with a touch of humor—about the doctors, and awaited his end as an imminent natural phenomenon. As fearless as he had been all his life, so he faced death humbly and quietly. He left the world without sentimentality or regrets."[73] He supposedly said to her, "I have accomplished what I came here to do."[74]

During the night, shortly before 1:00 a.m. on April 18, 1955, Albert Einstein departed this world. He said something in his mother tongue, but the night nurse did not understand German. The final words of one of the greatest men in the history of the world were incomprehensible.

A few hours later, a young American pathologist named Thomas Harvey came on duty, performed an autopsy, and cut Einstein's brain out of his skull. Einstein's life after death had begun.

Dear Posterity,

If you have not become more just, more peaceful, and generally more rational than we are (or were)—why, then, the Devil take you. Having, with all respect, given utterance to this pious wish,

I am (or was)

Yours,
Albert Einstein (1936)[75]

NOTES

1. HIS SECOND BIRTH: THE FATEFUL YEAR 1919

1. Calaprice, 237.
2. *The Times* (London), November 7, 1919.
3. Letter to Pauline Einstein, September 27, 1919; CP 9, 98.
4. Whitehead, 10–11.
5. *The New York Times*, November 10, 1919; quoted in Pais, *Einstein Lived Here*, 147.
6. *Der Abend*, November 3, 1919.
7. Letter to Pauline Einstein, November 5, 1919; CP 9, 82.
8. Letter to Hans Albert and Eduard Einstein, March 26, 1920; CP 9, 300.
9. *Berliner Morgenpost*, November 7, 1919.
10. *Berliner Illustrirte Zeitung*, November 2, 1919.
11. Letter from Sir Arthur Eddington, December 1, 1919; CP 9, 158.
12. Letter from Paul Ehrenfest, November 24, 1919; CP 9, 147.
13. Letter to Paul Ehrenfest, December 4, 1919; CP 9, 161.
14. *Berliner Illustrirte Zeitung*, December 14, 1919; quoted in Rosenkranz, *Albert Through the Looking Glass*, 132.
15. Moszkowski, 12–13.
16. Quoted in Calaprice, 41. (The English original of this book did not select this passage for inclusion; it was added to the German translation.)
17. Letter to Marcel Grossmann, September 12, 1920; CP 10, 271.
18. Letter to Max Planck, October 23, 1919; CP 9, 128.
19. Letter to Heinrich Zangger, early 1920; CP 9, 204.
20. 1934; Einstein Archives 31–160. Unpublished translation by Aaron Wiener.
21. Letter to Pauline Einstein et al., May 14, 1919; CP 9, 35.

2. HOW ALBERT BECAME EINSTEIN: THE PSYCHOLOGICAL MAKEUP OF A GENIUS

1. CP 1, 3.
2. CP 1, xviii.

3. Letter to C. Erlanger, March 16, 1929; Archives of the City of Ulm; quoted in Fölsing, 8.
4. March 16, 1929; Archives of the City of Ulm; quoted in Fölsing, 8.
5. Seelig, *Albert Einstein: Leben und Werk*, 399.
6. Quoted in Fölsing, 11.
7. Holton/Elkana, 157.
8. CP 1, xviii.
9. CP 1, xviii.
10. Schlipp, 9.
11. Schilpp, 9.
12. Reiser, 27.
13. Letter to Heinrich Zangger, June 6, 1919; CP 9, 94 (German edition only).
14. Hoffmann/Dukas, *Albert Einstein: Creator and Rebel*, 4.
15. Vallentin, 43.
16. Vallentin, 44.
17. Vallentin, 53.
18. Vallentin, 44.
19. Gardner, 25.
20. Gardner, 25.
21. Gardner, 26.
22. Moszkowski, 224.
23. Moszkowski, 225.
24. Gardner, 7–8.
25. Gardner, 31.
26. Gardner, 32.
27. Seelig, *Helle Zeit*, 72.
28. Letter to Marcel Grossmann, April 14, 1901; CP 1, 165.
29. Letter from Hermann Einstein to Jost Winteler, December 30, 1895; CP 1, 11.
30. Draft of a letter to Philipp Frank, 1940; quoted in Fölsing, 27.
31. Draft of a letter to Philipp Frank, 1940; quoted in Fölsing, 27.
32. Letter to Jost Winteler, July 6, 1901; CP 1, 177.
33. Gardner, 119.
34. Seelig, *Albert Einstein: A Documentary Biography*, 70–71.
35. Gardner, 128.
36. Letter from Max Born, October 13, 1920; CP 10, 291.
37. Letter from Michele Besso, September 18, 1932; Einstein/Besso, 286.
38. Probably 1932; Einstein Archives 31–101. Unpublished translation by Aaron Wiener.
39. Letter from Elsa Einstein, before May 20, 1920; CP 10, 165.
40. Vallentin, 85.
41. Vallentin, 85.
42. Katia Mann, 132.
43. Kessler, 155.
44. Letter to Heinrich Zangger, May 17, 1915; CP 8, 98.
45. Brian, 292.
46. Brian, 403.
47. Vallentin, 36.

48. Schilpp, 3.
49. Schilpp, 9.
50. Schilpp, 9.
51. Schilpp, 15.
52. Letter to Caesar Koch, Summer 1895; CP 1, 6.
53. Seelig, *Albert Einstein: A Documentary Biography*, 13.
54. Seelig, *Albert Einstein: A Documentary Biography*, 14.
55. David Reichinstein, 40.
56. Quoted in Herneck, *Einstein privat*, 89.
57. Letter to Pauline Winteler, June 7, 1897; CP 1, 33.
58. Letter to Mileva Marić, December 12, 1901; CP 1, 186.
59. Hoffmann/Dukas, *Albert Einstein: Creator and Rebel*, 254.
60. Holton/Elkana, 159.
61. Seelig, *Albert Einstein: A Documentary Biography*, 42.
62. *Ideas and Opinions*, 225.
63. *Ideas and Opinions*, 225.
64. Letter to Pauline Winteler, June 7, 1897; CP 1, 33.
65. *Out of My Later Years*, 5.
66. Letter to N. M. Butler, April 11, 1923; quoted in Fölsing, 475.
67. Letter to Paul Ehrenfest, October 24, 1916; CP 8, 256.
68. Infeld, *Leben mit Einstein*, 78.
69. Quoted in Brian, 230.
70. Quoted in Seelig, *Albert Einstein: A Documentary Biography*, 15.
71. Brian, 146.
72. Letter to Heinrich Zangger, May 17, 1915; CP 8, 98.
73. Brod, 221.
74. Brod, 13.
75. Brod, 146.
76. Brod, 89.
77. Brod, 156.
78. Infeld, *Leben mit Einstein*, 73.
79. Quoted in Highfield/Carter, 244.
80. Quoted in Hermann, 284.
81. Ernst Peter Fischer, 43.
82. Quoted in Herneck, *Einstein privat*, 76.
83. Brod, 78.
84. Brod, 13.
85. Brod, 14.
86. Brod, 89.
87. Brod, 89.
88. Vallentin, 38.
89. Brod, 85.
90. Brod, 149.
91. Quoted in Herneck, *Einstein privat*, 169.
92. Probably 1931; Einstein Archives 36–601.
93. Letter to Eduard Einstein, probably 1936/37; Einstein Archives 75–939.
94. Letter to Hans Albert Einstein, 1949; Einstein Archives 75–810.

95. Letter to Eduard Einstein, August 20, 1932; Einstein Archives 75–688.
96. Draft of a response to an inquiry by H. Freund, January 17, 1927; quoted in Rosenkranz, *Albert Through the Looking Glass*, 21.

3. "A NEW ERA!": FROM INDUSTRIALIST'S SON TO INVENTOR

1. *Schwabinger Gemeinde Zeitung*, March 2, 1889.
2. *Schwabinger Gemeinde Zeitung*, March 2, 1889.
3. CP 1, xvi.
4. Höchtl (1934); quoted in Hettler, 88.
5. Letter to Michele Besso, December 12, 1919; CP 9, 178.
6. Seelig, *Helle Zeit*, 12.
7. Letter to Hans Wohlwend, between August 15 and October 3, 1902; CP 5, 5.
8. Letter to Johannes Stark, December 14, 1908; CP 5, 95.
9. Letter to Michele Besso, December 26, 1911; CP 5, 242.
10. Letter to Michele Besso, February 4, 1912; CP 5, 258.
11. Moszkowski, 174.
12. Letter to Rudolph Goldschmidt, November 1928; Einstein Archives 31–071. Unpublished translation by Aaron Wiener.
13. Letter from Rudolph Goldschmidt, 1928. Einstein Archives 31–071. Unpublished translation by Aaron Wiener.
14. Paul G. Ehrhardt, August 1954; quoted in Fölsing, 448 (German edition only).
15. September 1, 1943; quoted in *Der Spiegel*, 33/2003, 128.
16. Letter to Hans Albert Einstein, March 31, 1928; Einstein Archives 75–753.
17. Letter to Hans Albert Einstein, May 12, 1928; Einstein Archives 75–755.
18. *Centralblatt für Elektrotechnik*, 1886, 605.
19. Archives of the City of Munich, 1884.
20. Quoted in Hettler, 58.
21. Hettler, 184.

4. OF DWARFS AND GIANTS: A BRIEF HISTORY OF SCIENCE, ACCORDING TO EINSTEIN

1. Einstein, "The Special Theory of Relativity," part 1, section 1 ("Physical Meaning of Geometrical Propositions"); Gribbin/Gribbin, 151.
2. Aaron Bernstein, vol. 3, 1.
3. Schilpp, 15.
4. Schilpp, 5.
5. Büchner, 96.
6. Aaron Bernstein, vol. 16, 3.
7. Aaron Bernstein, vol. 16, 265.
8. Büchner, 96.
9. Büchner, 111.
10. Büchner, 504.
11. Humboldt, 10.
12. Aaron Bernstein, vol. 1, 2.

13. Aaron Bernstein, vol. 3, 8.
14. Büchner, 13.
15. Bernal, 216.
16. Humboldt, 146.
17. *Sir Isaac Newton's "Mathematical Principles of Natural Philosophy" and His "System of the World,"* translated by Andrew Motte, vol. 2 (Berkeley. University of California Press, 1962), 545.
18. Schilpp, 19.
19. Quoted in Barnett, 31–32.
20. Hoffmann/Dukas, *Albert Einstein: Creator and Rebel,* 247.
21. Einstein Archives 31–049; quoted in Hoffmann/Dukas, *Albert Einstein: Creator and Rebel,* 141.
22. Aaron Bernstein, vol. 16, 1.
23. Aaron Bernstein, vol. 16, 4.
24. Aaron Bernstein, vol. 16, 10.
25. Aaron Bernstein, vol. 8, 138.
26. Aaron Bernstein, vol. 8, 129.
27. Aaron Bernstein, vol. 17, 92.
28. Aaron Bernstein, vol. 17, 93.
29. Aaron Bernstein, vol. 17, 95.
30. Aaron Bernstein, vol. 17, 91.
31. Aaron Bernstein, vol. 19, 22.
32. Aaron Bernstein, vol. 19, 5.
33. Calaprice, 36.
34. Aaron Bernstein, vol. 5, 101.
35. Aaron Bernstein, vol. 8, 131.
36. Quoted in Eberty, viii (German edition only).
37. Eberty, 21.
38. Eberty, 21.
39. Eberty, 22.
40. Eberty, 23–24.
41. Eberty, 23.
42. Eberty, 32.
43. Eberty, 32.
44. Eberty, 51.
45. Wells, 2.
46. Eberty, 76.
47. Eberty, 38.
48. Eberty, 67.
49. Eberty, 68–69.
50. Eberty, 79.

5. THE BURDEN OF INHERITANCE: EINSTEIN DETECTIVES IN ACTION

1. Quoted in Dukas/Hoffmann, *Albert Einstein: The Human Side,* 22.
2. Quoted in Calaprice, 7.

3. Letter to Adriaan D. Fokker, July 30, 1919; CP 9, 117 (German edition only).
4. Letter to Adriaan D. Fokker, July 30, 1919; CP 9, 117 (German edition only).
5. Letter to Adriaan D. Fokker, July 30, 1919; CP 9, 117 (German edition only).
6. Letter to Adriaan D. Fokker, July 30, 1919; CP 9, 117 (German edition only).
7. Travel diary, January 7, 1931; Einstein Archives 29–134. Unpublished translation by Josef Eisinger.
8. Quoted in Highfield/Carter, 211.
9. Highfield/Carter, 211.
10. Highfield/Carter, 279.
11. Holton, *Einstein, History, and Other Passions*, 56.
12. Holton, *Einstein, History, and Other Passions*, 57.
13. Highfield/Carter, 277.
14. Quoted in Abraham, 215.
15. Quoted in Abraham, 216.
16. 1936; Einstein Archives 31–178; translation in Highfield/Carter, 93–94.

6. "ELSA OR ILSE": THE PHYSICIST AND THE WOMEN

1. Letter to Mileva Marić, August 9(?), 1900; CP 1, 145.
2. Letter to Mileva Marić, May 1901; CP 1, 171.
3. Letter to Michele Besso, July 21, 1916; CP 8, 234.
4. Letter from Elsa Einstein to Antonina Luchaire-Vallentin, June 6, 1932; Archives of the Max Planck Society, Berlin.
5. Letter to Marie Winteler, April 21, 1896; CP 1, 12–13.
6. Letter to Mileva Marić, April 10, 1901; CP 1, 164.
7. Letter to Elsa Löwenthal, October 10, 1913; CP 5, 355.
8. Letter to Mileva Marić, September 13(?), 1900; CP 1, 149.
9. Letter to Elsa Löwenthal, April 30, 1912; CP 5, 292.
10. Unattributed quotations from individuals listed on page 436 are from interviews with the author.
11. Letter from Marie Winteler, November 1896; CP 1, 30.
12. Letter to Mileva Marić, September 28(?), 1899; CP 1, 136.
13. Letter from Mileva Marić, after October 20, 1897; CP 1, 34.
14. Letter to Mileva Marić, October 3, 1900; CP 1, 152.
15. Letter to Mileva Marić, August 30 or September 6, 1900; CP 1, 148.
16. Letter to Mileva Marić, December 17, 1901; CP 1, 186.
17. Letter to Pauline Winteler, May(?) 1897; CP 1, 32–33.
18. Letter to Mileva Marić, September 28(?), 1899; CP 1, 135–36.
19. Letter to Mileva Marić, July 29(?), 1900; CP 1, 141–42.
20. Letter from Pauline Einstein to Pauline Winteler, February 20, 1902; CP 1, 193.
21. Letter to Mileva Marić, July 29(?), 1900; CP 1, 142.
22. Letter to Mileva Marić, August 30 or September 6, 1900; CP 1, 148.
23. Seelig, *Albert Einstein: Leben und Werk*, 61.
24. Letter to Mileva Marić, September 13(?) 1900; CP 1, 149.
25. Letter to Mileva Marić, August 20, 1900; CP 1, 147.
26. Letter to Mileva Marić, June 28, 1902, or later; CP 5, 4.
27. Letter to Hans Mühsam, March 4, 1953; quoted in Seelig, *Helle Zeit*, 56.

28. Letter to Mileva Marić, August 20, 1900; CP 1, 147.
29. Letter to Mileva Marić, December 28, 1901; CP 1, 190.
30. Seelig, *Albert Einstein: A Documentary Biography*, 46.
31. Letter to Mileva Marić, September 13(?), 1900; CP 1, 149.
32. Letter to Mileva Marić, June 28, 1902, or later; CP 5, 4.
33. Letter to Mileva Marić, April 30, 1901; CP 1, 167.
34. Letter to Julia Niggli, August 6(?), 1899; CP 1, 129.
35. Letter from Mileva Marić to Helene Savić, December 11, 1900; CP 1, 154.
36. New York: Riverhead Books, 1999.
37. Einstein's words recounted in letter from Hedwig Born, October 9, 1944; Einstein/Born, 153.
38. Letter to Eduard Einstein, March 27, 1928(?); Einstein Archives 75–752.
39. Michelmore, 42.
40. Michelmore, 42.
41. Letter from Mileva Marić to Helene Savić, ca. March 20, 1903.
42. Letter to Mileva Marić, February 17(?), 1902; CP 1, 193.
43. Letter from Mileva Marić to Helene Savić, 1909; quoted in Highfield/Carter, 128.
44. Poem in the album of Anna Schmid, 1899; CP 1, 128.
45. Letter to Georg Meyer, June 7, 1909; CP 5, 127.
46. Letter to Michele Besso, November 17, 1909; CP 5, 140.
47. Letter from Mileva Einstein-Marić, October 4, 1911; CP 5, 210.
48. Highfield/Carter, 144.
49. Marianoff, 54.
50. Letter to Elsa Löwenthal, April 30, 1912; CP 5, 292.
51. Letter to Elsa Löwenthal, March 23, 1913; CP 5, 331.
52. Trbuhović-Gjurić, 122.
53. Letter to Elsa Löwenthal, after November 22, 1913; CP 5, 363.
54. Letter to Elsa Löwenthal, October 16, 1913; CP 5, 357.
55. Letter to Elsa Löwenthal, November 7, 1913; CP 5, 360.
56. Letter to Heinrich Zangger, July 7, 1915; CP 8, 110.
57. See letter to Elsa Löwenthal, October 16, 1913; CP 5, 358.
58. Letter to Elsa Löwenthal, after December 21, 1913; CP 5, 371–72.
59. Letter to Elsa Löwenthal, after December 21, 1913; CP 5, 372.
60. Letter to Elsa Löwenthal, before December 2, 1913; CP 5, 365.
61. Letter to Elsa Löwenthal, mid-January, 1914; CP 5, 376.
62. Letter to Elsa Löwenthal, December 27, 1913 to January 4, 1914; CP 5, 372.
63. Memorandum to Mileva Marić, ca. July 18, 1914; CP 8, 32–33.
64. Letter to Michele Besso, February 12, 1915; CP 8, 68.
65. Letter to Michele Besso, July 21, 1916; CP 8, 234.
66. Letter to Hans Albert Einstein, February 5, 1927; Einstein Archives 75–738.
67. Vallentin, 141–42.
68. Quoted in Grüning, 457.
69. Vallentin, 89.
70. Quoted in Highland/Carter, 195.
71. Quoted in Grüning, 42.
72. Letter to Elsa Löwenthal, after December 2, 1913; CP 5, 366.
73. Letter to Elsa Löwenthal, after November 22, 1913; CP 5, 363–64.

74. Letter from János Plesch to Peter Plesch, May 18, 1955, Notes Rec. Soc. Lond. 49 (2) 303–28 (1995), 309; quoted in Highfield/Carter, 206.
75. Quoted in Highfield/Carter, 206.
76. Marianoff, 189.
77. Highfield/Carter, 207.
78. Stern, 142.
79. Quoted in Highfield/Carter, 208.
80. Quoted in Highfield/Carter, 208.
81. Grüning, 159.
82. Grüning, 158.
83. Highfield/Carter, 210.
84. Highfield/Carter, 210.
85. Quoted in Renn/Schulmann, 68.
86. Letter to Vero Besso and Mrs. Bice, March 21, 1955; Einstein/Besso, 538.
87. Quoted in Highfield/Carter, 221, in German; the authors consider this saying untranslatable.
88. Letter from Ilse Einstein to Georg Nicolai, May 22, 1918; CP 8, 566.
89. Letter from Ilse Einstein to Georg Nicolai, May 22, 1918; CP 8, 564–66.
90. Letter to Ilse and Margot Einstein, August 17, 1919; CP 9, 133 (German edition only).
91. Letter to Fritz Haber, September 1920; quoted in Fölsing, 480.

7. THE MIRACULOUS PATH TO THE MIRACULOUS YEAR: EINSTEIN'S ANGELS

1. Trbuhović-Gjurić, 95.
2. Trbuhović-Gjurić, 94.
3. Letter to Mileva Marić, March 27, 1901; CP 1, 161.
4. Stachel, *Einstein from 'B' to 'Z'*, 39.
5. Letter to Mileva Marić, April 30, 1901; CP 1, 168.
6. Letter to Mileva Marić, May(?), 1901; CP 1, 171.
7. Seelig, *Albert Einstein: Leben und Werk*, 55.
8. Letter to Maja Einstein, 1898; CP 1, 123.
9. Letter to Alfred Stern, May 3, 1901; CP 1, 169.
10. Letter to Alfred Stern, May 3, 1901; CP 1, 169.
11. Seelig, *Albert Einstein: Leben und Werk*, 10.
12. Schilpp, 17.
13. Schilpp, 15.
14. Letter to Mileva Marić, February 16, 1898; CP 1, 123.
15. Seelig, *Helle Zeit*, 10.
16. Seelig, *Helle Zeit*, 10.
17. Quoted in Helferich, 228.
18. Seelig, *Helle Zeit*, 11.
19. Letter to Mileva Marić, April 4, 1901; CP 1, 162.
20. Vallentin, 54.
21. Letter to Johannes Stark, December 7, 1907; CP 5, 46.
22. Letter from Mileva Marić to Helene Savić, December 20, 1900; CP 1, 156.

23. Letter to Mileva Marić, May(?), 1901; CP1, 173.
24. Letter to Mileva Marić, July 7(?), 1901; CP 1, 176.
25. Letter to Mileva Marić, August 14(?), 1900; CP 1, 146.
26. Letter to Mileva Marić, December 17, 1901; CP 1, 187.
27. Letter to Michele Besso, January 22(?), 1903; CP 5, 7.
28. Letter to Mileva Marić, April 4, 1901; CP 1, 163
29. Letter to Mileva Marić, December 12, 1901; CP 1, 186.
30. Letter to Mileva Marić, April 15, 1901; CP 1, 166.
31. Letter to Marcel Grossmann, April 14, 1901; CP 1, 165.
32. Letter from Hermann Einstein to Wilhelm Ostwald, April 13, 1901; CP 1, 164–65.
33. Letter to Mileva Marić, July 7(?), 1901; CP 1, 176.
34. Letter to Mileva Marić, December 12, 1901; CP 1, 185.
35. Letter to Mileva Marić, December 12, 1901; CP 1, 185–86.
36. Letter to Conrad Habicht, February 4, 1902; CP 1, 190.
37. Letter to Marcel Grossmann, April 14, 1901; CP 1, 165.
38. City of Zurich police report, July 4, 1900; CP 1, 140.
39. Letter to Mileva Marić, February 4, 1902; CP 1, 191.
40. Letter to Mileva Marić, February 4, 1902; CP 1, 191.
41. Anzeiger für die Stadt Bern, February 5, 1902; CP 1, 192.
42. David Hume, An Enquiry Concerning Human Understanding, edited by Tom L. Beauchamp (Oxford and New York: Oxford University Press, 1999), section 4, part 1, 20.
43. Galison, 199.
44. Galison, 200.
45. Flückiger, 53.
46. Quoted in Seelig, Albert Einstein: A Documentary Biography, 58.
47. Quoted in Pais, Subtle Is the Lord, 47.
48. Letter to Michele Besso, January 22(?), 1903; CP 5, 7.
49. Letter to Mileva Einstein-Marić, September 19(?), 1903; CP 5, 14–15.
50. Letter to Conrad Habicht, late May 1905; CP 5, 19–20.

8. SQUARING THE LIGHT: WHY EINSTEIN HAD TO DISCOVER THE THEORY OF RELATIVITY

1. Quoted in Eberty, viii (German edition only).
2. Eberty, 23.
3. Aaron Bernstein, vol. 8, 131.
4. Letter to Marcel Grossmann, April 14, 1901; CP 1, 166.
5. Letter to Jakob Laub, May 19, 1909; CP 5, 121.
6. Letter to Mileva Marić, August 10(?), 1899; CP 1, 131.
7. Letter to Mileva Marić, September 10, 1899; CP 1, 133.
8. Draft of an article for Nature, after January 22, 1920; CP 7, 117.
9. CP 7, 117.
10. CP 2, 140.
11. See John Stachel, ed., Einstein's Miraculous Year, 111.
12. To Erika Oppenheimer, September 13, 1932; CP 2, 261 (German edition only).

13. Schilpp, 53.
14. *Physics Today*, 1982, 46.
15. Schilpp, 53.
16. CP 2, 141.
17. Hoffmann/Dukas, *Albert Einstein: Creator and Rebel*, 61.
18. CP 2, 141.
19. Gardner, 105.
20. Quoted in Fölsing, 189.
21. Hoffmann/Dukas, *Albert Einstein: Creator and Rebel*, 93.
22. Letter to Conrad Habicht, June 30–September 22, 1905; CP 5, 21.
23. Letter to Conrad Habicht, June 30–September 22, 1905; CP 5, 21.
24. Gaston Bachelard, "The Philosophic Dialectic of the Concepts of Philosophy," in Schilpp, 568.
25. Letter to Conrad Habicht, June 30–September 22, 1905; CP 5, 21.
26. Letter to Conrad Habicht, after July 20, 1905; CP 5, 21.

9. WHY IS THE SKY BLUE?: EINSTEIN—A CAREER

1. Calaprice, 259.
2. Einstein, "Religion and Science," *The New York Times Magazine*, November 9, 1930; quoted in *Ideas and Opinions*, 39.
3. Letter from Jakob Laub, January/March 1908; CP 5, 63.
4. Letter to Conrad Habicht, June 30–September 22, 1905; CP 5, 20.
5. CP 2, 123.
6. Evaluation by Alfred Kleiner, July 22–23, 1905; CP 5, 22.
7. CP 2, 104.
8. Letter to Maurice Solovine, April 27, 1906; CP 5, 25.
9. Letter to Maurice Solovine, April 27, 1906; CP 5, 25.
10. CP 2, 266 (German edition only).
11. Letter from Max Planck, July 6, 1907; CP 5, 31.
12. Letter to Johannes Stark, September 25, 1907; CP 5, 42.
13. Letter to Johannes Stark, November 1, 1907; CP 5, 44.
14. CP 2, 301.
15. CP 2, 310.
16. Letter to Heinrich Zangger, November 15, 1911; CP 5, 222.
17. Letter to Michele Besso, March 6, 1952; Einstein/Besso, 464.
18. Quoted in Fölsing, 217.
19. Letter to Marcel Grossmann, January 3, 1908; CP 5, 48–49.
20. To the educational council of the canton of Zurich, January 20, 1908; CP 5, 53.
21. Quoted in Flückiger, 115.
22. Quoted in Flückiger, 119.
23. Letter to Johannes Stark, December 14, 1908; CP 5, 96.
24. Quoted in Fölsing, 189.
25. Letter to Jakob Laub, July 30, 1908; CP 5, 81.
26. Letter to Jakob Laub, May 19, 1909; CP 5, 120.
27. Letter to Jakob Laub, May 19, 1909; CP 5, 120.
28. Letter to Jakob Laub, May 19, 1909; CP 5, 120.

29. Letter to Jakob Laub, May 19, 1909; CP 5, 120.
30. Quoted in Fölsing, 249.
31. Letter to Jakob Laub, May 19, 1909; CP 5, 120.
32. Letter to Jakob Laub, May 19, 1909; CP 5, 120.
33. Adolf Fisch; quoted in Seelig, *Albert Einstein: Leben und Werk*, 170.
34. Letter to Pauline Einstein, April 28, 1910; CP 5, 153.
35. Quoted in Fölsing, 271.
36. Letter to Marcel Grossmann, April 27, 1911; CP 5, 186.
37. Letter to Hans Tanner, April 24, 1911; CP 5, 186.
38. Letter to Alfred and Clara Stern, March 17, 1912; CP 5, 275.
39. Letter to Michele Besso, May 13, 1911; CP 5, 187–88.
40. Letter to Heinrich Zangger, November 15, 1911; CP 5, 222.
41. Letter to Michele Besso, December 26, 1911; CP 5, 241.
42. Quoted in Fölsing, 290–91.
43. Letter to Alfred and Clara Stern, February 2, 1912; CP 5, 255.
44. Proposal for Einstein's membership in the Prussian Academy of Sciences; CP 5, 337.
45. Letter to Heinrich Zangger, June 27, 1914; CP 8 Doc. 16a in CP 10, 12.
46. Letter to Elsa Löwenthal, July 14(?), 1913; CP 5, 341.
47. Letter to Elsa Löwenthal, July 19, 1913; CP 5, 343.
48. Letter to Jakob Laub, July 22, 1913; CP 5, 344.
49. Letter to the Prussian Academy of Sciences, December 7, 1913; CP 5, 369–70.
50. Quoted in Clark, 173.
51. Quoted in Fölsing, 184.
52. Letter to the Hurwitzes, May 4, 1914; CP 8, 13.
53. Letter to Heinrich Zangger, June 2, 1917; CP 8 Doc. 349a in CP 10, 55.
54. Dukas/Hoffmann, *Albert Einstein: The Human Side*, 73–74.
55. *Out of My Later Years*, 61.

10. "DEAR BOYS . . . YOUR PAPA": THE DRAMA OF THE BRILLIANT FATHER

1. Letter to Elsa Einstein, July 30, 1914; CP 8, 37.
2. Letter to Elsa Einstein, July 30, 1914; CP 8, 37.
3. Letter to Elsa Einstein, July 26, 1914; CP 8, 35.
4. Letter to Elsa Einstein, July 26, 1914; CP 8, 35.
5. Letter to Elsa Einstein, July 26, 1914; CP 8, 35.
6. Letter to Elsa Einstein, July 30, 1914; CP 8, 37.
7. Letter to Heinrich Zangger, after December 27, 1914; CP 8 Doc. 41a in CP 10, 14.
8. Letter to Heinrich Zangger, ca. April 10, 1915; CP 8, 88.
9. Letter to Mileva Einstein-Marić, ca. July 18, 1914; CP 8, 33.
10. Letter to Elsa Einstein, July 30, 1914; CP 8, 39.
11. Letter to Helene Savić, September 8, 1916; CP 8, 250.
12. Letter to Elsa Einstein, after August 3, 1914; CP 8, 40.
13. Letter to Elsa Einstein, July 30, 1914; CP 8, 38.
14. Letter to Elsa Einstein, July 30, 1914; CP 8, 38.

15. Letter to Elsa Einstein, after August 3, 1914; CP 8, 40.
16. Letter to Heinrich Zangger, June 27, 1914; CP 8 Doc. 16a in CP 10, 12.
17. Letter to Heinrich Zangger, May 28, 1915; CP 8, 101.
18. Letter to Heinrich Zangger, August 24, 1916; CP 8, 245.
19. Letter to Paul Ehrenfest, August 19, 1914; CP 8, 41.
20. CP 8, xli (German edition only).
21. Letter to Mileva Einstein-Marić, August 18, 1914; CP 8, 41.
22. Letter to Mileva Einstein-Marić, September 15, 1914; CP 8, 42.
23. Letter to Mileva Einstein-Marić, September 15, 1914; CP 8, 42.
24. Letter to Mileva Einstein-Marić, September 15, 1914; CP 8, 43.
25. Letter to Elsa Einstein, after July 26, 1914; CP 8, 36.
26. Letter to Mileva Einstein-Marić, December 12, 1914; CP 8, 48.
27. Letter to Mileva Einstein-Marić, January 12, 1915; CP 8, 58.
28. Letter to Mileva Einstein-Marić, January 12, 1915; CP 8, 58.
29. Letter from Hans Albert Einstein, before April 4, 1915; CP 8 Doc. 69b in CP 10, 16.
30. Letter to Heinrich Zangger, July 24–August 7, 1915; CP 8, 116.
31. Letter to Heinrich Zangger, July 7, 1915; CP 8, 109.
32. Letter from Hans Albert Einstein, June 28, 1915; CP 8 Doc. 91a in CP 10, 17. Einstein concluded, "The boy's soul is being systematically poisoned." Letter to Heinrich Zangger, before December 4, 1915; CP 8 Doc. 159a in CP 10, 20.
33. Letter to Hans Albert Einstein, before April 4, 1915; CP 8, 84.
34. Letter to Elsa Einstein, September 11, 1915; CP 8, 126.
35. Letter to Elsa Einstein, September 13, 1915; CP 8, 127.
36. Letter to Heinrich Zangger, July 24–August 7, 1915; CP 8, 115.
37. Letter to Heinrich Zangger, October 15, 1915; CP 8, 137.
38. Letter to Heinrich Zangger, October 15, 1915; CP 8, 137–38.
39. Letter to Hans Albert Einstein, before November 4, 1915; CP 8, 140.
40. Letter from Michele Besso, ca. October 30, 1915; CP 8, 139.
41. CP 8, n. 3, 49 (German edition only).
42. Letter from Michele Besso, ca. October 30, 1915; CP 8, 139.
43. Letter from Michele Besso, ca. October 30, 1915; CP 8, 139.
44. Letter to Mileva Einstein-Marić, November 15, 1915; CP 8, 146.
45. Letter from Mileva Einstein-Marić, November 5, 1915; CP 8, 141.
46. Letter to Hans Albert Einstein; November 15, 1915; CP 8, 146.
47. Letter to Hans Albert Einstein, before November 30, 1915; CP 8, 155.
48. Letter from Michele Besso, after November 30, 1915; CP 8, 156.
49. Letter to Mileva Einstein-Marić, December 1, 1915; CP 8, 156.
50. Letter to Mileva Einstein-Marić, December 1, 1915; CP 8, 157.
51. Letter to Mileva Einstein-Marić, December 10, 1915; CP 8, 160.
52. Letter to Hans Albert Einstein, December 18, 1915; CP 8, 162.
53. Letter to Mileva Einstein-Marić, February 6, 1916; CP 8, 189.
54. Letter to Mileva Einstein-Marić, March 12, 1916; CP 8, 200.
55. Letter to Mileva Einstein-Marić, March 12, 1916; CP 8, 200.
56. Letter to Mileva Einstein-Marić, April 1, 1916; CP 8, 206.
57. Letter to Mileva Einstein-Marić, April 1, 1916; CP 8, 207.
58. Letter to Mileva Einstein-Marić, April 8, 1916; CP 8, 208.

59. Postcard to Elsa Einstein, April 12, 1916; CP 8, 209.
60. Letter to Mileva Einstein-Marić, before May 8, 1918; CP 8, 554.
61. Letter to Hans Albert Einstein, March 16, 1916; CP 8, 203.
62. Letter to Hans Albert Einstein, September 19, 1927; Einstein Archives 75–745.
63. Letter to Elsa Einstein, October 19, 1920; CP 10, 293.
64. Letter to Elsa Einstein, April 21, 1916; CP 8, 210–11.
65. Letter to Michele Besso, July 14, 1916; CP 8, 230.
66. Letter to Heinrich Zangger, July 11, 1916; CP 8 Doc. 232a in CP 10, 24.
67. Letter from Michele Besso, July 17, 1916; CP 8, 234.
68. Letter to Michele Besso, July 21, 1916; CP 8, 234.
69. Letter to Michele Besso, July 21, 1916; CP 8, 235.
70. Letter to Hans Albert Einstein, July 25, 1916; CP 8, 237.
71. Letter to Heinrich Zangger, July 25, 1916; CP 8, 237–38.
72. Letter to Heinrich Zangger, July 25, 1916; CP 8, 238.
73. Letter to Michele Besso, August 24, 1916; CP 8, 244.
74. Letter to Michele Besso, September 6, 1916; CP 8, 246.
75. Letter to Hans Albert Einstein, September 26, 1916; CP 8, 251.
76. Letter to Hans Albert Einstein, October 13, 1916; CP 8, 252–53.
77. Letter to Hans Albert Einstein, after October 31, 1916; CP 8, 259.
78. Letter from Michele Besso, December 5, 1916; CP 8, 270.
79. Letter to Hans Albert Einstein, after January 8, 1917; CP 8, 278.
80. Letter to Michele Besso, March 9, 1917; CP 8, 293.
81. Letter to Michele Besso, March 9, 1917; CP 8, 293.
82. Letter to Michele Besso, March 9, 1917; CP 8, 292.
83. Letter to Michele Besso, March 9, 1917; CP 8, 293.
84. Letter to Mileva Einstein-Marić, March 6, 1926; Einstein Archives 75–652.
85. Letter to Heinrich Zangger, February 16, 1917; CP 8 Doc. 299a in CP 10, 43.
86. Letter to Hans Albert Einstein, August 4, 1948; Einstein Archives 75–836.
87. Letter to Heinrich Zangger, February 16, 1917; CP 8 Doc. 299a in CP 10, 43.
88. Letter to Michele Besso, October 21, 1932; Einstein/Besso, 290.
89. Letter to Heinrich Zangger, March 10, 1917; CP 8, 298.
90. Letter to Heinrich Zangger, December 6, 1917; CP 8, 412.
91. Letter to Michele Besso, March 9, 1917; CP 8, 292.
92. Letter to Elsa Löwenthal, after August 11, 1913; CP 5, 348.
93. Letter to Michele Besso, March 9, 1917; CP 8, 292.
94. Letter to Heinrich Zangger, July 29, 1917; CP 8, 361.
95. Letter to Michele Besso and Anna Besso-Winteler, August 1, 1917; CP 8, 363.
96. Letter from Heinrich Zangger, May 20, 1917; CP 8, 332.
97. Letter to Hans Albert Einstein, December 9, 1917; CP 8, 417.
98. Letter to Hans Albert Einstein, December 24, 1917; CP 8, 424.
99. Letter to Hans Albert Einstein, January 25, 1918; CP 8, 449.
100. Letter to Paul Ehrenfest, March 22, 1919; CP 9, 9.
101. Letter to Heinrich Zangger, November 26, 1915; CP 8, 151.
102. Letter to Mileva Einstein-Marić, January 31, 1918; CP 8, 455.
103. Letter to Mileva Einstein-Marić, January 31, 1918; CP 8, 456.
104. Letter from Mileva Einstein-Marić, February 9, 1918; CP 8 Doc. 461a in CP 10, 86.

105. Letter from Mileva Einstein-Marić, after February 6, 1918; CP 8, 465.
106. Letter to Michele Besso, January 5, 1918; CP 8, 436.
107. Letter to Hans Albert Einstein, January 25, 1918; CP 8, 449.
108. Letter to Hans Albert Einstein, January 25, 1918; CP 8, 449.
109. Letter from Heinrich Zangger to Michele Besso, March 4, 1918; CP 8, n. 4, 665 (German edition only).
110. Letter to Anna Besso-Winteler, after March 4, 1918; CP 8, 490.
111. Letter to Anna Besso-Winteler, after March 4, 1918; CP 8, 489.
112. Letter to Anna Besso-Winteler, after March 4, 1918; CP 8, 489–90.
113. Letter from Anna Besso-Winteler, after March 4, 1918; CP 8, 490–91.
114. Letter to Mileva Einstein-Marić, June 4, 1918; CP 8, 579.
115. Letter to Michele Besso, before June 28, 1918; CP 8, 598.
116. Letter to Mileva Einstein-Marić, March 17, 1918; CP 8, 497.
117. Letter to Mileva Einstein-Marić, before May 8, 1918; CP 8, 554.
118. Letter to Mileva Einstein-Marić, May 23, 1918; CP 8, 566.
119. Letter to Michele Besso, July 9, 1918; CP 8, 610.
120. Letter to Michele Besso, before June 28, 1918; CP 8, 599.
121. Letter to Eduard Einstein, before June 28, 1918; CP 8, 599.
122. Letter to Max Born, after June 29, 1918; CP 8, 600.
123. Letter to Hans Albert Einstein, after June 29, 1918; CP 8, 600–601.
124. Letter to Michele Besso, July 29, 1918; CP 8, 613.
125. Letter to Heinrich Zangger, before August 11, 1918; CP 8, 622.
126. Michelmore, 79.
127. Letter to Eduard Einstein, April 17, 1926; Einstein Archives 75–654.
128. Letter from Heinrich Zangger, before August 11, 1918; CP 8, 623.
129. Letter to Mileva Einstein-Marić, ca. November 9, 1918; CP 8, 689.
130. Deposition in divorce proceedings, Berlin, December 23, 1918; CP 8, 713.
131. Letter to Hans Albert Einstein, June 13, 1919; CP 9, 50.
132. Letter to Pauline Einstein, August 9, 1919; CP 9, 72.
133. Letter to Pauline Einstein, August 9, 1919; CP 9, 72.
134. Letter to Mileva Einstein-Marić, October 15, 1919; CP 9, 115.
135. Letter to Paul Ehrenfest, March 1, 1920; CP 9, 282.
136. Letter to Mileva Einstein-Marić, November 16, 1918; CP 9, 140.
137. Letter to Hans Albert and Eduard Einstein, December 15, 1920; CP 10, 334.
138. Letter to Mileva Einstein-Marić, July 20, 1938; Einstein Archives 75–949.
139. Letter to Mileva Einstein-Marić, December 5, 1919; CP 9, 163.
140. Letter to Mileva Einstein-Marić, November 16, 1919; CP 9, 140.
141. Letter to Hans Albert Einstein, February 27, 1920; CP 9, 279.
142. Letter to Hans Albert Einstein, February 27, 1920; CP 9, 279.
143. Letter to Hans Albert Einstein and Eduard Einstein, March 26, 1920; CP 9, 300.
144. Letter to Hans Albert Einstein, April 5, 1920; CP 9, 306.
145. Letter to Hans Albert Einstein, July 4, 1920; CP 10, 204.
146. Letter to Mileva Einstein-Marić, July 23, 1920; CP 10, 213.
147. Letter to Eduard Einstein, August 1, 1920; CP 10, 226.
148. Letter to Mileva Einstein-Marić, 1921(?); Einstein Archives 75–723.
149. Letter from Eduard Einstein to Mileva Einstein-Marić, 1921(?); Einstein Archives 75–724.

150. Letter to Mileva Einstein-Marić, August 28, 1921; Einstein Archives 75–721.
151. Letter to Eduard Einstein, July 15, 1923; Einstein Archives 75–627.
152. Letter to Mileva Einstein-Marić, August 14, 1925; Einstein Archives 75–963.
153. Letter to Maja Winteler-Einstein, September 14, 1925; Einstein Archives 29–402.
154. Letter to Mileva Einstein-Marić, 1925; Einstein Archives 75–719.
155. Letter to Mileva Einstein-Marić, 1925; Einstein Archives 75–719.
156. Letter from Mileva Einstein-Marić, n.d.; Einstein Archives (unnumbered).
157. Letter from Mileva Einstein-Marić, n.d.; Einstein Archives (unnumbered).
158. Letter from Mileva Einstein-Marić, n.d.; Einstein Archives (unnumbered).
159. Letter from Mileva Einstein-Marić, n.d.; Einstein Archives (unnumbered).
160. Letter to Mileva Einstein-Marić, August 14, 1925; Einstein Archives 75–963.
161. Letter to Mileva Einstein-Marić, May 5, 1928; Einstein Archives 75–754.
162. Letter to Hans Albert Einstein, February 23, 1927; Einstein Archives 75–739.
163. Letter to Hans Albert Einstein, September 7, 1927; Einstein Archives 75–657.
164. Letter to Eduard Einstein, February 23, 1927; Einstein Archives 75–748.
165. Letter to Mileva Einstein-Marić, Summer 1929; Einstein Archives 75–776.
166. Letter to Eduard Einstein, April 17, 1926; Einstein Archives 75–654.
167. Letter to Eduard Einstein, July 27, 1932; Einstein Archives 75–670.
168. Letter to Eduard Einstein, December 23, 1927; Einstein Archives 75–748.
169. Letter to Eduard Einstein, December 23, 1927; Einstein Archives 75–748.
170. Letter to Eduard Einstein, July 10, 1929(?); Einstein Archives 75–782.
171. Letter to Eduard Einstein, before December 2, 1931; Einstein Archives 75–984.
172. Letter to Eduard Einstein, October 8, 1932; Einstein Archives 75–668.
173. Letter to Eduard Einstein, before March 14, 1929; Einstein Archives 75–779.
174. Letter to Eduard Einstein, n.d.; Einstein Archives 75–993.
175. Letter to Eduard Einstein, 1935(?); Einstein Archives 75–662.
176. Letter to Eduard Einstein, January 24, 1930; Einstein Archives 75–992.
177. Letter to Eduard Einstein, February 5, 1930; Einstein Archives 75–990.
178. Letter to Eduard Einstein, n.d.; Einstein Archives 75–988.
179. Letter to Mileva Einstein-Marić, 1926(?); Einstein Archives 75–656.
180. Letter to Maja Winteler-Einstein, September 14, 1925; Einstein Archives 29–410.
181. Letter to Mileva Einstein-Marić, 1926(?); Einstein Archives 75–655.
182. Letter to Mileva Einstein-Marić, 1926(?); Einstein Archives 75–651.
183. Letter to Mileva Einstein-Marić, 1926(?); Einstein Archives 75–656.
184. Letter to Mileva Einstein-Marić, January 11, 1928; Einstein Archives 75–697.
185. Letter to Mileva Einstein-Marić, October 15, 1926; Einstein Archives 75–658.
186. Letter to Mileva Einstein-Marić, 1927(?); Einstein Archives 75–676.
187. Letter to Hans Albert Einstein, February 1927; Einstein Archives 75–738.
188. Quoted in Herneck, *Einstein privat*, 50.
189. Letter to Hans Albert Einstein, March 31, 1928; Einstein Archives 75–753.
190. Letter to Michele Besso, January 5, 1929; Einstein/Besso, 241.
191. Letter to Hans Albert Einstein, October 6, 1932; Einstein Archives 75–787.
192. Vallentin, 93.
193. Letter to Eduard Einstein, August 30, 1932; Einstein Archives 75–686.
194. Vallentin, 196.
195. Letter to Eduard Einstein, September 7, 1932; Einstein Archives 75–666.
196. Letter from Hans Albert Einstein, 1932(?); Einstein Archives 75–684.

197. Letter to Hans Albert Einstein, November 5, 1932; Einstein Archives 75–685.
198. Letter to Mileva Einstein-Marić, April 29, 1933; Einstein Archives 75–678.
199. Letter to Hans Albert Einstein, May 30, 1933; Einstein Archives 75–663.
200. Letter to Mileva Einstein-Marić, June 15, 1933; Einstein Archives 75–962.
201. Letter to Hans Albert Einstein, January 4, 1937; Einstein Archives 75–926.
202. Quoted in Seelig, *Helle Zeit*, 45.
203. Letter to Hans Albert and Frieda Einstein, January 7, 1939; Einstein Archives 75–904.
204. Letter to Hans Albert and Frieda Einstein, 1941(?); Einstein Archives 75–906.
205. Letter from Hans Albert Einstein to Mileva Einstein-Marić, July 26, 1942; Einstein Archives 75–813.
206. Letter from Hans Albert Einstein to Mileva Einstein-Marić, July 23, 1942; Einstein Archives 75–956.
207. Letter to Hans Albert Einstein, June 24, 1947; Einstein Archives 75–808.
208. Letter to Hans Albert Einstein, n.d.; Einstein Archives 75–831.
209. Letter to Mileva Einstein-Marić, September 7, 1947; Einstein Archives 75–845.
210. Letter to Hans Albert Einstein, June 7, 1948; Einstein Archives 75–958.
211. Letter to Hans Albert Einstein, June 14, 1948; Einstein Archives 75–835.
212. Letter to Hans Albert Einstein, August 4, 1948; Einstein Archives 75–836.
213. Letter to Hans Albert Einstein, December 26, 1948; Einstein Archives 75–837.
214. Letter to Hans Albert Einstein, March 2, 1949; Einstein Archives 75–835.
215. Letter to Hans Albert and Frieda Einstein, March 18, 1949; Einstein Archives 75–829.
216. Letter from Maja Winteler-Einstein to Eduard Einstein, November 19, 1944; Einstein Archives 75–806.
217. Letter from Maja Winteler-Einstein to Theresia Mutzenbecher, July 15, 1946; quoted in Highfield/Carter, 249.
218. Letter to Hans Albert and Frieda Einstein, June 27, 1951; Einstein Archives 75–794.
219. Letter to Hans Albert Einstein, December 28, 1954; Einstein Archives 75–917.
220. Letter to Hans Albert Einstein, May 11, 1954; Einstein Archives 75–918; quoted in Highfield/Carter, 258.
221. Letter to Mileva Einstein-Marić, 1928(?); Einstein Archives 75–783.
222. Letter to Mileva Einstein-Marić, June 4, 1932; Einstein Archives 75–672.
223. Letter from Michele Besso, September 18, 1932; Einstein/Besso, 286.
224. Letter to Eduard Einstein, October 8, 1932; Einstein Archives 75–668.
225. Letter to Maja Winteler-Einstein, n.d.; Einstein Archives 29–415.
226. Letter from Maja Winteler-Einstein to Theresia Mutzenbecher, April 20, 1934; quoted in Highfield/Carter, 240–41.
227. Letter to Eduard Einstein, May 25, 1937; Einstein Archives 75–933.
228. Letter from Michele Besso, June 19, 1937; Einstein/Besso, 316.
229. Letter from Carl Seelig to Albert Einstein; quoted in Highfield/Carter, 257.
230. Quoted in Huonker, 222.
231. Quoted in Huonker, 223.
232. Quoted in Huonker, 223.
233. Letter to Mileva Einstein-Marić, 1934; Einstein Archives 75–969.
234. Letter to Michele Besso, November 11, 1940; Einstein/Besso, 352.

235. Letter to Mileva Einstein-Marić, December 22, 1946; Einstein Archives 75–846.
236. Quoted in Huonker, 223.
237. Letter to Carl Seelig, January 4, 1954; quoted in Highfield/Carter, 257.
238. Quoted in Huonker, 223.
239. Quoted in Huonker, 223f.

11. ANATOMY OF A DISCOVERY: HOW EINSTEIN FOUND THE GENERAL THEORY OF RELATIVITY

1. Calaprice, 8.
2. Quoted in Infeld, *Leben mit Einstein*, 81.
3. Quoted in Bührke, 90.
4. Draft of an article for *Nature*, after January 22, 1920; CP 7, 136.
5. Draft of an article for *Nature*, after January 22, 1920; CP 7, 136.
6. Draft of an article for *Nature*, after January 22, 1920; CP 7, 136.
7. Quoted in Seelig, *Helle Zeit*, 27.
8. Letter to Heinrich Zangger, July 24–August 7, 1915; CP 8, 116.
9. Letter to Erwin Freundlich, September 30, 1915; CP 8, 133.
10. Letter to Erwin Freundlich, September 30, 1915; CP 8, 133.
11. Letter to David Hilbert, November 7, 1915; CP 8, 141.
12. Letter from David Hilbert, November 13, 1915; CP 8, 144.
13. Letter to David Hilbert, November 15, 1915; CP 8, 146.
14. Minutes of the proceedings of the Prussian Academy of Sciences, May 4, 1915; CP 6, 98.
15. Letter to Michele Besso, November 17, 1915; CP 8, 148.
16. Letter to Paul Ehrenfest, January 17, 1916; CP 8, 179.
17. Letter to Heinrich Zangger, October 9, 1915; CP 8 Doc. 161a in CP 10, 21.
18. Quoted in Frank, 179.
19. Letter to Heinrich Zangger, November 26, 1915; CP 8, 151.
20. Letter to David Hilbert, December 20, 1915; CP 8, 163.
21. Letter from Willem de Sitter, November 4, 1920; CP 10, 303.
22. Draft of an article for *Nature*, after January 22, 1920; CP 7, 130.
23. Draft of an article for *Nature*, after January 22, 1920; CP 7, 149.

12. LAMBDA LIVES: EINSTEIN, "CHIEF ENGINEER OF THE UNIVERSE"

1. Quoted in *GEO* 11 (2002), 61.
2. Letter to Karl Schwarzschild, January 9, 1916; CP 8, 176.
3. Letter to Paul Ehrenfest, February 4, 1917; CP 8, 282.
4. Letter from Willem de Sitter, November 1, 1916; CP 8, 261.
5. Letter to Heinrich Zangger, February 1, 1917; CP 8 Doc. 291a in CP 10, 41.
6. Letter from Willem de Sitter, April 18, 1917; CP 8, 317.
7. Letter from Willem de Sitter, April 1, 1917; CP 8, 313.
8. Letter to Willem de Sitter, April [13]–14, 1917; CP 8, 316.
9. Letter from Willem de Sitter, April 18, 1917; CP 8, 317.
10. *Nature*, May 9, 1931; quoted in Bührke, 149.

13. SPACETIME QUAKES: THE THEORY OF RELATIVITY PUT TO THE TEST

1. Calaprice, 226.
2. Letter to J. Cattell, December 18, 1936; Einstein Archives 65–603.

14. HIS BEST FOE: EINSTEIN, GERMANY, AND POLITICS

1. Letter to Mileva Einstein-Marić, May 15, 1915; CP 8, 97.
2. Clara Haber to Richard Abegg, April 25, 1909; quoted in Stern, 77.
3. Quoted in Stern, 121.
4. Quoted in Stern, 119.
5. Quoted in Moszkowski, 173.
6. Quoted in Grüning, 179.
7. Letter to Elsa Löwenthal, after December 2, 1913; CP 5, 366.
8. Letter to Hendrik A. Lorentz, August 1, 1919; CP 9, 68.
9. Mid-October 1914; CP 6, 28–29.
10. Letter to Paul Ehrenfest, August 19, 1914; CP 8, 41.
11. Letter to Paul Ehrenfest, August 19, 1914; CP 8, 41.
12. Letter to Paul Ehrenfest, early December 1914; CP 8, 47.
13. Letter to Heinrich Zangger, ca. April 10, 1915; CP 8, 87.
14. Letter to Heinrich Zangger, May 28, 1915; CP 8, 101.
15. Letter to Hendrik A. Lorentz, January 23, 1915; CP 8, 63.
16. Letter to Paul Ehrenfest, early December 1914; CP 8, 46.
17. Manuscript for the Goethe League, October 23–November 11, 1915; CP 6, 96.
18. Letter to Heinrich Zangger, December, 1917; CP 8, 412.
19. Letter to Romain Rolland, August 22, 1917; CP 8, 368.
20. Lecture notes, October 11, 1918–February 1919; CP 7, 90 (German edition only).
21. Speech manuscript, November 13, 1918; CP 7, 76.
22. Letter to Max Born, January 27, 1920; CP 9, 236.
23. *Ideas and Opinions*, 178.
24. Quoted in Grüning, 245.
25. Quoted in Vallentin, 119.
26. Letter to Michele Besso, December 4, 1918; CP 8, 703.
27. Letter to Michele Besso, December 4, 1918; CP 8, 703.
28. Letter to Arnold Sommerfeld, December 6, 1918; CP 8, 705.
29. Letter to Arnold Sommerfeld, December 6, 1918; CP 8, 705–6.
30. Letter to Max Born, June 4, 1919; Einstein/Born, 11.
31. Quoted in the *Frankfurter Allgemeine Zeitung*, October 5, 1981.
32. Letter to Paul Winteler and Maja Winteler-Einstein, November 11, 1918; CP 8, 693.
33. Letter to Michele Besso, December 4, 1918; CP 8, 703.
34. Letter to Heinrich Zangger, December 24, 1919; CP 9, 198.
35. Letter to Heinrich Zangger, December 24, 1919; CP 9, 198.
36. Holton, *Einstein, History, and Other Passions*, 8.
37. Quoted in Moszkowski, 15.

38. Letter to Wilhelm Wien, October 17, 1916; CP 8, 255.
39. *Dt. Phys. Ges. Verhandlungen*, December 30, 1918; CP 7, 78.
40. *Die Naturwissenschaften*, November 29, 1918; CP 7, 74.
41. *Die Naturwissenschaften*, November 29, 1918; CP 7, 74–75.
42. Letter to Hendrik A. Lorentz, November 15, 1919; CP 9, 139.
43. Letter to Heinrich Zangger, December 15 or 22, 1919; CP 9, 186.
44. Letter to Heinrich Zangger, December 15 or 22, 1919; CP 9, 185.
45. *8-Uhr-Abendblatt*, February 13, 1920; CP 7, 152.
46. *8-Uhr-Abendblatt*, February 13, 1920; CP 7, 152.
47. *8-Uhr-Abendblatt*, February 13, 1920; CP 7, 152.
48. Letter to Heinrich Zangger, February 13, 1920; CP 9, 261.
49. Letter to Aurel Stodola, March 31, 1919; CP 9, 15.
50. *Morus*, August 27, 1920; quoted in Grundmann, 162. Unpublished translation by Aaron Wiener.
51. *Berliner Tageblatt*, August 27, 1920; CP 7, 197.
52. *Berliner Tageblatt*, August 31, 1920.
53. *Berliner Tageblatt*, August 27, 1920; CP 7, 197.
54. Letter from Paul Ehrenfest, September 2, 1920; CP 10, 254.
55. Letter from Paul Ehrenfest, August 28, 1920; CP 10, 245.
56. Letter from Hedwig Born, September 8, 1920; CP 10, 263.
57. Letter to Max and Hedwig Born, September 9, 1920; CP 10, 265.
58. Letter to Paul Ehrenfest, before September 9, 1920; CP 10, 265.
59. *Tägliche Rundschau*, August 26, 1920.
60. *Berliner Tageblatt*, August 31, 1920.
61. Quoted in Fölsing, 464.
62. Letter to Marcel Grossmann, February 27, 1920; CP 9, 276.
63. Letter to Max and Hedwig Born, September 9, 1920; CP 10, 265.
64. *Deutsche Allgemeine Zeitung*, September 25, 1920, morning edition; CP 7, 109, n. 52 (German edition only).
65. *Berliner Tageblatt*, September 24, 1920; CP 7, 358 (German edition only).
66. *Deutsche Zeitung*, quoted in CP 7, n. 63, 110 (German edition only).
67. *Berliner Tageblatt*, September 25, 1920, morning edition; CP 7, n. 57, 110 (German edition only).
68. *Staatsbürger-Zeitung*, January 9, 1921.
69. Letter to Max Planck, July 6, 1922; quoted in Grundmann, 176.
70. Letter from Max Planck to Max von Laue, July 9, 1922; quoted in Hermann, 281.
71. Quoted in CP 7, n. 76, 113 (German edition only).
72. Grundmann, 182.
73. Grundmann, 184.
74. Quoted in Grundmann, 237.
75. *Daily Mail*; quoted in Grundmann, 200.
76. Quoted in Grundmann, 214.
77. *Vossische Zeitung*, April 6, 1922.
78. *Berliner Tageblatt*, April 12, 1922.
79. Quoted in Grundmann, 214.
80. Quoted in Grundmann, 220.

81. Letter to Elsa Löwenthal, after December 2, 1913; CP 5, 366.
82. Quoted in Grundmann, 231.
83. Quoted in Grundmann, 274.
84. Quoted in Grundmann, 276.
85. Letter to Mileva Einstein-Marić, July 20, 1938; Einstein Archives 75–949.
86. Quoted in Grundmann, 278.
87. Quoted in Grundmann, 281.
88. Letter from Max Planck, November 10, 1923; quoted in Fölsing, 345.
89. Letter to Michele Besso, May 24, 1924; Einstein/Besso, 202.
90. Quoted in Grundmann, 264.
91. Quoted in Grundmann, 260.
92. Travel diary, April 17, 1925. Unpublished translation by Josef Eisinger.
93. Letter to M. Pierre Comert, March 21, 1923; quoted in Grundmann, 296.
94. Nathan/Norden, 111.
95. Letter to Michele Besso, October 21, 1932; Einstein/Besso, 290.
96. To the children at Mopr Reform School, March 27, 1929; quoted in Grundmann, 409.
97. Quoted in Grundmann, 410.
98. Speech given on August 22, 1930; quoted in Grundmann, 381.
99. Nathan/Norden, 113.
100. July 20, 1930; quoted in Pais, *Einstein Lived Here*, 175.
101. Quoted in Grüning, 259.
102. Quoted in Fölsing, 635.
103. Letter from Romain Rolland to Stefan Zweig, September 15, 1933; quoted in Grundmann, 433.
104. November 4, 1931; quoted in Calaprice, 261.
105. Quoted in Grüning, 198–99.
106. Speech to the International League for Sexual Reforms, September 6, 1929; quoted in Grüning, 305–6.
107. Letter to Alfons Goldschmidt, August 21, 1931; quoted in Grüning, 357.
108. Letter to Henri Barbusse, April 20, 1932; quoted in Grüning, 386–87.
109. Quoted in Grüning, 254.
110. Quoted in Grüning, 198.
111. Quoted in Nathan/Norden, 190.
112. Quoted in Nathan/Norden, 199–200.
113. Quoted in Nathan/Norden, 202.
114. Quoted in Nathan/Norden, 229.
115. Letter from Romain Rolland to Stefan Zweig, September 15, 1933; quoted in Grundmann, 361.
116. *Kölnische Zeitung*, March 30, 1933.
117. Travel diary, December 6, 1931; Einstein Archives 29–136. Unpublished translation by Josef Eisinger.
118. Letter to Max Planck, July 17, 1931; quoted in Grüning, 353.
119. Letter to the Prussian Academy of Sciences, March 23, 1933; quoted in Fölsing, 661.
120. Minutes of the plenary session of the Prussian Academy of Sciences, March 30, 1933; quoted in Grundmann, 167.

121. Prussian Academy of Sciences press release, April 1, 1933; quoted in *Ideas and Opinions*, 206.
122. Letter from Max von Laue to Heinrich von Ficker, July 11, 1947; quoted in Kirsten/Treder, 274.
123. *Deutsche Tageszeitung*, April 1, 1933.
124. Letter from Max von Laue to Heinrich von Ficker, July 11, 1947; quoted in Kirsten/Treder, 273.
125. Letter to Mileva Einstein-Marić, April 29, 1933; Einstein Archives 75–678.
126. *Grossdeutscher Pressedienst*, February 4, 1934; quoted in Grundmann, 434.
127. Letter to Maja Winteler-Einstein, 1933; Einstein Archives 79–416.
128. Quoted in Bucky/Weakland, 137.
129. Letter to Max Planck, April 6, 1933; quoted in Nathan/Norden, 217–18.
130. Letter to Fritz Haber, August 8, 1933; quoted in Fölsing, 668.
131. Minutes of the plenary session of the Prussian Academy of Sciences, May 11, 1933; Fölsing, 665.
132. *The New York Times*, October 4, 1933.
133. Speech on October 3, 1933; quoted in Nathan/Norden, 237–38.
134. Speech on October 3, 1933; quoted in Nathan/Norden, 238.
135. Hermann, 407.
136. Quoted in Grüning, 286.
137. Letter to Fritz Haber, August 8, 1933; quoted in Fölsing, 668.

15. "I AM NOT A TIGER": EINSTEIN, THE HUMAN SIDE

1. Letter to Max and Hedwig Born, September 9, 1920; CP 10, 265.
2. Letter to Maja Winteler-Einstein, 1926; Einstein Archives 29–403.
3. *The New York Times*, October 29, 1930; quoted in Pais, *Einstein Lived Here*, 181.
4. Vallentin, 13.
5. Quoted in Grüning, 145.
6. Vallentin, 14.
7. Quoted in Grüning, 265.
8. Vallentin, 91.
9. Quoted in Grüning, 35.
10. Quoted in Grüning, 36.
11. Quoted in Grüning, 155.
12. Vallentin, 185.
13. Quoted in Grüning, 450.
14. Quoted in Hermann, *Jahresbericht Preussischer Kulturbesitz* VIII (1970), 90.
15. Quoted in Grüning, 33.
16. Vallentin, 26.
17. Letter to Mileva Einstein-Marić, 1930(?); Einstein Archives 75–985.
18. Quoted in Grüning, 169.
19. Quoted in Grüning, 170.
20. Quoted in Grüning, 173.
21. Quoted in Grüning, 169.
22. Vallentin, 26.
23. Quoted in Grüning, 141.

24. Quoted in Herneck, *Einstein privat*, 27.
25. Quoted in Herneck, *Einstein privat*, 58.
26. Quoted in Herneck, *Einstein privat*, 114.
27. Quoted in Grüning, 239.
28. Letter to Maja Winteler-Einstein, 1928/1929; Einstein Archives 29–406.
29. Quoted in Grundmann, 371.
30. Quoted in Grüning, 239.
31. Quoted in Grüning, 211.
32. Letter to Mileva Einstein-Marić, July 4, 1929; Einstein Archives 75–784.
33. Quoted in Grüning, 75.
34. Quoted in Grüning, 42.
35. Quoted in Grüning, 54.
36. Quoted in Grüning, 55.
37. Quoted in Grüning, 375.
38. *Berliner Tageblatt*, May 14, 1929.
39. Quoted in Grüning, 123.
40. Quoted in Grüning, 78.
41. Quoted in Grüning, 130.
42. Holton, *Einstein, History, and Other Passions*, 7.
43. Gardner, 127.
44. Quoted in Grüning, 241.
45. Quoted in Grüning, 237.
46. Quoted in Grüning, 241.
47. Proust, 62.
48. Thomas Mann, 64.
49. Faulkner, 171.
50. Faulkner, 104.
51. Quoted in Grüning, 152.
52. Quoted in Grüning, 206.
53. Quoted in Grüning, 461.
54. Quoted in Herneck, *Einstein privat*, 68.
55. Quoted in Herneck, *Einstein privat*, 129.
56. Bucky/Weakland, 139.
57. Quoted in Calaprice, 148.
58. Quoted in Seelig, *Albert Einstein: A Documentary Biography*, 114.
59. Quoted in Rosenkranz, *Albert Through the Looking Glass*, 103.
60. Quoted in Calaprice, 151.
61. Quoted in Seelig, *Helle Zeit*, 37.
62. Preface to the Caputh guest book, May 4, 1930; Einstein Archives 31–067. Unpublished translation by Aaron Wiener.
63. Quoted in Grüning, 216.
64. Quoted in Grüning, 312.
65. Quoted in Grüning, 475.
66. Letter to Hans Albert Einstein, June 5, 1929; Einstein Archives 75–777.
67. Quoted in Grüning, 515.

16. A JEW NAMED ALBERT: HIS GOD WAS A PRINCIPLE

1. Travel diary, February 3, 1923; Einstein Archives 29–129 to 29–131. Unpublished translation by Josef Eisinger.
2. Quoted in Fölsing, 531.
3. Travel diary, February 13, 1923; Einstein Archives 29–129 to 29–131. Unpublished translation by Josef Eisinger.
4. Letter to Maurice Solovine, Whitsun 1923; *Letters to Solovine*, 59.
5. *Jüdische Rundschau*, June 21, 1921; CP 7, 235.
6. Letter to Hedwig Born, September 8, 1916; CP 8, 249.
7. Letter from Paul Ehrenfest to Tatiana Ehrenfest, February 25, 1912; CP 5, n. 3, 254 (German edition only).
8. CP 1, xx.
9. CP 1, xx.
10. CP 1, xx.
11. Reiser, 30.
12. Letter to Paul Nathan, April 3, 1920; CP 9, 304.
13. Quoted in Hettler, 155.
14. Schilpp, 3.
15. Schilpp, 5.
16. Schilpp, 3–7.
17. Letter to Mileva Marić, March 27, 1901; CP 1, 160.
18. Letter from Mileva Marić to Helene Savić, November/December, 1901; CP 1, 183.
19. Letter to Marcel Grossmann, January 3, 1908; CP 5, 49.
20. Quoted in Fölsing, 250.
21. Quoted in Fölsing, 271.
22. Nachama, 90.
23. Nachama, 90.
24. Quoted in Charpa/Grunwald, 133.
25. Nachama, 129.
26. Nachama, 129.
27. *The Times* (London) November 28, 1919; Frank, 144.
28. Letter to Paul Ehrenfest, December 4, 1919; CP 9, 162.
29. *Berliner Tageblatt*, December 30, 1919, morning edition; CP 7, 110.
30. Letter to Max Born, before November 9, 1919; CP 9, 137.
31. *Jüdische Rundschau*, June 21, 1921; CP 7, 234.
32. *Jüdische Rundschau*, June 21, 1921; CP 7, 236.
33. Grüning, 261.
34. *Israelitisches Wochenblatt für die Schweiz*, September 24, 1920; CP 7, 159.
35. *Israelitisches Wochenblatt für die Schweiz*, September 24, 1920; CP 7, 159.
36. Letter to the Jewish Community of Berlin, December 22, 1920; CP 10, 338.
37. *Ideas and Opinions*, 188.
38. *Ideas and Opinions*, 171.
39. *Ideas and Opinions*, 171.
40. *Ideas and Opinions*, 188.
41. Stern, 68.
42. Vallentin, 222.

43. Vallentin, 224.
44. Letter to Max Born, March 22, 1934; Einstein/Born, 121–22.
45. Letter to Michele Besso, May 28, 1921; Einstein/Besso, 163.
46. Letter from Fritz Haber, March 9, 1921; quoted in Fölsing, 497.
47. Letter to Fritz Haber, March 9, 1921; quoted in Fölsing, 495.
48. Letter to Maurice Solovine, March 8, 1921; *Letters to Solovine*, 41.
49. Letter to Michele Besso, May 28, 1921; quoted in Fölsing, 504.
50. *Jüdische Rundschau*, July 1, 1921; CP 7, 244.
51. Travel diary, November 10, 1922; Einstein Archives 29–129 to 29–131. Unpublished translation by Josef Eisinger.
52. Grüning, 321.
53. Letter to Edward Freed, July 11, 1932; quoted in Grüning, 321.
54. Quoted in Grundmann, 432.
55. Quoted in Vallentin, 224.
56. Letter to Michele Besso, August 8, 1938; Einstein/Besso, 321.
57. Einstein Archives 31–324. Unpublished translation by Josef Eisinger.
58. *Out of My Later Years*, 249.
59. Speech before the National Labor Committee for Palestine, April 17, 1938; Calaprice, 133–34.
60. Calaprice, 135.
61. Radio broadcast for a conference of the United Jewish Appeal, November 27, 1949; Einstein Archives 58–904; *Out of My Later Years*, 275.
62. Letter to Michele Besso, September 4 (or 11), 1929; Einstein/Besso, 255.
63. Letter to Abba Eben, November 18, 1952; Einstein Archives 48–943; Calaprice, 138.
64. Letter to Margot Einstein, n.d., Calaprice, 139.
65. Einstein, "Religion and Science," *The New York Times Magazine*, November 9, 1930; *Ideas and Opinions*, 39.
66. Einstein, "Religion and Science," *The New York Times Magazine*, November 9, 1930; *Ideas and Opinions*, 38.
67. Holton, "Einstein's Third Paradise," 26–34.
68. *Ideas and Opinions*, 11.
69. *Ideas and Opinions*, 187.
70. *Ideas and Opinions*, 11.
71. *Ideas and Opinions*, 186.
72. Reiser, 223.
73. Infeld, *Leben mit Einstein*, 56.
74. Gardner, 129.
75. Calaprice, 209.
76. Pais, *Einstein Lived Here*, 122.
77. Moszkowski, 46.
78. Letter to Eduard Einstein, n.d., Einstein Archives 75–987.
79. Jammer, 49.
80. *The New York Times*, April 25, 1929; Clark, 413.
81. Spinoza, 104.
82. Seelig, *Helle Zeit*, 90.
83. Seelig, *Helle Zeit*, 58.

17. THE END JUSTIFIES THE DOUBTS: EINSTEIN AND QUANTUM THEORY

1. CP 2, 87.
2. Calaprice, 260.
3. Anton Zeilinger, *Einsteins Spuk* (Munich: Bertelsmann, 2005); *Einsteins Schleier* (Munich: C. H. Beck, 2003).
4. Quoted in Görnitz, 125.
5. Letter to Mileva Marić, March 1899; CP 1, 126.
6. Letter to Mileva Marić, April 4, 1901; CP 1, 162.
7. Quoted in Zeilinger, 16.
8. Quoted in Kragh, 62.
9. Quoted in Kragh, 63, 66.
10. CP 2, n. 97, 147 (German edition only).
11. CP 2, 147 (German edition only).
12. CP 2, 148 (German edition only).
13. CP 3, n. 36, xxii–xxiii (German edition only).
14. Letter to Jakob Laub, March 16, 1910; CP 5, 149.
15. Letter to Michele Besso, May 13, 1911; CP 5, 187.
16. November 3, 1911; CP 3, 426.
17. Letter to Heinrich Zangger, November 15, 1911; CP 5, 222.
18. Letter to Hendrik A. Lorentz, November 23, 1911; CP 5, 228.
19. CP 3, xxviii (German edition only).
20. Frank, 98.
21. CP 5, 337–38.
22. Hoffmann/Dukas, *Albert Einstein: Creator and Rebel*, 173.
23. Quoted in Kragh, 53.
24. Hoffmann/Dukas, *Albert Einstein: Creator and Rebel*, 176.
25. Letter to Michele Besso, August 11, 1916; CP 8, 243.
26. CP 6, 232.
27. Letter to Michele Besso, September 6, 1916; CP 8, 246.
28. Letter from Niels Bohr, June 24, 1920; CP 10, 200.
29. Letter to Max and Hedwig Born, January 27, 1920; CP 9, 237.
30. Max Born's appended commentary to letter to Max and Hedwig Born, April 29, 1924, Einstein/Born, 119 (German edition only).
31. Letter to Max and Hedwig Born, April 29, 1924; Einstein/Born, 82.
32. Kragh, 159.
33. Letter from Max Born, July 15, 1925; Einstein/Born, 84.
34. Letter from Max Born, July 15, 1925; Einstein/Born, 84.
35. Letter to Max Born, December 4, 1926; Einstein/Born, 91.
36. Max Born, 1926; quoted in Fölsing, 584.
37. Quoted in Görnitz, 137f.
38. Quoted in Calaprice, 209.
39. Quoted in Barnett, 115.
40. Paul Ehrenfest, November 3, 1927; quoted in Fölsing, 589.
41. Paul Ehrenfest, November 3, 1927; quoted in Fölsing, 589.
42. Letter to Max Born, September 7, 1944; Einstein/Born, 149.
43. Letter to Michele Besso, December 12, 1951; Einstein/Besso, 453.

44. Quoted in Ernst Peter Fischer, 177.
45. Quoted in Ernst Peter Fischer, 178.
46. Letter from Wolfgang Pauli to Werner Heisenberg, June 15, 1935; quoted in Fölsing, 697.
47. Quoted in Hoffmann/Dukas, *Albert Einstein: Creator and Rebel*, 234.
48. Quoted in Zeilinger, 173.

18. OF THE MAGNITUDE OF FAILURE: THE QUEST FOR THE UNIFIED THEORY

1. Letter to Michele Besso, January 5, 1929; Einstein/Besso, 240.
2. Letter to Michele Besso, January 5, 1929; Einstein/Besso, 240.
3. Quoted in Hoffmann/Dukas, *Albert Einstein: Creator and Rebel*, 234.
4. Quoted in Hoffmann/Dukas, *Albert Einstein: Creator and Rebel*, 225.
5. Letter to Michele Besso, January 5, 1929; Einstein/Besso, 240.
6. Quoted in Hoffmann/Dukas, *Albert Einstein: Creator and Rebel*, 227.
7. Letter from Wolfgang Pauli, December 19, 1929; quoted in Fölsing, 606.
8. *Les Prix Nobel en 1921–1922*, edited by Carl Gustaf Santesson (Stockholm: Nobel Foundation, 1923), 9.
9. *Daily Chronicle*, January 26, 1929; quoted in Fölsing, 605.
10. Calaprice, 118.
11. Letter from Elsa Einstein to Hedwig Born, September 13, 1930; quoted in Fölsing, 615.
12. Letter to Michele Besso, August 8, 1938; Einstein/Besso, 527.
13. Quoted in Fölsing, 716–17.
14. Letter to Michele Besso, August 8, 1938; Einstein/Besso, 527.

19. FROM BARBARIA TO DOLLARIA: EINSTEIN'S AMERICA

1. Travel diary, December 29, 1930; Einstein Archives 29–135. Unpublished translation by Josef Eisinger.
2. Travel diary, December 29, 1930; Einstein Archives 29–135. Unpublished translation by Josef Eisinger.
3. *Princeton Alumni Weekly*, May 11, 1921; quoted in Calaprice, 326.
4. *The New York Times*, April 3, 1921; quoted in Pais, *Einstein Lived Here*, 154.
5. Frank, 179.
6. *The New York Times*, April 3, 1921; quoted in Pais, *Einstein Lived Here*, 154.
7. Quoted in Brian, 122.
8. Quoted in Brian, 124.
9. *Chicago Daily Tribune*, May 13, 1921.
10. Quoted in Brian, 129.
11. *Cleveland Press*, May 25, 1921.
12. Calaprice, 228.
13. *The New York Times*, July 11, 1921.
14. *The New York Times*, July 11, 1921.
15. *Berliner Tageblatt*, July 1921; quoted in Pais, *Einstein Lived Here*, 156.
16. A. *Einstein: How I See the World*, PBS video, 1995.

17. Travel diary, December 29, 1930. Unpublished translation by Josef Eisinger.
18. Travel diary, January 7, 1931. Unpublished translation by Josef Eisinger.
19. Letter from Hedwig Born, February 22, 1931; Einstein/Born, 110.
20. Travel diary, January 7, 1931; Einstein Archives 29–135. Unpublished translation by Josef Eisinger.
21. Letter to the Lebach family, January 16, 1931.
22. Nathan/Norden, 122.
23. Travel diary, January 8, 1931; Einstein Archives 29–142. Unpublished translation by Josef Eisinger.
24. Travel diary, January 8, 1931; Einstein Archives 29 142. Unpublished translation by Josef Eisinger.
25. Quoted in Jerome, 22.
26. Travel diary, May 15, 1931; Einstein Archives 29–142. Unpublished translation by Josef Eisinger.
27. Travel diary, before June 15, 1931; Einstein Archives 29–142.
28. Report of the German General Consulate in New York, March 21, 1931; Kirsten/Treder, 237.
29. Calaprice, 45.
30. Letter to Michele Besso, June 5, 1924; Einstein/Besso, 204.
31. Letter to Paul Ehrenfest, April 3, 1932; quoted in Fölsing, 648.
32. Quoted in Jerome, 7.
33. *The New York Times*, December 6, 1932; quoted in Sayen, 7–8.
34. Quoted in Jerome, 11.
35. Quoted in Jerome, 11.
36. Quoted in Jerome, 11.
37. Quoted in Jerome, 11.
38. *Ideas and Opinions*, 7–8.
39. Letter to Eduard Einstein, February 5, 1930; Einstein Archives 75–990.
40. Quoted in Brian, 252.
41. Letter to Eduard Einstein, December 1933; Einstein Archives 75–665.
42. *Ideas and Opinions*, 205.
43. *The Crisis*, February 1932; quoted in Nathan/Norden, 158.
44. Letter to Queen Elisabeth of Belgium, November 20, 1933; quoted in Calaprice, 48.
45. Letter to Maja Winteler-Einstein, 1937; Einstein Archives 29–424.
46. Quoted in Fölsing, 682–83.
47. Letter to Maurice Solovine, April 10, 1938; *Letters to Solovine*, 87.
48. Letter to Maja Winteler-Einstein, August 8, 1935; Einstein Archives 29–417.
49. Letter to Maja Winteler-Einstein, 1936; Einstein Archives 29–419.
50. Infeld, *Leben mit Einstein*, 65.
51. Infeld, *Leben mit Einstein*, 55.
52. Infeld, *Leben mit Einstein*, 55.
53. Infeld, *Leben mit Einstein*, 54.
54. Hoffmann/Dukas, *Albert Einstein: Creator and Rebel*, 231.
55. Brian, 293.
56. Infeld, *Leben mit Einstein*, 73.
57. 1938; Einstein Archives 31–216. Bucky/Weakland, 137.

58. All examples in the "curiosity file" that are cited here are from Pais, *Einstein Lived Here*, 89–95.
59. Letter to Michele Besso, October 10, 1938; Einstein/Besso, 330.
60. Letter to Maurice Solovine, April 10, 1938; *Letters to Solovine*, 87.
61. Letter to Max Born, n.d., Einstein/Born, 125.
62. Letter to Max Born, n.d., Einstein/Born, 128.
63. Vallentin, 292.
64. Quoted in Seelig, *Helle Zeit*, 54.
65. Letter to Eduard Einstein, May 25, 1937; Einstein Archives 75–933.
66. Letter to Maja Winteler-Einstein, 1937; Einstein Archives 29–425.
67. Letter to Hans Albert Einstein, August 30, 1937; Einstein Archives 75–424.
68. Letter to Michele Besso, October 10, 1938; Einstein/Besso, 330.
69. Letter to Michele Besso, August 8, 1938; Einstein/Besso, 321.
70. Letter to Maja Winteler-Einstein, June 1938; Einstein Archives 29–425.
71. Infeld, *Leben mit Einstein*, 77.
72. In *The Tablet*, May 14, 1938; quoted in Jerome, 224.

20. "PEOPLE ARE A BAD INVENTION": EINSTEIN, THE
ATOMIC BOMB, MCCARTHY, AND THE END
1. Quoted in Pais, *Einstein Lived Here*, 217.
2. Vallentin, 271.
3. Letter to Franklin D. Roosevelt, August 2, 1939; Einstein Archives 33–088; quoted in Nathan/Norden, 294–95.
4. Letter to Franklin D. Roosevelt, March 3, 1940; quoted in Nathan/Norden, 299.
5. Moszkowski, 36.
6. Moszkowski, 33.
7. Moszkowski, 34.
8. Moszkowski, 37.
9. Kragh, 258.
10. Jerome, 39.
11. Letter to Maja Winteler-Einstein, 1937; Einstein Archives 29–424.
12. Jerome, 39.
13. Letter to Hans Albert Einstein, 1943(?); Einstein Archives 75–832.
14. Thomas Mann to Agnes E. Meyer, January 24, 1941, in Erika Mann, ed., *Letters of Thomas Mann 1889–1955*, 279.
15. *The New York Times*, December 30, 1941; quoted in Pais, *Einstein Lived Here*, 220.
16. *The New York Times*, December 30, 1941; quoted in Pais, *Einstein Lived Here*, 220.
17. Quoted in Pais, *Einstein Lived Here*, 203.
18. *Out of My Later Years*, 265.
19. Letter to Niels Bohr, December 12, 1944; quoted in Fölsing, 719.
20. Nathan/Norden, 304.
21. Quoted in Kragh, 269.
22. Quoted in Jerome, 58.
23. Quoted in Jerome, 59.

24. *Out of My Later Years*, 200.
25. Nathan/Norden, 347.
26. Quoted in Jerome, 56.
27. Quoted in Jerome, 56.
28. Vallentin, 278.
29. Quoted in Brian, 420.
30. Letter to Hans Albert Einstein, September 2, 1945; Einstein Archives 75–790.
31. Quoted in Pais, *Einstein Lived Here*, 231.
32. Quoted in Pais, *Einstein Lived Here*, 227.
33. Vallentin, 278.
34. Letter to Hans Albert Einstein, February 18, 1946 or 1947; Einstein Archives 75–790.
35. Letter to Hans Albert Einstein, August 25, 1947; Einstein Archives 75–844.
36. Vallentin, 286.
37. Quoted in Nathan/Norden, 475.
38. Quoted in Pais, *Einstein Lived Here*, 230.
39. *The New York Times*, September 15, 1945.
40. *Out of My Later Years*, 200.
41. Quoted in Jerome, 93.
42. Vallentin, 294–96.
43. Quoted in Jerome, 171.
44. Letter to Michele Besso, December 12, 1951; Einstein/Besso, 453.
45. Letter to Max Born, May 12, 1952; Einstein/Born, 192.
46. Seelig, *Albert Einstein: A Documentary Biography*, 216.
47. *Weltwoche*, no. 20/04, 51.
48. Letter to Hans Albert Einstein, February 3, 1936; Einstein Archives 75–976.
49. Letter to Mileva Einstein-Marić, July 7, 1937; Einstein Archives 75–938.
50. Quoted in Jerome, 239.
51. Quoted in Jerome, 240.
52. Quoted in Jerome, 240.
53. Quoted in Jerome, 240.
54. Grundmann, 615.
55. Quoted in *Weltwoche*, no. 20/04, 51.
56. Quoted in *The New York Times*, April 24, 2004.
57. "Liebe Hanne! Einstein's Poems to Johanna Fantova," translated by Alfred Engel, in *Princeton University Library Chronicle* 65, no. 1 (Autumn 2003): 84. All other remarks to Fantova quoted below are from her unpublished diary, Firestone Library of Princeton University, unless otherwise noted.
58. Quoted in Calaprice, 186.
59. Letter to Hans Albert Einstein, December 28, 1954; Einstein Archives 75–917.
60. Letter to Hans Albert Einstein, n.d.; Einstein Archives 75–800.
61. Quoted in Calaprice, 363.
62. Letter to Otto Hahn, January 28, 1949; quoted in Fölsing, 728.
63. Letter to Arnold Sommerfeld, December 14, 1945; quoted in Fölsing, 727.
64. Letter to Theodor Heuss, January 16, 1951; quoted in Fölsing, 728.
65. Quoted in Johanna Fantova, introduction to "Gespräche mit Einstein," in *Princeton University Library Chronicle* 65, no. 1 (Autumn 2003): 58.

66. *The New York Times*, April 24, 2004.
67. Letter to Maurice Solovine, February 27, 1955; Einstein, *Letters to Solovine*, 159.
68. Letter to Maurice Solovine, April 3, 1953; Einstein, *Letters to Solovine*, 143.
69. Letter from Michele Besso, October 27, 1949; Einstein/Besso, 420.
70. Letter to Vero Besso and Mrs. Bice, March 21, 1955; Einstein/Besso, 538.
71. Quoted in Nathan/Norden, 634.
72. Quoted in Calaprice, 61.
73. Quoted in Einstein/Born, 234.
74. Seelig, *Helle Zeit*, 86.
75. Rosenkranz, *Albert Through the Looking Glass*, 137.

BIBLIOGRAPHY

WORKS BY ALBERT EINSTEIN

COLLECTED WORKS

The Collected Papers of Albert Einstein. Translated by Anna Beck et al. Princeton, NJ: Princeton University Press, 1987– . The first ten volumes have been issued to date; each volume has a hardbound German edition accompanied by a softbound English translation of selected passages. Quotations from this edition are abbreviated in the notes as CP followed by the volume and page and refer to the English-language volumes unless otherwise indicated.

CORRESPONDENCE

with Max Born: *The Born-Einstein Letters.* Translated by Irene Born. New York: Walker, 1971. Abbreviated in notes as Einstein/Born.

with Michele Besso: *Correspondence 1903–1955.* Translated by Pierre Speziali. Paris: Hermann, 1972. Abbreviated in notes as Einstein/Besso.

with Maurice Solovine: *Letters to Solovine.* Translated by Wade Baskin. New York: Philosophical Library, 1987. Abbreviated in notes as *Letters to Solovine.*

CO-AUTHORED BOOK

with Leopold Infeld: *The Evolution of Physics.* New York: Simon & Schuster, 1938.

ADDITIONAL WRITINGS BY ALBERT EINSTEIN

Out of My Later Years. New York: Philosophical Library, 1950.

Ideas and Opinions. New translations by Sonja Bargmann. New York: Crown Publishers, 1954.

ARCHIVES CONSULTED

Einstein Archives, Hebrew University, Jerusalem

Einstein Papers Project, California Institute of Technology, Pasadena

Archives of the Max Planck Society, Berlin

Nachlass of Ernst Gehrcke, Max Planck Institute for the History of Science, Berlin
Press Archives of Spiegel-Verlag, Hamburg
Press Archives of Gruner & Jahr, Hamburg
Newspaper Archives of the State Library, Berlin

INTERVIEWS

Dr. Markus Aspelmeyer, Vienna; Dr. Peter Aufmuth, Hannover; Prof. John Beckman, Tenerife; Erika Britzke, Caputh; Alice Calaprice, Princeton; Evelyn Einstein, Berkeley; Paul Einstein, Ulm; Prof. Gerd Grasshoff, Bern; Dr. Thomas Harvey, Princeton; Prof. Dieter Hoffmann, Berlin; Prof. Michel Janssen, Minneapolis; Prof. Diana Kormos-Buchwald, Pasadena; Dr. Christoph Lehner, Berlin; Ze'ev Rosenkranz, Jerusalem; Dr. Tilman Sauer, Pasadena; Prof. Robert Schulmann, Bethesda, Maryland; Prof. Ursula Staudinger, Berlin; Prof. Thomas Thiemann, Waterloo; Prof. Hans-Jürgen Treder, Potsdam; Philip Walther, Vienna; Milena Wazeck, Berlin; Anton Zeilinger, Vienna.

SOURCES CONSULTED

Abraham, Carolyn. *Possessing Genius*. New York: St. Martin's Press, 2003.

Atkins, Peter. *Galileo's Finger*. New York: Oxford University Press, 2003.

Auerbach, Felix. *Raum und Zeit, Materie und Energie*. Leipzig: Dürrsche Buchhandlung, 1921.

Balibar, Françoise. *Einstein: Decoding the Universe*. New York: Harry N. Abrams, 2001.

Barnett, Lincoln. *The Universe and Dr. Einstein*. 2nd ed. New York: Harper, 1957.

Bartusiak, Marcia. *Einstein's Unfinished Symphony: Listening to the Sounds of Space-Time*. Washington, DC: Joseph Henry Press, 2000.

Baumgart, I. et al., eds. *Albert Einstein in Berlin 1913–1933*. Part 2. Berlin: Akademie Verlag, 1979.

Bernal, J. D. *Science in History*. 3rd ed. Vol. 1: *The Emergence of Science*. Harmondsworth, England: Penguin, 1969.

Bernstein, Aaron. *Naturwissenschaftliche Volksbücher*. Berlin: Verlag von Franz Duncker, 1869.

Bernstein, Jeremy. *Einstein*. New York: Viking Press, 1973.

Bodanis, David. *E=mc²: A Biography of the World's Most Famous Equation*. New York: Walker, 2000.

Born, Max. *Die Relativitätstheorie Einsteins*. Berlin, 1921. Reprint, Berlin: Springer, 2003.

Brecht, Bertolt. *Life of Galileo*. Translated by John Willett. London: Methuen, 2001.

Brian, Denis. *Einstein: A Life*. New York: John Wiley, 1996.

Brod, Max. *The Redemption of Tycho Brahe*. Translated by Felix Warren Crosse. London: Knopf, 1928.

Büchner, Ludwig. *Kraft und Stoff*. Leipzig: T. Thomas, 1888.

Bucky, Peter A., and Allen G. Weakland. *The Private Albert Einstein*. Kansas City, MO: Andrews McNeel Publishing, 1993.

Bührke, Thomas. *Albert Einstein*. Munich: dtv, 2004.

Calaprice, Alice, ed. *The New Quotable Einstein*. Princeton, NJ: Princeton University Press, 2005.

Calder, Nigel. *Einstein's Universe*. New York: Viking Press, 1979.

Charpa, Ulrich, and Armin Grunwald. *Albert Einstein*. Frankfurt: Campus Verlag, 2001.

Clark, Ronald W. *Einstein: The Life and Times*. New York: World Publishing Co., 1971.

Cohen, Bernard I. *The Birth of a New Physics*. Garden City, NY: Doubleday, 1960.

———. *Revolution in Science*. Cambridge, MA: Harvard University Press, 1985.

Deutsch, David. *The Fabric of Reality*. New York: Allan Lane, 1997.

Dukas, Helen, and Banesh Hoffmann, eds. *Albert Einstein, the Human Side: New Glimpses from His Archives*. Princeton, NJ: Princeton University Press, 1979.

Eberty, Felix. *Stars and the Earth, or, Thoughts upon Space, Time, and Eternity*. 3rd ed. Boston: Crosby, Nichols, Lee and Co., 1860.

Epstein, Lewis C. *Relativity Visualized*. San Francisco: Insight Press, 1984.

Faulkner, William. *The Sound and the Fury*. New York: Random House, 1984.

Ferris, Timothy. *The Whole Shebang*. New York: Simon & Schuster, 1997.

Feyerabend, Paul. *Against Method*. 3rd ed. New York: Verso, 1993.

Fischer, Ernst Peter. *Einstein: Ein Genie und sein überfordertes Publikum*. Berlin: Springer-Verlag, 1996.

Fischer, Klaus. *Einstein*. Freiburg: Panorama Wissenschaftsbücher, 1999.

Flückiger, Max. *Albert Einstein in Bern*. Bern: Paul Haupt, 1974.

Fölsing, Albrecht. *Albert Einstein: A Biography*. Translated by Ewald Osers. New York: Viking, 1997.

Frank, Philipp. *Einstein: His Life and Times*. New York: Knopf, 1953.

Friedman, Alan J., and Carol C. Donley. *Einstein as Myth and Muse*. New York: Cambridge University Press, 1985.

Fritzsch, Harald. *An Equation that Changed the World: Newton, Einstein, and the Theory of Relativity*. Translated by Karin Heusch. Chicago: University of Chicago Press, 1994.

Galison, Peter. *Einstein's Clocks*. New York: W. W. Norton, 2003.

Gardner, Howard. *Creating Minds*. New York: Basic Books, 1994.

Geier, Manfred. *Kants Welt*. Reinbek: Rowohlt, 2005.

Gleick, James. *Isaac Newton*. New York: Pantheon Books, 2003.

Gloy, Karen. *Das Verständnis der Natur*. 2 vols. Munich: C. H. Beck, 1995–96.

Goenner, Hubert. *Einsteins Relativitätstheorien*. Munich: C. H. Beck, 2005.

Görnitz, Thomas. *Quanten sind anders*. Heidelberg: Spektrum Verlag, 1999.

Gott, J. Richard. *Time Travel in Einstein's Universe*. Boston: Houghton Mifflin, 2001.

Greene, Brian. *The Elegant Universe: Superstrings, Hidden Dimensions, and the Quest for the Ultimate Theory*. New York: W. W. Norton, 2003.

Gribbin, John, and Mary Gribbin. *Annus Mirabilis: 1905, Albert Einstein, and the Theory of Relativity*. New York: Chamberlain Brothers, 2005.

Grundmann, Siegfried. *Einsteins Akte*. Berlin: Springer-Verlag, 1998.

Grüning, Michael. *Ein Haus für Albert Einstein*. Berlin: Verlag der Nation, 1990.

Hawking, Stephen. *Black Holes and Baby Universes and Other Essays*. New York: Bantam, 1993.

———. *The Illustrated A Brief History of Time*. New York: Bantam, 1996.

Hejlek, Ossi. *Albert Einstein für Einsteiger*. Vol. 2 of *Wissen mit Pfiff*. Vienna: Boehlau, 1999.

Helferich, Christoph. *Geschichte der Philosophie*. Stuttgart: Metzler, 2001.

Hermann, Armin. *Einstein: Der Weltweise und sein Jahrhundert*. Munich: Piper, 1994.

Herneck, Friedrich. *Einstein und sein Weltbild*. Berlin: Morgenbuch Verlag, 1976.

———. *Einstein privat*. Berlin: Morgenbuch Verlag, 1990.

———. *Albert Einstein*. Leipzig: B. G. Teubner, 1975.

Hettler, Nicolaus. "Die Elektrotechnische Firma J. Einstein u. Cie. in München 1876–1894." PhD diss. University of Stuttgart, 1996.

Highfield, Roger, and Paul Carter. *The Private Lives of Albert Einstein*. Boston: Faber and Faber, 1993.

Hoffmann, Banesh. *Relativity and Its Roots*. New York: Scientific American Books, 1983.

———, and Helen Dukas. *Albert Einstein, Creator and Rebel*. New York: Viking Press, 1972.

Holton, Gerald. *Einstein, History, and Other Passions*. Woodbury, NY: AIP Press, 1995.

———. "Einstein's Third Paradise." *Daedalus*, Fall 2003: 26–34.

———, and Yehuda Elkana, eds. *Albert Einstein: Historical and Cultural Perspectives*. Princeton, NJ: Princeton University Press, 1982.

Horgan, John. *The End of Science*. Reading, MA: Addison-Wesley, 1996.

Humboldt, Alexander von. *Kosmos: Entwurf einer physischen Weltbeschreibung*. Stuttgart: Cotta, 1850.

Huonker, Thomas. *Diagnose: "moralisch defekt."* Zurich: Orell Füssli, 2003.

Infeld, Leopold. *Albert Einstein: His Work and Its Influence on Our World*. New York: Scribner, 1950.

———. *Leben mit Einstein*. Vienna: Europa-Verlag, 1969.

Jammer, Max. *Einstein and Religion: Physics and Theology*. Princeton, NJ: Princeton University Press, 2002.

Jerome, Fred. *The Einstein File*. New York: St. Martin's Press, 2002.

Kaku, Michio. *Einstein's Cosmos*. New York: W. W. Norton, 2004.

Karamanolis, Stratis. *Albert Einstein: Mythos und Realität*. Munich: Karamanolis, 1991.

Kessler, Charles, ed. and trans. *Berlin in Lights: The Diaries of Count Harry Kessler (1918–1937)*. New York: Grove Press, 1999.

Kirsten, Christa, and Hans-Jürgen Treder, eds. *Albert Einstein in Berlin 1913–1933*. Part 1. Berlin: Akademie Verlag, 1979.

Kragh, Helge. *Quantum Generations*. Princeton, NJ: Princeton University Press, 2001.

Kuhn, Thomas. *The Structure of Scientific Revolutions*. 3rd ed. Chicago: University of Chicago Press, 1996.

Kuznetsov, B. G. *Einstein*. Translated by V. Talmy. Moscow: Progress Publishers, 1965.

Levenson, Thomas. *Einstein in Berlin*. New York: Bantam, 2004.

Lightman, Alan. *Einstein's Dreams*. New York: Warner Books, 1994.

Livio, Mario. *The Accelerating Universe*. New York: John Wiley, 2000.

Maddox, John. *What Remains to Be Discovered*. New York: Free Press, 1999.

Mann, Erika, ed. *Letters of Thomas Mann, 1889–1955*. Selected and translated by Richard and Clara Winston. Berkeley: University of California Press, 1975.

———, and Klaus Mann. *Escape to Life*. Boston: Houghton Mifflin, 1939.

Mann, Katia. *Meine ungeschriebenen Memoiren*. Frankfurt: S. Fischer Verlag, 2000.

Mann, Thomas. *The Magic Mountain*. Translated by John E. Woods. New York: Alfred A. Knopf, 1995.

Marianoff, Dimitri. *Einstein: An Intimate Study of a Great Man*. Garden City, NY: Doubleday, 1944.

Melcher, Horst. *Einstein wider Vorurteile und Denkgewohnheiten*. Cologne: Aulis, 1978.

Michelmore, Peter. *Einstein: Profile of the Man*. New York: Dodd, Mead, 1962.

Miller, Arthur I. *Einstein, Picasso: Space, Time, and the Beauty that Causes Havoc*. New York: Basic Books, 2001.

Moszkowski, Alexander. *Conversations with Einstein*. Translated by Henry L. Brose. New York: Horizon Press, 1971.

Nachama, Andreas et al., eds. *Juden in Berlin*. Berlin: Henschel Verlag, 2001.

Nathan, Otto, and Heinz Norden, eds. *Albert Einstein on Peace*. New York: Simon & Schuster, 1961.

Overbye, Dennis. *Einstein in Love*. New York: Viking, 2000.

Pais, Abraham. *Subtle Is the Lord: The Science and Life of Albert Einstein*. New York: Oxford University Press, 1982.

———. *Einstein Lived Here*. New York: Oxford University Press, 1994.

Panek, Richard. *The Invisible Century: Einstein, Freud, and the Search for Hidden Universes*. New York: Viking Press, 2004.

Parker, Barry. *Einstein: The Passions of a Scientist*. New York: Prometheus Books, 2003.

Paterniti, Michael. *Driving Mr. Albert: A Trip Across America with Einstein's Brain*. New York: Dial Press, 2000.

Pickover, Clifford A. *Time: A Traveler's Guide*. New York: Oxford University Press, 1998.

Proust, Marcel. *Swann's Way*. Vol. 1 of *In Search of Lost Time*. Translated by Lydia Davis. New York: Viking, 2003.

Pyenson, Lewis. *The Young Einstein: The Advent of Relativity*. Boston: Adam Hilger, 1985.

Radkau, Joachim. *Das Zeitalter der Nervosität*. Darmstadt: Econ Taschenbücher, 1998.

Reichinstein, David. *Albert Einstein: A Picture of His Life and His Conception of the World*. Translated by M. Juers and D. Sigmund. Prague: Stella Publishing House, Ltd., 1934.

Reiser, Anton [pseudonym of Rudolf Kayser]. *Albert Einstein: A Biographical Portrait*. New York: Albert & Charles Boni, 1930.

Renn, Jürgen. *Auf den Schultern von Riesen und Zwergen: Einsteins unvollendete Revolution*. Weinheim: Wiley-VCH, 2006.

——— et al. "Albert Einstein: Alte und neue Kontexte in Berlin." Speech in Berlin, 1997.

———, and Robert Schulmann, eds. *Albert Einstein and Mileva Marić: The Love Letters*. Translated by Shawn Smith. Princeton, NJ: Princeton University Press, 1992.

Rosenkranz, Ze'ev. *Albert Through the Looking Glass: The Personal Papers of Albert Einstein*. Jerusalem: The Jewish National and University Library, 1998.

———. *The Einstein Scrapbook*. Baltimore: Johns Hopkins University Press, 2002.

Sayen, Jamie. *Einstein in America*. New York: Crown Publishers, 1985.

Schilpp, Paul Arthur, ed. *Albert Einstein: Philosopher-Scientist*. 3rd ed. La Salle, IL: Open Court, 1970.

Seelig, Carl. *Albert Einstein: A Documentary Biography*. Translated by Mervyn Savill. London: Staples Press, Ltd., 1956.

——. *Albert Einstein: Leben und Werk eines Genies unserer Zeit*. Gütersloh: Bertelsmann Lesering, 1960.

——, ed. *Helle Zeit—dunkle Zeit*. Zurich: Europa, 1956.

Serres, Michel, ed. *Eléments d'histoire des sciences*. Paris: Bordas, 1989.

Sexl, Lore, and Anne Hardy. *Lise Meitner*. Reinbek: Rowohlt, 2002.

Silver, Brian L. *The Ascent of Science*. Oxford: Oxford University Press, 2000.

Smith, Peter D. *Einstein*. London: Haus Publishers, 2003.

Sowell, Thomas. *The Einstein Syndrome*. New York: Basic Books, 2002.

Spinoza, Benedict de [Baruch]. *The Ethics and Other Works*. Edited and translated by Edwin Curley. Princeton, NJ: Princeton University Press, 1994.

Stachel, John. *Einstein from 'B' to 'Z'*. Boston: Birkhäuser, 2002.

——, ed. *Einstein's Miraculous Year: Five Papers That Changed the Face of Physics*. Princeton, NJ: Princeton University Press, 1998.

Steiner, Frank. *Albert Einstein*. Berlin: Springer, 2005.

Stern, Fritz. *Einstein's German World*. Princeton, NJ: Princeton University Press, 1999.

Trbuhović-Gjurić, Desanka. *Im Schatten Albert Einsteins*. Bern: Paul Haupt, 1988.

Treder, Hans-Jürgen. *Einstein in Potsdam*. Leipzig: Akademie-Verlag, 1986.

Vallentin, Antonina. *The Drama of Albert Einstein*. Translated by Moura Budberg. Garden City, NY: Doubleday, 1954.

Weissensteiner, Friedrich. *Die Frauen der Genies*. Munich: Piper, 2004.

Weizsäcker, Carl Friedrich von. *Zeit und Wissen*. Munich: dtv, 1995.

Wells, H. G. *The Time Machine*. Cambridge, MA: Robert Bentley, Inc., 1955.

Whitehead, Alfred North. *Science and the Modern World*. New York: Macmillan, 1925.

Wickert, Johannes. *Einstein*. Reinbek: Rowohlt, 1991.

Wolfson, Richard. *Simply Einstein*. New York: W. W. Norton, 2003.

Zackheim, Michele. *Einstein's Daughter*. New York: Riverhead Books, 1999.

Zeilinger, Anton. *Einsteins Schleier*. Munich: C. H. Beck, 2003.

ACKNOWLEDGMENTS

I am deeply grateful to the Einstein Archives at the Hebrew University in Jerusalem and the Einstein Papers Project at the California Institute of Technology in Pasadena, California. This book would not have been possible without the collegial support of the staff at both of these institutions, who made unpublished source material available to me.

I would also like to thank the Max Planck Society, especially Department 1 of the Max Planck Institute for the History of Science in Berlin (director: Prof. Jürgen Renn), which kindly granted me the status of visiting researcher and provided expert guidance and technical support, and the staff of the Max Planck Society Archives (director: Prof. Echart Henning), also in Berlin.

I wish to extend my special thanks to the Berlin historian and journalist Jörg von Bilavsky for sharing his expertise and offering cordial and creative suggestions for improvement, and to Siegfried Grundmann, also a Berlin historian, whose comprehensive research provided me with many valuable leads and passages.

A heartfelt thank-you goes also to the staff of Rowohlt Publishers, who accompanied me on my exciting and sometimes difficult path through Einstein's life, in particular my editor Uwe Naumann, as well as Barbara Wenner of the Graf & Graf Agency in Berlin.

I would like to thank everyone who read (and improved) the manuscript as a whole or in part, especially Mathias Greffrath, Prof. Dieter Hoffmann, Dr. Arno Nehlsen, Harald Schumann, and particularly Dr. Hania Luczak—as well as Dr. Markus Aspelmeyer, Dr. Peter Aufmuth,

Prof. John Beckman, Prof. Diana Kormos-Buchwald, Prof. Michel Janssen, Dr. Christoph Lehner, Christian Ludwig, Prof. Jürgen Renn, Prof. Robert Schulmann, Prof. Ursula Staudinger, and Prof. Thomas Thiemann.

I am most grateful to all the readers who gave me so much advice and sent me new source material after the publication of the German edition, which was invaluable in shaping the American edition of this book, especially Prof. Peter Plesch in Newcastle under Lyme (UK), who allowed me to include previously unpublished pages from the diary of his father, Dr. János Plesch.

I would like to thank my American publisher for making this book accessible to a wider, English-speaking readership. Special thanks go to Eric Chinski, my meticulous, ever-supportive, and refreshingly inquisitive editor at FSG.

Last but not least, I would like to express my deep admiration for my American translator, Shelley Frisch. Her English-language version of my book has surpassed my wildest dreams.

INDEX

216–17; of general theory of relativity, 216–30
GEO 600, 243–46, 251
Gerber, Paul, 262
German Democratic Republic (GDR), 290
German League for Human Rights, 258
German Natural Scientists, Cologne, 158
German Physics Society, 256, 268, 269
Germany, 20, 31, 50; atomic bomb and, 379, 380; Einstein's emigration from, 199, 284–88, 359, 394; Einstein's views on, 73, 256–59, 267, 270–88, 290, 320, 363–64, 379, 384, 389, 393–94, 400; Nazi, 7, 199, 276, 282–88, 300, 308–309, 367, 377–78, 383–84; 1918 Revolution, 257, 259; post–World War I, 11–12, 15, 271–88; Weimar Republic, 15, 76, 274–88; World War I, 172, 254–59, 274, 277; World War II, 383–88
Gestapo, 377, 378, 383, 394
Giedion, Sigfried, *Space, Time, and Architecture*, 301
Gilbert, William, *Magnete, De*, 54
Glasgow, 244
Global Positioning System (GPS), 248
Gödel, Kurt, 371
Goethe, Johann Wolfgang von, 235, 256, 303, 398; *Faust*, 372
Goethe League, Berlin, 256
Goldschmidt, Rudolph, 42, 43
Goldstein, Rabbi Herbert S., 324
Gorki, Maxim, 282
Göttingen, 222, 223
Grass, Günther, *Tin Drum, The*, 27
gravitation, 59, 60, 156, 209, 233–35, 325, 332; acceleration and, 208–16; confirmation of gravitation waves, 243–51; cosmology and, 233–35; Newton's law of, 210–12, 226, 247; role in general theory of relativity, 209–30, 243–51

gravitational lenses, 248–49
Gravity Probe B, 249–50
Great Britain, 9, 10, 14, 15, 166, 267, 271–72, 287, 338
Gropius, Walter, 298, 301
Grossmann, Marcel, 16, 111, 112, 113–14, 115, 132, 154, 157, 163, 216, 217, 218, 220, 313; death of, 376
Grundman, Siegfried, 270, 397
gyrocompass, 43, 254

H

Haber, Clara, 252, 253
Haber, Fritz, 15, 107, 162, 168–69, 252–54, 259, 278, 288, 314, 317; death of, 288; Einstein and, 253–54, 286, 288, 295, 319; as father of chemical warfare, 252, 253, 288, 379; Judaism of, 317, 318; Nazis and, 286, 288
Habicht, Conrad, 42, 118, 120–21, 123, 124, 125, 148, 149, 150, 152, 188
Habicht, Paul, 42
Hague, 252
Hahn, Otto, 148, 381
Haller, Friedrich, 118, 122
Halley, Edmund, 62
Halley's comet, 62
Hannover, 243, 245
Harding, Warren G., 361
Harvard University, 25, 78, 144
Harvey, Thomas, 3–5, 83, 206; Einstein's brain taken by, 4–6, 77, 83, 403
Hauptmann, Benvenuto, 303
Hauptmann, Gerhart, 254, 303, 306; *Weavers, The*, 303
Hawking, Stephen, 239; *Brief History of Time, A*, 239

M